Lecture Notes on Acoustics and Noise Control

Hejie Lin • Turgay Bengisu
Zissimos P. Mourelatos

Lecture Notes on Acoustics and Noise Control

 Springer

Hejie Lin
General Motors
Warren, MI, USA

Turgay Bengisu
Oakland University
Rochester, MI, USA

Zissimos P. Mourelatos
Mechanical Engineering
Oakland University
Rochester, MI, USA

ISBN 978-3-030-88215-0 ISBN 978-3-030-88213-6 (eBook)
https://doi.org/10.1007/978-3-030-88213-6

This Springer imprint is published by the registered company Springer Nature Switzerland AG
The registered company address is: Gewerbestrasse 11, 6330 Cham, Switzerland

Preface

Objectives of the Book

Lecture Notes on Acoustics and Noise Control provides a mathematical backbone for acoustics and noise control. This mathematical foundation is comprised of straightforward mathematical derivations and formulations of sound waves. All formulations are built from the acoustic wave equation based on the kinetic theory of gases. The mathematical formulations are condensed into a minimalistic yet complete acoustic foundation for a one-semester course. The formulations covered in this book can be used as reliable references for researchers and engineers working in the field of acoustics and noise control.

Style

This book is written in a course notes format for a one-semester course of "acoustics and noise control." This course is offered to senior undergraduates and beginning graduate students without a prior background of acoustics or noise control. The materials in this book are compiled from a series of noise, vibration, and harshness (NVH) courses taught by the authors of this book during the past 14 years at Oakland University.

Straightforward derivations of formulations are presented in the course notes. When the derivation of formulas is the main objective of a section, the final formulas are placed at the beginning of that section to serve as a compass of derivation. The properties of sound can be observed and understood through the derivation of the formulation. Class examples and homework are included to support and reinforce the derivations. Through the derivations, examples, and homework, students can understand the physics behind the formulas.

Two significant projects in numerical analysis of sound propagation are included in this course. The first project revolves around numerically calculating the sound pressure induced by a vibrating plate. This project requires students to understand the formulations for acoustic point sources and apply them to the real-world situations. The first project concludes the fundamental acoustics theory in the first half of the course (Chaps 1, 2, 3, 4, 5 and 6). The second project requires numerical calculation of the sound power transmission in pipes. This project requires students to understand the formula for power transmission in pipes as filters and apply them to real-world problems. The second project concludes the filter design in the second half of the course (Chaps. 7, 8, 9, 10, 11 and 12).

Prerequisites

Students should have a basic knowledge of practical experience with calculus and differential equations.

The Big Picture

The first half of this book covers the fundamentals of acoustic wave formulations. Chapter 1 reviews *comp*lex numbers and introduces four equivalent forms of complex numbers for harmonic waves. Understanding these four equivalent forms of complex numbers is crucial for understanding the mathematical formulations of acoustics and noise control. Chapters 2, 3, and 4 derive and solve the plane acoustic wave equation. Chapters 5 and 6 derive and solve the spherical wave equation. At the end of Chap. 6, Project #1 is included to reinforce and consolidate the learning from Chaps. 1 to 6 (the first half of the course). This project requires students to numerically calculate sound pressure using the point source formulation obtained in Chap. 6.

The second half of this book covers applications of acoustics in the field of noise control. Chapters 7 and 8 explain and formulate the sound resonance in rectangular cavities and sound propagation in wave guides. Chapter 9 introduces the concept of weighted sound pressure levels. Chapter 10 introduces the basic formulations for noise control of room acoustics. Chapters 11 and 12 cover the theory of three basic acoustic filters: the high-pass filter, the low-pass filter, and the band-stop (Helmholtz) filters.

At the end of Chapter 11, Project 2A is included to numerically calculate sound power transmission in pipelines using the formulas developed in this chapter. At the end of Chapter 12, Project 2B is included to model low-pass filters, high-pass filters, and band-stop filters as pipes with side branches. In this project, sound power transmission of filters will be calculated.

Warren, MI, USA   Hejie Lin
Rochester, MI, USA   Turgay Bengisu
Rochester, MI, USA   Zissimos P. Mourelatos

Contents

Chapter 1
Complex Numbers for Harmonic Functions

Sound waves are typically described as simple harmonic motions. Simple harmonic motion can be extended to describe any motion (of the sound wave) by adding a series of simple harmonic motions such as a Fourier series. Simple harmonic motions can be formulated using either trigonometric functions or complex exponential functions. Even though we are more familiar with trigonometric functions such as a cosine or a sine function, the complex exponential functions are commonly used due to their compact and elegant form for derivation and integration in vibration and acoustic analysis.

Since trigonometric functions and complex exponential functions can be used interchangeably in the analysis of vibrations and acoustics, the ability to convert a harmonic function between trigonometric functions and complex exponential functions is essential for the analysis of either vibrations or acoustics. For this reason, conversions between trigonometric functions and complex exponential functions are introduced at the beginning of this course to provide a mathematical tool for analyzing acoustics throughout this course.

This chapter will introduce the four equivalent forms for simple harmonic motions. After you are familiar with the four equivalent forms, you will be able to express any simple harmonic function using either a trigonometric function or a complex exponential function using the four equivalent forms. Furthermore, you will be able to convert any simple harmonic function between the four equivalent forms.

This chapter can be treated as an independent chapter from the rest of the chapters in this book. The materials in this chapter can be used for any vibration-related course to provide students the mathematical skills to handle harmonic functions in vibration analysis.

© The Author(s), under exclusive license to Springer Nature Switzerland AG 2021
H. Lin et al., *Lecture Notes on Acoustics and Noise Control*,
https://doi.org/10.1007/978-3-030-88213-6_1

1.1   Review of Complex Numbers

This section and the next section cover the basic formulas of complex numbers in either high school- or college-level mathematics courses. The goal of these two sections is to refresh your knowledge of complex numbers for harmonic functions. You can skim through these two sections if you already know these formulas.

So what are complex numbers? The polynomial equation:

$$X^2 + 1 = 0$$

does not have a solution that can be represented by real numbers. Instead, the solution of this polynomial equation can be represented by the complex numbers:

$$X = \pm\sqrt{-1}$$

The square root of -1 is not a real number and has been defined as an imaginary number.

Hence, a combination of a real and an imaginary number is called a complex number.

For example:

$$\mathbf{Z} = a + jb$$

where a and b are real numbers, j is the imaginary number, and the bold face character indicates that $\mathbf{Z}$ is a complex number.

The following are some properties of complex numbers:

- The square of the imaginary number j is -1. That is:

$$j^2 = -1$$

- The absolute value of a complex number is called its modulus. That is:

$$|\mathbf{Z}| = Z = \sqrt{a^2 + b^2}$$

- The complex conjugate of $\mathbf{Z} = a + jb$ is defined as:

$$\mathbf{Z}^* = a - jb$$

- The addition of a complex conjugate function pair is a real number as shown:

$$\mathbf{Z} + \mathbf{Z}^* = a + jb + a - jb = 2a$$

The star symbol (*) indicates the complex conjugate of $\mathbf{Z}$.

1.2   Complex Numbers in Polar Form

A complex number can be presented in both rectangular and polar coordinates as:

$$Z = A_r + jA_i = \sqrt{A_r^2 + A_i^2}\left(\frac{A_r}{\sqrt{A_r^2 + A_i^2}} + j\frac{A_i}{\sqrt{A_r^2 + A_i^2}}\right) = A(\cos(\theta) + j\sin(\theta))$$

where:

$$A = \sqrt{A_r^2 + A_i^2}; \theta = \tan^{-1}\left(\frac{A_i}{A_r}\right); A_r = A\cos(\theta); A_i = A\sin(\theta)$$

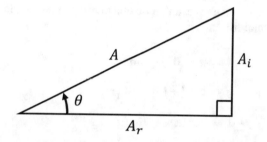

Note that the complex number in polar coordinates (shown above) is formulated using trigonometric functions such as cosine and sine functions. The complex number in polar coordinates can be further expressed in a complex exponential format using Euler's formula. Euler's formula shows the relationship between the trigonometric functions and complex exponential functions:

$$e^{\pm j\theta} = \cos(\theta) \pm j\sin(\theta)$$

Based on Euler's formula, a complex number can be presented as a complex exponential function in polar form as:

$$Z = A(\cos(\theta) + j\sin(\theta)) = Ae^{j\theta}$$

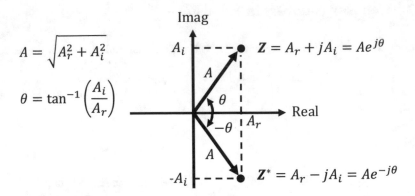

$$A = \sqrt{A_r^2 + A_i^2}$$

$$\theta = \tan^{-1}\left(\frac{A_i}{A_r}\right)$$

Based on the equation above, a real or imaginary number can be expressed in the complex exponential format as:

$$1 = \cos(0) + j\sin(0) = e^{j(0)}$$

$$j = \cos\left(\frac{\pi}{2}\right) + j\sin\left(\frac{\pi}{2}\right) = e^{j\left(\frac{\pi}{2}\right)}$$

$$-1 = \cos(\pi) + j\sin(\pi) = e^{j(\pi)}$$

$$-j = \cos\left(-\frac{\pi}{2}\right) + j\sin\left(-\frac{\pi}{2}\right) = e^{j\left(-\frac{\pi}{2}\right)}$$

1.3   Four Equivalent Forms to Represent Harmonic Waves

A summary of the four equivalent forms for simple harmonic motion is shown in this section. The derivations of these four equivalent forms are shown in the following sections.

Four equivalent forms can be used to describe simple harmonic motion. The following is a summary of four equivalent forms used to describe simple harmonic motions:

(Form 1)	$A_c \cos(\omega t) - A_s \sin(\omega t)$	Real trigonometric function
(Form 2)	$A \cos(\omega t + \phi)$	Real trigonometric function
(Form 3)	$\frac{1}{2}\left[Ae^{j(\omega t + \phi)} + Ae^{-j(\omega t + \phi)}\right]$	Complex conjugate function pair
(Form 4)	$\frac{1}{2}\left[(A_c + jA_s)e^{j\omega t} + (A_c - jA_s)e^{-j\omega t}\right]$	Complex conjugate function pair

where A, A_c, and A_s are real numbers and have the relationships shown in the following table. In the figure below, A is a positive number; A_c and A_s are either positive or negative numbers. Therefore, the phase angle ϕ is between $-\pi$ and π, and

$\phi = \tan^{-1}\left(\frac{A_s}{A_c}\right)$ can be computed by the function atan2(A_s, A_c) in most programming codes.

Note that Form 1 and Form 2 are formulated using *real* numbers (R). Form 3 and Form 4 are formulated using *complex* numbers (C). In addition, Form 1 and Form 4 are formulated using *implicit phase* (IP) of ϕ. Form 2 and Form 3 are formulated using *explicit phase* (EP) of ϕ. According to these properties, the four forms above can be rearranged into the following table:

Real Trigonometric Function	Complex Conjugate Function Pair
$A_c\cos(\omega t) - A_s\sin(\omega t)$ Form 1: R_{IP}	$\frac{1}{2}\left[(A_c + jA_s)e^{j\omega t} + (A_c - jA_s)e^{-j\omega t}\right]$ Form 4: C_{IP}
$A\cos(\omega t + \phi)$ Form 2: R_{EP}	$\frac{1}{2}\left[Ae^{j(\omega t+\phi)} + Ae^{-j(\omega t+\phi)}\right]$ Form 3: C_{EP}

R: Real Trigonometric Function IP: Implicit Phase of ϕ
C: Complex Conjugate Function EP: Explicit Phase of ϕ

where:

Geometric Relationship	
Geom 1	$A = \sqrt{A_c^2 + A_s^2}$
Geom 2	$\phi = \tan^{-1}\left(\dfrac{A_s}{A_c}\right)$
Geom 3	$A_c = A\cos(\phi)$
Geom 4	$A_s = A\sin(\phi)$

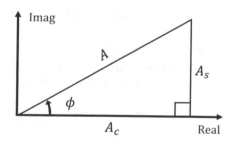

Note that Form 3 can also be expressed as:

(Form 3) $\frac{1}{2}\left[Ae^{j(\omega t+\phi)} + Ae^{-j(\omega t+\phi)}\right] = \frac{1}{2}\left[Ae^{j\omega t}e^{j\phi} + Ae^{-j\omega t}e^{-j\phi}\right]$ (Form 3.5)

Also, Form 3 and Form 4 are often expressed with a $Re[]$ function which takes the real part of the complex number as Form 3' and Form 4'.

(Form 3) $\frac{1}{2}\left[Ae^{j(\omega t+\phi)} + Ae^{-j(\omega t+\phi)}\right] = Re\left[Ae^{j(\omega t+\phi)}\right]$ (Form 3')
(Form 4) $\frac{1}{2}\left[(A_c + jA_s)e^{j\omega t} + (A_c - jA_s)e^{-j\omega t}\right] = Re\left[(A_c + jA_s)e^{j\omega t}\right]$ (Form 4')

Instead of using a complex conjugate pair to ensure a real number (Form 3 and Form 4), Form 3' and Form 4' use a brute force to take the real part of a complex number to get the real number.

The pros and cons of using $Re[\ ]$ functions (Form 3' and Form 4') are listed below:

Pros:
Form 3' and Form 4' do not require showing the whole complex conjugate pair. They only need to show one complex number, which is half of the complex conjugate pair. As a result, formulations using Form 3' and Form 4' are more compact than ones using Form 3 and Form 4.

Cons:
$Re[\ ]$ functions (Form 3' and Form 4') may mislead us into thinking that the real solution is just ignoring the imaginary part of the complex solution. To be mathematically correct, whenever $Re[\ ]$ functions are used, we should know that the real solution is the result of the addition of two complex solutions as in Form 3 and Form 4.

In conclusion, even though using $Re[\ ]$ functions (Form 3' and Form 4') might not be mathematically correct, Form 3' and Form 4' are widely used due to their compact formulation showing only half of a complex conjugate pair. Therefore, we need to know how to convert Form 3' and Form 4' back to Form 3 and Form 4. For this purpose, converting the $Re[\ ]$ functions to Form 3 and Form 4 is demonstrated in Example 1.2. Also, exercises to convert the $Re[\ ]$ functions to Form 3 and Form 4 are provided in the Homework Exercises section.

1.4 Mathematical Identity

Four mathematical identities will be used in deriving the four equivalent forms for representing simple harmonic waves. Math 1 is Euler's formula, and Math 2 will be explained and derived. Math 3 and Math 4 will be shown below for reference:

Mathematical Identity	
Math 1	$e^{\pm j\theta} = \cos(\theta) \pm j\,\sin(\theta)$ [Euler's Formula]
Math 2	$\cos(\theta)=\frac{1}{2}\left[e^{j\theta} + e^{-j\theta}\right]$; $\sin(\theta)=\frac{1}{2j}\left[e^{j\theta} - e^{-j\theta}\right]$
Math 3	$\cos(a \pm b) = \cos(a)\cos(b) \mp \sin(a)\sin(b)$
Math 4	$e^{a\pm b} = e^a e^{\pm b}$

Math 1 is Euler's formula. It defines a complex exponential function, $e^{j\theta}$, in polar coordinates using cosine and sine functions:

$$e^{\pm j\theta} = \cos(\theta) \pm j\sin(\theta) \tag{1}$$

which is shorthand for

$$e^{j\theta} = \cos(\theta) + j\sin(\theta) \tag{2}$$

$$e^{-j\theta} = \cos(\theta) - j\sin(\theta) \tag{3}$$

Math 2 provides a bridge between real numbers, $\cos(\theta)$, and complex numbers, $e^{j\theta}$. The cosine function in Eq.(4) can be derived from Euler's formulas by adding Eq.(2) and Eq.(3). The sine function in Eq.(5) can be derived from Euler's formulas by subtracting Eq.(3) from Eq.(2):

$$\cos(\theta) = \frac{1}{2}\left[e^{j\theta} + e^{-j\theta}\right] \tag{4}$$

$$\sin(\theta) = \frac{1}{2j}\left[e^{j\theta} - e^{-j\theta}\right] \tag{5}$$

Note that, in Eq.(4), both sides of the equation are real numbers. The left-hand side of the equation, $\cos(\theta)$, is a real number. The right-hand side of the equation, $\frac{1}{2}\left[e^{j\theta} + e^{-j\theta}\right]$, is also a real number because the addition of a complex conjugate pair is a real number. Even though both sides of Eq.(4) are real, the formats are totally different. The left-hand side of the equation is formulated as a real trigonometric function, $\cos(\theta)$. The right-hand side of the equation is formulated as complex exponential functions, $\frac{1}{2}\left[e^{j\theta} + e^{-j\theta}\right]$. Therefore, Math 2 is a bridge between real numbers and complex numbers.

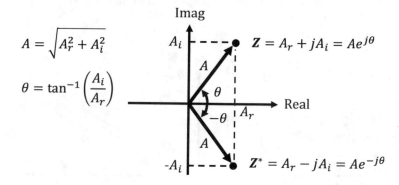

Math 3 is one of the trigonometric identities.

Math 4 is a property of exponential functions, valid for both real and complex a and b.

1.5   Derivation of Four Equivalent Forms

1.5.1   Obtain Form 2 from Form 1

Form 1 represents a simple harmonic motion using a cosine function with amplitude A_c and a sine function with amplitude A_s as:

$$x(t) = A_c \cos(\omega t) - A_s \sin(\omega t) \qquad\qquad \text{(Form 1:}R_{IP}\text{)}$$

$$= \sqrt{A_c^2 + A_s^2} \left[\frac{A_c}{\sqrt{A_c^2 + A_s^2}} \cos(\omega t) - \frac{A_s}{\sqrt{A_c^2 + A_s^2}} \sin(\omega t) \right]$$

$$= A \left[\frac{A_c}{A} \cos(\omega t) - \frac{A_s}{A} \sin(\omega t) \right]$$

Let:

$$A \equiv \sqrt{A_c^2 + A_s^2}; \frac{A_c}{A} \equiv \cos(\phi); \frac{A_s}{A} \equiv \sin(\phi); \text{and } \phi = \tan^{-1}\left(\frac{A_s}{A_c} \right)$$

Therefore, Form 1 becomes:

$$x(t) = A[\cos(\phi)\cos(\omega t) - \sin(\phi)\sin(\omega t)]$$

Geometric Relationship	
Geom 1	$A = \sqrt{A_c^2 + A_s^2}$
Geom 2	$\phi = \tan^{-1}\left(\dfrac{A_s}{A_c}\right)$
Geom 3	$A_c = A\cos(\phi)$
Geom 4	$A_s = A\sin(\phi)$

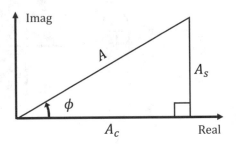

Use Math 3, $\cos(a + b) = \cos(a)\cos(b) - \sin(a)\sin(b)$, to transfer Form 1 to Form 2 as:

$$x(t) = A\cos(\omega t + \phi) \quad \text{(Form 2:}R_{EP}\text{)}$$

Note that in Form 2, A is a combined amplitude of the cosine function, and ϕ is the phase shift of the simple harmonic motion.

1.5.2   Obtain Form 3 from Form 2

Form 2 represents simple harmonic motion using one cosine function with a phase shift. Form 3 will represent the same simple harmonic motion using complex exponential functions.

Use Math 2, $\cos(\theta) = \frac{1}{2}\left[e^{j\theta} + e^{-j\theta}\right]$ as a bridge to transfer Form 2 to Form 3 as:

$$x(t) = \frac{1}{2}\left[Ae^{j(\omega t + \phi)} + Ae^{-j(\omega t + \phi)}\right] \quad \text{(Form 3:}C_{EP})$$

Note that, in Form 3, the right-hand side of the equation is a real number because the imaginary part is zero due to the addition of the complex conjugate pair, $e^{j(\omega t + \phi)}$ and $e^{-j(\omega t + \phi)}$.

1.5.3   Obtain Form 4 from Form 3

In Form 3, the constant phase ϕ is combined with the time variable ωt as one part of the complex exponential function, $e^{-j(\omega t + \phi)}$. In Form 4, the constant phase ϕ will be separated from the time variable ωt to become a constant coefficient of $e^{j\omega t}$.

Use Math 4, $e^{a+b} = e^a \cdot e^b$, to transfer Form 3 to get Form 3.5 as:

$$x(t) = \frac{1}{2}\left[Ae^{j\omega t}e^{j\phi} + Ae^{-j\omega t}e^{-j\phi}\right] \quad \text{(Form 3.5)}$$

Note that the above form is called Form 3.5 because phase ϕ has been separated from the time variable ωt. However, $e^{j\phi}$ and $e^{-j\phi}$ are still not expressed in Form 4 as $[A_c + jA_s]$ and $[A_c - jA_s]$.

Use Math 1, Euler's formula, to modify Form 3.5 as:

$$x(t) = \frac{1}{2}\left\{A[\cos(\phi) + j\sin(\phi)]e^{j\omega t} + A[\cos(\phi) - j\sin(\phi)]e^{-j\omega t}\right\}$$

Introducing Geom 3, $A_c = A\cos(\phi)$, and Geom 4, $A_s = A\sin(\phi)$, to the above equation yields:

$$x(t) = \frac{1}{2}\left\{[A_c + jA_s]e^{j\omega t} + [A_c - jA_s]e^{-j\omega t}\right\} \quad \text{(Form 4:}C_{IP})$$

Form 4 can also be derived from Form 1 using Math 2. The derivation is straightforward and is not shown here.

1.6 Visualization and Numerical Validation of Form 1 and Form 2

Even though the equivalency between Form 1 and Form 2 was mathematically proven in Sect. 1.5.1, the geometric equivalency between them is not obvious. The geometric equivalency between them can be visualized and observed by plotting each term of these two forms.

Before we plot Form 1 and Form 2, let's take a look at the geometric difference between these two forms. Form 1 is the result of an addition of two simple harmonic motions with two independent amplitudes A_c and A_s. Contrast to Form 1, Form 2 is just a harmonic motion with an amplitude A but with a phase ϕ as relisted below:

(Form 1) $A_c \cos(\omega t) - A_s \sin(\omega t)$ Real trigonometric function
(Form 2) $A \cos(\omega t + \phi)$ Real trigonometric function

where A, A_c, A_s, and ϕ are real numbers and can be related using the following formula:

$$A = \sqrt{A_c^2 + A_s^2}; \phi = \tan^{-1}\left(\frac{A_s}{A_c}\right); A_c = A\cos(\phi); A_s = A\sin(\phi)$$

The equivalency of Form 1 and Form 2 can be observed by plotting them. Also, the equivalency of Form 1 and Form 2 can be numerically validated by adding the cosine function, $A_c \cos(\omega t)$, and sine function, $-A_s \sin(\omega t)$, to get $A \cos(\omega t + \phi)$ as shown in the figure below.

For the purposes of demonstration, the following parameters:

$$\omega = 2\pi \left[\frac{rad}{sec}\right]$$

$$A = 2$$

$$\phi = \frac{-1}{6}\pi[rad]$$

are used to plot Form 1 and Form 2 for a time duration of 2 seconds with a time increment of 1/8 of a period T as shown in the four figures below:

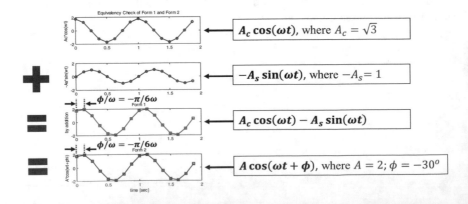

The four figures above show that the results from the addition of a cosine and sine functions (Form 1) are the same as a cosine function (Form 2) with a phase ϕ.

Students are encouraged to regenerate these four figures with their preferred tools such as MATLAB, Python, or Excel spreadsheet. After regenerating these four figures, they can change the values of the three parameters ω, A, and ϕ to three arbitrary numbers for generating a new set of four figures. The results calculated from Form 1 and Form 2 should be the same for any given parameters ω, A, and ϕ. This serves a numerical validation of equivalency between Form 1 and Form 2.

The exercise aims to gain deeper insight into the equivalency of Form 1 and Form 2 by visualization and numerical validation. A MATLAB code for generating the above four figures is provided in Sect. 1.10.

Example 1.1

A simple harmonic motion is expressed in Form 1 by a cosine and a sine function as:

$$x(t) = \sqrt{3}\cos(5t) + \sin(5t)$$

Express this simple harmonic motion in Form 2, Form 3, and Form 4 to complete the four equivalent forms as shown below:

(Form 1)	$A_c \cos(\omega t) - A_s \sin(\omega t)$	Real trigonometric function
(Form 2)	$A \cos(\omega t + \phi)$	Real trigonometric function
(Form 3)	$\frac{A}{2}\left[e^{j(\omega t+\phi)} + e^{-j(\omega t+\phi)}\right]$	Complex conjugate function pair
(Form 4)	$\frac{1}{2}\left[(A_c + jA_s)e^{j\omega t} + (A_c - jA_s)e^{-j\omega t}\right]$	Complex conjugate function pair

where A, A_c, A_s, and ϕ are real numbers and can be related using the following formula:

$$A = \sqrt{A_c^2 + A_s^2}; \phi = \tan^{-1}\left(\frac{A_s}{A_c}\right); A_c = A\cos(\phi); A_s = A\sin(\phi)$$

(a) Solve this problem *without using the formula*.
(b) Solve this problem *using the formula*.

Example 1.1 Solution

Solving the problem without using the formula

Form **1** is given as:

$$x(t) = \sqrt{3}\cos(5t) + \sin(5t) \qquad\qquad (\text{Form } 1:R_{IP})$$

$$= \sqrt{\left(\sqrt{3}\right)^2 + (1)^2}\left(\frac{\sqrt{3}}{\sqrt{\left(\sqrt{3}\right)^2 + (1)^2}}\cos(5t) + \frac{1}{\sqrt{\left(\sqrt{3}\right)^2 + (1)^2}}\sin(5t)\right)$$

$$= 2\left(\frac{\sqrt{3}}{2}\cos(5t) + \frac{1}{2}\sin(5t)\right)$$

Let:

$$\cos(\phi) = \frac{\sqrt{3}}{2} \text{ and } \sin(\phi) = \frac{-1}{2}$$

then:

$$\phi = \tan^{-1}\left(\frac{-1}{\sqrt{3}}\right) = -30^o = -\frac{1}{6}\pi \text{ [rad]or } \frac{11}{6}\pi \text{ [rad]}$$

Geometric Relationship	
Geom 1	$A = \sqrt{A_c^2 + A_s^2}$
Geom 2	$\phi = \tan^{-1}\left(\frac{A_s}{A_c}\right)$
Geom 3	$A_c = A\cos(\phi)$
Geom 4	$A_s = A\sin(\phi)$

$A_c = \sqrt{3}$

$\phi = -30^o$

$= -\pi/6$

$A = 2$

$A_s = -1$

Therefore:

$$x(t) = 2\left[\cos(5t)\cos\left(-\frac{\pi}{6}\right) - \sin(5t)\sin\left(-\frac{\pi}{6}\right)\right]$$
$$= 2\cos\left(5t - \frac{\pi}{6}\right) \qquad\qquad\qquad \text{(Form 2:}R_{EP})$$

Use Math 2 to get the Form 3:

$$x(t) = \left[e^{j\left(5t-\frac{\pi}{6}\right)} + e^{-j\left(5t-\frac{\pi}{6}\right)}\right] \quad \text{(Form 3:}C_{EP})$$

For Form 4:

$$x(t) = \frac{1}{2}\left[2e^{-j\left(\frac{\pi}{6}\right)}e^{j(5t)} + 2e^{j\left(\frac{\pi}{6}\right)}e^{-j(5t)}\right] \qquad\qquad \text{(Form 3.5)}$$
$$= \frac{1}{2}\left\{2\left[\cos\left(\frac{\pi}{6}\right) - j\sin\left(\frac{\pi}{6}\right)\right]e^{j(5t)} + 2\left[\cos\left(\frac{\pi}{6}\right) + j\sin\left(\frac{\pi}{6}\right)\right]e^{-j(5t)}\right\}$$
$$= \frac{1}{2}\left[(\sqrt{3} - j)e^{j(5t)} + (\sqrt{3} + j)e^{-j(5t)}\right] \qquad \text{(Form 4:}C_{IP})$$

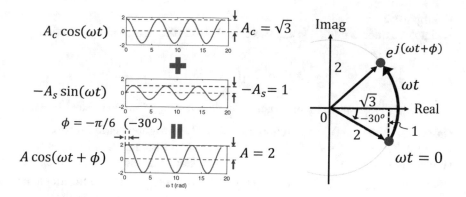

(b) Solving the problem using the formula

Compare the given simple harmonic motion (Form 1) with Form 1 formulas:

$x(t)$ $= \sqrt{3} \cos (5t) + \sin (5t)$ (Form 1:R_{IP})
(Form 1) $A_c \cos (\omega t) - A_s \sin (\omega t)$ Real trigonometric function

Therefore:

$$\omega = 5$$
$$A_c = \sqrt{3}$$
$$A_s = -1$$

then:

$$A = \sqrt{A_c^2 + A_s^2} = \sqrt{\left(\sqrt{3}\right)^2 + (-1)^2} = 2$$

$$\phi = \tan^{-1}\left(\frac{A_s}{A_c}\right) = \tan^{-1}\left(\frac{-1}{\sqrt{3}}\right) = -\frac{\pi}{6} \text{ [rad]or } \frac{11}{6} \pi \text{ [rad]}$$

Therefore, we have:

$x(t)$ $= A \cos (\omega t + \phi)$
      $= 2 \cos \left(5t - \frac{\pi}{6}\right)$ (Form 2:R_{EP})
$x(t)$ $= \frac{1}{2} A \left[e^{j(\omega t + \phi)} + e^{-j(\omega t + \phi)}\right]$
      $= \left[e^{j\left(5t - \frac{\pi}{6}\right)} + e^{-j\left(5t - \frac{\pi}{6}\right)}\right]$ (Form 3:C_{EP})
$x(t)$ $= \frac{1}{2}\left\{(A_c + jA_s)e^{j\omega t} + (A_c - jA_s)e^{-j\omega t}\right\}$
      $= \frac{1}{2}\left[(\sqrt{3} - j)e^{j(5t)} + (\sqrt{3} + j)e^{-j(5t)}\right]$ (Form 4:C_{IP})

Example 1.2

A simple harmonic motion is expressed as the real part of a complex function:

$$x(t) = Re\left[(-2 + j3)e^{j4t}\right]$$

Express this simple harmonic motion in the following four equivalent forms as:

(Form 1) $A_c \cos(\omega t) - A_s \sin(\omega t)$ Real trigonometric function
(Form 2) $A \cos(\omega t + \phi)$ Real trigonometric function
(Form 3) $\frac{A}{2}\left[e^{j(\omega t+\phi)} + e^{-j(\omega t+\phi)}\right]$ Complex conjugate function pair
(Form 4) $\frac{1}{2}\left[(A_c + jA_s)e^{j\omega t} + (A_c - jA_s)e^{-j\omega t}\right]$ Complex conjugate function pair

where A, A_c, A_s, and ϕ can be related using the following formula:

$$A = \sqrt{A_c^2 + A_s^2}; \quad \phi = \tan^{-1}\left(\frac{A_s}{A_c}\right); \quad A_c = A\cos(\phi); A_s = A\sin(\phi)$$

(a) Solve this problem *without using the formula*.
(b) Solve this problem *using the formula*.

Example 1.2 Solution
(a) **Solving the problem without using the formula**

The given simple harmonic motion is formulated as the real part of a complex function. A real number can be expressed as half of the summation of a complex conjugate pair. Therefore, we can express the given simple harmonic motion (real part of a complex function) as half of the summation of a complex conjugate pair (Form 4) as:

$$
\begin{aligned}
x(t) &= Re\left[(-2 + j3)e^{j4t}\right] && \text{(Form 4')}\\
&= \tfrac{1}{2}\left[(-2 + j3)e^{j4t} + (-2 - j3)e^{-j4t}\right] && \text{(Form 4:}C_{IP})\\
&= \tfrac{1}{2}\left[(-2 + j3)e^{j4t} + cc\right]
\end{aligned}
$$

Let:

$$\cos\alpha = \frac{-2}{\sqrt{13}} \text{ and } \sin\alpha = \frac{3}{\sqrt{13}}$$

Then:

$$\alpha = \tan^{-1}\left(\frac{3}{-2}\right) = 2.159 \text{ [rad]}$$

and

$x(t)$		
	$= \frac{1}{2}\left[\sqrt{13}(e^{j\alpha})e^{\,j4t} + cc\right]$	
	$= \frac{1}{2}\left[\sqrt{13}e^{\,j2.159}e^{\,j4t} + cc\right]$	(Form 3.5)
	$= \frac{1}{2}\left[\sqrt{13}e^{\,j(4t+2.159)} + cc\right]$	
	$= \frac{1}{2}\left[\sqrt{13}e^{\,j(4t+2.159)} + \sqrt{13}e^{-j(4t+2.159)}\right]$	(Form 3:C_{EP})
	$= Re\left(\sqrt{13}e^{\,j(4t+2.159)}\right)$	(Form 3')

Form 2 can be derived from Form 3 as:

$$x(t) = \sqrt{13}\cos(4t + 2.159) \quad \text{(Form 2:}R_{EP})$$

Form 1 can be derived from Form 2 as:

$$
\begin{aligned}
x(t) &= \sqrt{13}(\cos 4t \cdot \cos(2.159) - \sin 4t \cdot \sin(2.159)) \\
&= \sqrt{13}(\cos 4t \cdot \cos(\alpha) - \sin 4t \cdot \sin(\alpha)) \\
&= -2\cos 4t - 3\sin 4t \qquad\qquad\qquad\text{(Form 1:}R_{IP})
\end{aligned}
$$

(b) Solving the problem using the formula

The given simple harmonic motion is formulated as the real part of a complex function. A real number can be expressed as half of the summation of a complex conjugate pair. Therefore, we can express the given simple harmonic motion (real part of a complex function) as half of the summation of a complex conjugate pair (Form 4) as:

$$
\begin{aligned}
x(t) &= Re\left[(-2 + j3)e^{\,j4t}\right] \qquad\qquad \text{(Form 4')} \\
&= \frac{1}{2}\left[(-2 + j3)e^{\,j4t} + (-2 - j3)e^{-j4t}\right] \quad \text{(Form 4:}C_{IP})
\end{aligned}
$$

Compare the above simple harmonic motion (Form 4) with Form 4 formulas:

$$x(t) = \frac{1}{2}\left\{[(-2 + j3)e^{\,j4t}] + (-2 - j3)e^{-j4t}\right\} \qquad\qquad \text{(Form 4:}C_{IP})$$

(Form 4) $\frac{1}{2}[(A_c + jA_s)e^{j\omega t} + (A_c - jA_s)e^{-j\omega t}]$ Complex conjugate function pair

Therefore, we have:

$$\omega = 4$$
$$A_c = -2$$
$$A_s = 3$$

then:

$$A = \sqrt{A_c^2 + A_s^2} = \sqrt{(-2)^2 + (3)^2} = \sqrt{13}$$

$$\phi = \tan^{-1}\left(\frac{A_s}{A_c}\right) = \tan^{-1}\left(\frac{3}{-2}\right) = 2.159 \text{ [rad]}$$

Therefore, we have:

$$
\begin{aligned}
x(t) \ &= A_c \cos{(\omega t)} - A_s \sin{(\omega t)} \\
&= -2 \cos{(4t)} - 3 \sin{(4t)} && \text{(Form 1:} R_{IP}\text{)} \\
x(t) \ &= A \cos{(\omega t + \phi)} \\
&= \sqrt{13} \cos{(4t + 2.159)} && \text{(Form 2:} R_{EP}\text{)} \\
x(t) \ &= \tfrac{1}{2} A \left[e^{j(\omega t + \phi)} + e^{-j(\omega t + \phi)} \right] \\
&= \tfrac{1}{2} \left[\sqrt{13} e^{j(4t + 2.159)} + \sqrt{13} e^{-j(4t + 2.159)} \right] && \text{(Form 3:} C_{EP}\text{)}
\end{aligned}
$$

1.7 Space-Time Harmonic Functions Expressed in Four Equivalent Forms

Four Equivalent forms of the time harmonic function $p(t)$ can be extended to express space-time harmonic motions $p_\pm(x, t)$ by simply adding one term, $\mp kx$, after ωt as:

(Form 1:R_{IP}) $A_c \cos{(\omega t \mp kx)} - A_s \sin{(\omega t \mp kx)}$
(Form 2:R_{EP}) $A \cos{(\omega t \mp kx + \phi)}$
(Form 3:C_{EP}) $\frac{A}{2} \left[e^{j(\omega t \mp kx + \phi)} + e^{-j(\omega t \mp kx + \phi)} \right]$
(Form 4:C_{IP}) $\frac{1}{2} \left[(A_c + jA_s) e^{j(\omega t \mp kx)} + (A_c - jA_s) e^{-j(\omega t \mp kx)} \right]$

Note that Form 1 and Form 2 are formulated using *r*eal numbers (R). Form 3 and Form 4 are formulated using *c*omplex numbers (C). In addition, Form 2 and Form 3 are formulated using *e*xplicit *p*hase (EP) of phase ϕ. Form 1 and Form 4 are formulated using *i*mplicit *p*hase (IP) of ϕ. According to these properties, the four forms above can be rearranged into the following table:

Real Trigonometric Function	Complex Conjugate Function Pair
$A_{\pm c} \cos{(\omega t \mp kx)}$ $-A_{\pm s} \sin{(\omega t \mp kx)}$ Form 1: R_{IP}	$\frac{1}{2} [(A_{\pm c} + jA_{\pm s}) e^{j(\omega t \mp kx)}$ $+ (A_{\pm c} - jA_{\pm s}) e^{-j(\omega t \mp kx)}]$ Form 4: C_{IP}
$A_\pm \cos{(\omega t \mp kx + \phi_\pm)}$ Form 2: R_{EP}	$\frac{1}{2} \left[A_\pm e^{j(\omega t \mp kx + \phi_\pm)} + A_\pm e^{-j(\omega t \mp kx + \phi_\pm)} \right]$ Form 3: C_{EP}

where $A_\pm$, $A_{\pm c}$, $A_{\pm s}$, and $\phi_\pm$ are real numbers related by the following geometry relationships:

Geometric Relationship	
Geom 1	$A_\pm = \sqrt{A_{\pm c}^2 + A_{\pm s}^2}$
Geom 2	$\phi_\pm = \tan^{-1}\left(\dfrac{A_{\pm s}}{A_{\pm c}}\right)$
Geom 3	$A_{\pm c} = A_\pm \cos(\phi_\pm)$
Geom 4	$A_{\pm s} = A_\pm \sin(\phi_\pm)$

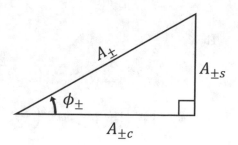

Example 1.3

A space-time harmonic function $p(x,t)$ is given as:

$$p(x,t) = Re\left[(-2+j3)e^{\,j(4t-5x)}\right]$$

Express this harmonic motion in the four equivalent forms as:

(Form 1:R_{IP}) $A_c \cos(\omega t \mp kx) - A_s \sin(\omega t \mp kx)$
(Form 2:R_{EP}) $A \cos(\omega t \mp kx + \phi)$
(Form 3:C_{EP}) $\frac{A}{2}\left[e^{\,j(\omega t \mp kx + \phi)} + e^{-j(\omega t \mp kx + \phi)}\right]$
(Form 4:C_{IP}) $\frac{1}{2}\left[(A_c + jA_s)e^{\,j(\omega t \mp kx)} + (A_c - jA_s)e^{-j(\omega t \mp kx)}\right]$

A, A_c, A_s, and ϕ are real numbers related using the following formula:

$$A = \sqrt{A_c^2 + A_s^2}; \quad \phi = \tan^{-1}\left(\frac{A_s}{A_c}\right); \quad A_c = A\cos(\phi); \quad A_s = A\sin(\phi)$$

Example 1.3 Solution

Compare the given harmonic motion (in Form 4:C_{IP}) with the Form 4 formulas:

$x(t)$ $= \frac{1}{2}\left[(-2+j3)e^{\,j(4t-5x)} + (-2-j3)e^{-j(4t-5x)}\right]$ (Form 4)
(Form 4) $\frac{1}{2}\left[(A_c + jA_s)e^{\,j(\omega t - kx)} + (A_c - jA_s)e^{-j(\omega t - kx)}\right]$

Therefore, we have:

$$\omega = 4$$

$$k = 5$$

$$A_c = -2$$

$$A_s = 3$$

Then:

$$A = \sqrt{A_c^2 + A_s^2} = \sqrt{(-2)^2 + (3)^2} = \sqrt{13}$$

$$\phi = \tan^{-1}\left(\frac{A_s}{A_c}\right) = \tan^{-1}\left(\frac{3}{-2}\right) = 2.159 \text{ [rad]} \text{ or } -4.124 \text{ [rad]}$$

Therefore, we have:

$x(t)$ $= A \cos(\omega t - kx + \phi)$

$\quad = \sqrt{13} \cos(4t - 5x + 2.159)$ (Form 2)

$x(t)$ $= \frac{1}{2}A\left[e^{j(\omega t - kx + \phi)} + e^{-j(\omega t - kx + \phi)}\right]$

$\quad = \frac{1}{2}\left[\sqrt{13}e^{j(4t - 5x + 2.159)} + \sqrt{13}e^{-j(4t - 5x + 2.159)}\right]$ (Form 3)

$x(t)$ $= A_c \cos(\omega t - kx) - A_s \sin(\omega t - kx)$

$\quad = -2 \cos(4t - 5x) - 3 \sin(4t - 5x)$ (Form 1)

1.8 Homework Exercises

Four equivalent forms as shown can be used to describe simple harmonic vibration:

(Form 1) $A_c \cos(\omega t) - A_s \sin(\omega t)$ Real trigonometric function
(Form 2) $A \cos(\omega t + \phi)$ Real trigonometric function
(Form 3) $\frac{1}{2}\left[Ae^{j(\omega t + \phi)} + Ae^{-j(\omega t + \phi)}\right]$ Complex conjugate function pair
(Form 4) $\frac{1}{2}\left[(A_c + jA_s)e^{j\omega t} + (A_c - jA_s)e^{-j\omega t}\right]$ Complex conjugate function pair

where A, A_c, A_s, and ϕ are real numbers and can be related using the following formula:

$$A = \sqrt{A_c^2 + A_s^2}; \phi = \tan^{-1}\left(\frac{A_s}{A_c}\right); A_c = A \cos(\phi); A_s = A \sin(\phi)$$

Exercise 1.1
Express the following simple harmonic motion (Form 2) in the rest of the four equivalent forms:

(Form 2) $x(t) = 10 \cos(2\beta t - 105°)$

(a) Solve this problem *without using the formula*.
(b) Solve this problem *using the formula*.

(Answers): (You must show all of your work for full credit!)

(Form 1) $x(t) = -2.588 \cos(2\beta t) + 9.659 \sin(2\beta t)$
(Form 3) $x(t) = \frac{1}{2}\left[10e^{\,j(2\beta t - 1.83)} + cc\right]$
(Form 4) $x(t) = \frac{1}{2}\left[(-2.588 - j9.659)e^{\,j2\beta t} + cc\right]$

Exercise 1.2

Express following simple harmonic motion in the four equivalent forms:

$$x(t) = \text{Re}\left[(-1+j)e^{\left(\,j3t-45^\circ\right)}\right]$$

(a) Solve this problem *without using the formula.*
(b) Solve this problem *using the formula.*

(Answers): (You must show all of your work for full credit!)

(Form 1) $x(t) = -e^{-\frac{\pi}{4}}\cos(3t) - e^{-\frac{\pi}{4}}\sin(3t)$
(Form 2) $x(t) = \sqrt{2}\,e^{-\frac{\pi}{4}}\cos\left(3t + \frac{3\pi}{4}\right)$
(Form 3) $x(t) = \frac{1}{2}\left[\sqrt{2}e^{-\frac{\pi}{4}}e^{\,j\left(3t+\frac{3\pi}{4}\right)} + cc\right]$
(Form 4) $x(t) = \frac{1}{2}\left[(-e^{-\frac{\pi}{4}} + je^{-\frac{\pi}{4}})e^{\,j3t} + cc\right]$

Exercise 1.3

Express the following simple harmonic motion (Form 1) in the rest of the four equivalent forms:

(Form 1) $x(t) = -\cos(3t) - 2\sin(3t)$

(a) Solve this problem *without using the formula.*
(b) Solve this problem *using the formula.*

(Answers): (You must show all of your work for full credit!)

(Form 2) $x(t) = \sqrt{5}[\cos(3t + 2.03)]$
(Form 3) $x(t) = \frac{\sqrt{5}}{2}\left[\left(e^{\,j(3t+2.03)}\right) + \left(e^{\,j(3t+2.03)}\right)^*\right]$
(Form 4) $x(t) = \frac{1}{2}\left[(-1+j2)e^{\,j(3t)} + cc\right]$

Exercise 1.4

Express the following simple harmonic motion in the four equivalent forms:

$$x(t) = \cos 5t + 4\sqrt{2}\sin\left(5t - \frac{\pi}{4}\right)$$

(a) Solve this problem *without using the formula.*
(b) Solve this problem *using the formula.*

(Answers): (You must show all of your work for full credit!)

(Form 1) $-3 \cos 5t + 4 \sin 5t$

(Form 2) $5(\cos(5t - 2.214))$

(Form 3) $\frac{1}{2}\left[5e^{\,j(5t-2.214)} + 5e^{-j(5t-2.214)}\right]$ or $\frac{1}{2}\left[5e^{\,j(5t-2.214)} + cc\right]$

(Form 4) $\frac{1}{2}\left[(-3 - j4)e^{\,j5t} + (-3 + j4)e^{-j5t}\right]$ or $\frac{1}{2}\left[(-3 - j4)e^{\,j5t} + cc\right]$

Exercise 1.5

Express the following harmonic motion in the four equivalent forms:

$$x(x, t) = \cos(5t - 7x) + 4\sqrt{2}\sin\left(5t - 7x - \frac{\pi}{4}\right)$$

(Answers): (You must show all your work for full credit!)

(Form 1) $-3 \cos(5t - 7x) + 4 \sin(5t - 7x)$

(Form 2) $5(\cos(5t - 7x - 2.214))$

(Form 3) $\frac{1}{2}\left[5e^{\,j(5t-7x-2.214)} + 5e^{-j(5t-7x-2.214)}\right]$

(Form 4) $\frac{1}{2}\left[(-3 - j4)e^{\,j(5t-7x)} + (-3 + j4)e^{-j(5t-7x)}\right]$

Exercise 1.6

A forward traveling wave $p_+(x, t)$ with phase shift is given as:

$$p_+(x, t) = Re\left[\sqrt{13}\,e^{\,j(4t-5x-4.124)}\right]$$

Express this harmonic motion in the following four equivalent forms:

(Form 1) $p_\pm(x, t) = A_{\pm c} \cos(\omega t \mp kx) - A_{\pm s} \sin(\omega t \mp kx)$

(Form 2) $p_\pm(x, t) = A_\pm \cos(\omega t \mp kx + \phi_\pm)$

(Form 3) $p_\pm(x, t) = \frac{1}{2}A_\pm\left[e^{\,j(\omega t \mp kx + \phi_\pm)} + e^{-j(\omega t \mp kx + \phi_\pm)}\right]$

(Form 4) $p_\pm(x, t) = \frac{1}{2}\left[(A_{\pm c} + jA_{\pm s})e^{\,j(\omega t \mp kx)} + (A_{\pm c} - jA_{\pm s})e^{-j(\omega t \mp kx)}\right]$

where $A_{\pm c}, A_{\pm s}, A_\pm$, and $\phi_\pm$ are real numbers and can be related using the following formula:

$$A_\pm = \sqrt{A_{\pm c}^2 + A_{\pm s}^2};\ \ \phi_\pm = \tan^{-1}\left(\frac{A_{\pm s}}{A_{\pm c}}\right);$$

$$A_{\pm c} = A_\pm \cos(\phi_\pm); A_{\pm s} = A_\pm \sin(\phi_\pm)$$

(Answers): (You must show all your work for full credit!)

(Form 1) $p_+(x, t) = -2 \cos(4t - 5x) - 3 \sin(4t - 5x)$

(Form 2) $p_+(x, t) = \sqrt{13} \cos(4t - 5x - 4.124)$

(Form 3) $p_+(x, t) = \frac{1}{2}\left[\sqrt{13}e^{\,j(4t-5x-4.124)} + \sqrt{13}e^{-j(4t-5x-4.124)}\right]$

(Form 4) $p_+(x,t) = \frac{1}{2}\left\{\left[(-2+j3)e^{j(4t-5x)}\right] + (-2-j3)e^{-j(4t-5x)}\right\}$

Exercise 1.7

Express the following simple harmonic motion in the four equivalent forms:

$$x(t) = \cos 5t - 4\sqrt{2}\sin\left(5t + \frac{3\pi}{4}\right)$$

(Answers): (You must show all your work for full credit!)

(Form 1) $= -3\cos 5t + 4\sin 5t$
(Form 2) $= 5\cos(5t - 2.214)$
(Form 3) $= \frac{1}{2}\left[5e^{j(5t-2.214)} + 5e^{-j(5t-2.214)}\right]$
(Form 4) $= \frac{1}{2}\left[(-3-j4)e^{j5t} + (-3+j4)e^{-j5t}\right]$

Exercise 1.8

Express the following harmonic motion in the four equivalent forms:

$$x(x,t) = Re\left[(-4+j3)e^{j45^\circ}e^{j(2t-5x)}\right]$$

(Answers): (You must show all your work for full credit!)

(Form 1) $\frac{-7\sqrt{2}}{2}\cos(2t - 5x) + \frac{\sqrt{2}}{2}\sin(2t - 5x)$
(Form 2) $5\cos(2t - 5x - 3)$
(Form 3) $\frac{5}{2}\left[e^{j(2t-5x-3)} + e^{-j(2t-5x-3)}\right]$
(Form 4) $\frac{1}{4}\left[(-7\sqrt{2} - j\sqrt{2})e^{j(2t-5x)} + (-7\sqrt{2} + j\sqrt{2})e^{-j(2t-5x)}\right]$

1.9   References of Trigonometric Identities

A collection of trigonometric identities is shown in this section for ease of reference. The derivations of most of these formulas are not difficult but are prone to small mistakes. The goal of this collection is to increase confidence and reduce mistakes when working with trigonometric functions and complex exponential functions for harmonic motion.

1.9.1   Trigonometric Identities of a Single Angle

Trigonometric identities of a single angle can be found by the unit circle:

$$\cos(\theta) = \cos(-\theta)$$
$$\sin(\theta) = -\sin(-\theta)$$

Some identities of trigonometric functions can be found by playing around with the unit circle:

$$\sin(\theta) = -\sin(\theta - \pi) = -\sin(\theta + \pi)$$
$$\cos(\theta) = -\cos(\theta - \pi) = -\cos(\theta + \pi)$$

The sine of any angle is equal to the cosine of its complementary angle. The relationship between the sine and cosine of complementary angles is:

$$\cos(\theta) = \sin\left(\frac{\pi}{2} - \theta\right) = -\sin\left(\theta - \frac{\pi}{2}\right) = \sin\left(\theta + \frac{\pi}{2}\right)$$

$$\sin(\theta) = \cos\left(\frac{\pi}{2} - \theta\right) = \cos\left(\theta - \frac{\pi}{2}\right)$$

Euler's formula:

$$e^{\pm j\theta} = \cos(\theta) \pm j \sin(\theta)$$

or:

$$e^{j\theta} = \cos(\theta) + j \sin(\theta);$$
$$e^{-j\theta} = \cos(\theta) - j \sin(\theta)$$

The reverse of Euler's formula:

$$\cos(\theta) = \frac{1}{2}\left(e^{j\theta} + e^{-j\theta}\right) = \mathrm{Re}\left[e^{j\theta}\right]$$

$$\sin(\theta) = \frac{1}{2j}\left(e^{j\theta} - e^{-j\theta}\right) = -\frac{1}{2}\left(je^{j\theta} - je^{-j\theta}\right) = -\mathrm{Re}\left[je^{j\theta}\right]$$

Sine and cosine functions can be expressed in complex exponential functions as:

$$\begin{aligned}
\cos(\theta) &= \tfrac{1}{2}\left(e^{j\theta} + e^{-j\theta}\right) = Re\left[e^{j\theta}\right]\\
\sin(\theta) &= \cos\left(\tfrac{\pi}{2} - \theta\right) = \tfrac{1}{2}\left(e^{j\left(\frac{\pi}{2}-\theta\right)} + cc\right) = Re\left[e^{j\left(\frac{\pi}{2}-\theta\right)}\right]\\
&= \cos\left(\theta - \tfrac{\pi}{2}\right) = \tfrac{1}{2}\left(e^{j\left(\theta-\frac{\pi}{2}\right)} + cc\right) = Re\left[e^{j\left(\theta-\frac{\pi}{2}\right)}\right]\\
\sin(\theta) &= \tfrac{1}{2j}\left(e^{j\theta} - e^{-j\theta}\right) = -\tfrac{1}{2}\left(je^{j\theta} - je^{-j\theta}\right) = -Re\left[je^{j\theta}\right]\\
&= \tfrac{1}{2}\left(je^{-j\theta} - je^{j\theta}\right) = Re\left[je^{-j\theta}\right]
\end{aligned}$$

1.9.2 Trigonometric Identities of Two Angles

Some trigonometric identities of two angles are:

$$\cos(a \pm b) = \cos(a)\cos(b) \mp \sin(a)\,\sin(b)$$
$$\sin(a \pm b) = \sin(a)\cos(b) \pm \cos(a)\,\sin(b)$$

Reverse:

$$\cos(a)\cos(b) = \frac{1}{2}[\cos(a+b) + \cos(a-b)]$$

$$\sin(a)\,\sin(b) = \frac{1}{2}[\cos(a-b) - \cos(a+b)]$$

$$\sin(a)\cos(b) = \frac{1}{2}[\sin(a+b) + \sin(a-b)]$$

$$\cos(a)\sin(b) = \frac{1}{2}[\sin(a+b) - \sin(a-b)]$$

1.10 A MATLAB Code for Visualization of Form 1 and Form 2

```
function ANC_PRJ011_PlotTwoEquivalentForms
clear all
%-----------------------------%
% Section 1: Define Variables and Parameters
%-----------------------------%
% define the parameters of the wave function
% f(t) = Ac cos(wt) - As sin(wt)     (Form 1)
%      = A cos(wt + phi)             (Form 2)
Ac=3^(1/2);  % the coefficient of the cosine function in Form 1
As=-1;       % the coefficient of the sine function in Form 1
w=2*pi/1;    % the angular frequency [rad/sec] in Form 1 and Form 2
% define the parameters of time
% Use the period (T) to determine the time increment (TimeIncrement)
% where period T = 2 * pi / w
nDataPerPeriod= 8;   % number of data per period (T)
nPeriodPerFigure=2;  % total number of periods for each figure
%-----------------------------%
% Section 2: Setup
%-----------------------------%
% get period (T) and time array
T=2*pi/w;                    % period of wave [sec]
TimeIncrement=T/nDataPerPeriod;      % time increment of data [sec]
```

```
nTime=nDataPerPeriod*nPeriodPerFigure;  % total number of time data
TimeArray=(0:1:nTime-1)'*TimeIncrement; % [sec]
%————————————————————————————%
% Section 3: Calculation
%————————————————————————————%
% calculate Form 1 and Form 2
for kTime=1:1:nTime
  TimeTemp=TimeArray(kTime,1); % time used in this for-loop
  % the first term in Form 1: Ac*cos(wt)
  Term1Cosine(kTime,1)=Ac*cos(w*TimeTemp);
  % the second term in Form 1: -As*sin(wt)
  Term2Sine(kTime,1)=-As*sin(w*TimeTemp);
  % addition of the two terms in Form 1: [Ac*cos(wt)]+[-As*sin(wt)]
  Form1(kTime,1)=...
     Term1Cosine(kTime,1)+Term2Sine(kTime,1);
  % get A and phi (Form 2) from Ac and As (Form 1):
  % p(t) = Ac cos(wt) - As sin(wt)     (Form 1)
  %      = A cos(wt + phi)         (Form 2)
  % where Ac=A*cos(phi); As=A*sin(phi)
  [A,phi]=get_A_and_phi(Ac, As);
  Form2(kTime,1)=A*cos(w*TimeTemp+phi);
end
%————————————————————————————%
% Section 4: Plotting
%————————————————————————————%
% figures setup
figure(110); set(gcf,'Position',[100 100 600 800])
% plot term 1 of Form 1
subplot(4,1,1)
plot(TimeArray,Term1Cosine,'ro-','lineWidth',2)
title('Equivalency Check of Form 1 and Form 2')
ylabel('Ac*cos(wt)')
ylim([-2 2])
% plot term 2 of Form 1
subplot(4,1,2)
plot(TimeArray,Term2Sine,'ro-','lineWidth',2)
ylabel('-As*sin(wt)')
ylim([-2 2])
% plot the combined terms in Form 1
subplot(4,1,3)
plot(TimeArray,Form1,'bs-','lineWidth',2)
title('Form 1')
ylabel('by addition')
ylim([-2 2])
% plot the formula of Form 2
subplot(4,1,4)
h=plot(TimeArray,Form2,'rs-'); set(h,'LineWidth',2);
title('Form 2')
xlabel('time [sec]')
ylabel('A*cos(wt+phi)')
ylim([-2 2])
% save figure
saveas(gcf,'Figure_ANC_PRJ011_TwoEquivalentForms','emf')
```

```
end
%────────────────────────────────────%
% Section 5: Functions
%──────────────────────%
function [A, phi]=get_A_and_phi(Ac, As)
A = (Ac^2+As^2)^(1/2);
phi= atan2(As,Ac);
end
```

Chapter 2
Derivation of Acoustic Wave Equation

In the analysis of acoustics, sound waves are typically formulated as the additions of series of harmonic functions such as cosine and sine functions. Each harmonic function is a solution of an acoustic wave equation under the given boundary conditions.

Because all the solutions of sound waves are calculated from the acoustic wave equation, the acoustic wave equation is the most fundamental equation for the analysis of acoustics. The focus of this chapter will be on the derivation of the acoustic wave equation. The following is a summary of this chapter.

The one-dimensional acoustic wave equation (shown below) will be derived in detail in this chapter from the fundamental principles of physics:

$$\frac{\partial^2}{\partial x^2} p(x,t) = \frac{1}{c^2} \frac{\partial^2}{\partial t^2} p(x,t)$$

where c is the speed of sound and $p(x,t)$ is sound pressure as a function of space coordinate x and time t.

The acoustic wave equation (shown above) is derived by combining the three fundamental principles of physics: (I) Newton's laws of motion, (II) equation of continuity, and (III) equation of state. Each fundamental principle constructs a relationship between two physical quantities of a media: (i) sound pressure, (ii) air particle flow velocity, and (iii) air density as shown in the figure below.

H. Lin et al., *Lecture Notes on Acoustics and Noise Control*, https://doi.org/10.1007/978-3-030-88213-6_2

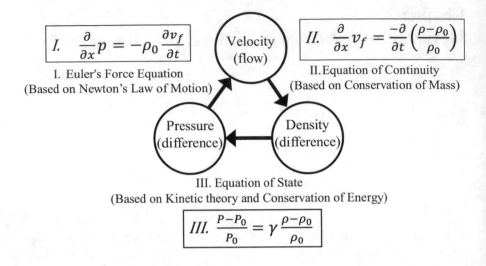

The derivation of these three fundamental principles (I–III) will be covered in this chapter and is summarized as follows: (I) Euler's force equation is based on Newton's law of motion; (II) equation of continuity is based on conservation of mass; and (III) equation of state is based on the kinetic theory of gases and conservation of energy.

After the acoustic wave equation is derived, three formulas for calculating the speed of sound are derived near the end of this chapter from Sects. 2.5.1 to 2.5.3. Three formulas for calculating the speed of sound are summarized below:

$$c = \sqrt{\frac{\gamma P_o}{\rho_o}}$$

$$c = \sqrt{\frac{B}{\rho_o}}$$

$$c = \sqrt{\gamma \frac{RT}{M}}$$

where:

c is the speed of the sound in a transmission media.
γ is the ratio of specific heats of a transmission media.
P_o is the pressure of a transmission media.
ρ_o is the density of a transmission media.
B is the bulk modulus of a transmission media.
γ is the ratio of specific heats of a transmission media.

R is the ideal gas constant.

T is the temperature in Kelvin of a transmission media.

M is the molar mass of a transmission media.

There are two goals in deriving the acoustic wave equation: (1) to understand how sound waves propagate through the air as a result of the collisions of air particles and (2) to understand the physical meaning of each tern in the acoustic wave equation.

2.1 Euler's Force Equation

Euler's force equation shows the relationship between sound pressure (pressure difference) and molecular flow velocity. Euler's force equation in Cartesian coordinates is:

$$\left(\widehat{e}_x\frac{\partial}{\partial x}+\widehat{e}_y\frac{\partial}{\partial y}+\widehat{e}_z\frac{\partial}{\partial z}\right)p(x,y,z,t)=-\rho_0\frac{\partial}{\partial t}\overrightarrow{v}_f(x,y,z,t)$$

where p is sound pressure (pressure difference); x, y, and z are coordinates; t is time; ρ_0 is the average molecular density; and $\overrightarrow{v}_f$ is the flow velocity of air molecules.

Euler's force equation in vector format is independent of coordinate systems. Euler's force equation in vector format is:

$$\nabla p(x,y,z,t)=-\rho_0\frac{\partial}{\partial t}\overrightarrow{v}_f(x,y,z,t)$$

where the gradient operator ∇ in Cartesian coordinate is a vector with bases $\widehat{e}_x,\widehat{e}_y$, and e_z:

$$\nabla\equiv\widehat{e}_x\frac{\partial}{\partial x}+\widehat{e}_y\frac{\partial}{\partial y}+\widehat{e}_z\frac{\partial}{\partial z}$$

and the flow velocity of air molecules $\overrightarrow{v}_f$ in Cartesian coordinate is a vector with bases $\widehat{e}_x,\widehat{e}_y$, and $\widehat{e}_z$:

$$\overrightarrow{v}_f(x,y,z,t)\equiv\widehat{e}_xv_{fx}(t)+\widehat{e}_yv_{fy}(t)+\widehat{e}_zv_{fz}(t)$$

For a one-dimensional plane wave, let $v_{fy}=0$, $v_{fz}=0$, and $v_{fx}=v_f$; hence, Euler's force equation in Cartesian coordinate can be simplified as:

$$\frac{\partial}{\partial x}p = -\rho_0\frac{\partial v_f}{\partial t}$$

The Euler's force equation can be derived from Newton's second law of motion. To simplify the derivation of Euler's force equation, we will use the one-dimensional plane wave in Cartesian coordinate as an example:

Molecular flow velocity: $v_f \left[\frac{distance}{unit\ time}\right]$

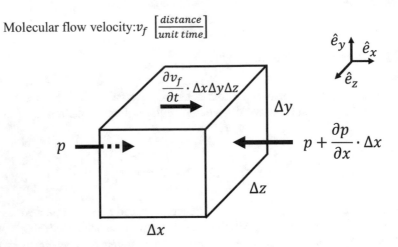

According to Newton's law of motion, the force $\vec{f}$ applied on the above cubic will cause an acceleration $\vec{a}$ on this cubic with three-dimensional lengths of Δx, Δy, and Δz. The free body diagram of the cubic is shown in the figure above.

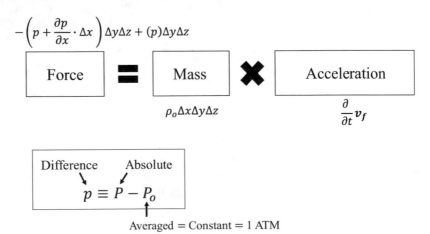

The governing equation of the cubic can be expressed as:

$$-\left(p + \frac{\partial p}{\partial x} \cdot \Delta x\right)\Delta y \Delta z + (p)\Delta y \Delta z = \rho_0 \Delta x \Delta y \Delta z \cdot \frac{\partial}{\partial t} v_f$$

It can be simplified as:

$$\left(\frac{\partial p}{\partial x} \cdot \Delta x\right)\Delta y \Delta z = -\rho_0 \Delta x \Delta y \Delta z \cdot \frac{\partial}{\partial t} v_f$$

When the above equation is compared to Newton's law of motion, the left-hand side of the equation is the force ($\frac{\partial p}{\partial x} \cdot \Delta x \Delta y \Delta z$) due to the pressure difference, and the right-hand side of the equation is the mass ($\rho_0 \Delta x \Delta y \Delta z$) multiplied by the acceleration ($\frac{\partial}{\partial t} v_f$).

Simplifying the equation above by eliminating $\Delta x \Delta y \Delta z$ yields the one-dimensional Euler's force equation in Cartesian coordinates:

$$\frac{\partial p}{\partial x} = -\rho_0 \frac{\partial}{\partial t} v_f$$

The one-dimensional Euler's force equation above can be extended to three-dimensional Euler's force as:

$$\nabla p(x, y, z, t) = -\rho_0 \frac{\partial}{\partial t} \vec{v}_f(x, y, z, t)$$

where both $\nabla p(x, y, z, t)$ and $\frac{\partial}{\partial t} \vec{v}_f(x, y, z, t)$ are vectors.

2.2   Equation of Continuity

The equation of continuity shows the relationship between molecular flow velocity and molecular density. The equation of continuity in Cartesian coordinates is:

$$\frac{\partial}{\partial t}\left(\frac{\rho - \rho_0}{\rho_0}\right) = -\left(\hat{e}_x \frac{\partial}{\partial x} + \hat{e}_y \frac{\partial}{\partial y} + \hat{e}_z \frac{\partial}{\partial z}\right) \cdot \vec{v}_f$$

where $\vec{v}_f$ is the flow velocity of air molecules; variables x, y, and z are space coordinates; t is time; ρ_0 is the averaged molecular density; and ρ is the instantaneous molecular density.

The equation of continuity in vector format is independent of coordinate systems. The equation of continuity in vector format is:

$$\frac{\partial}{\partial t}\left(\frac{\rho - \rho_0}{\rho_0}\right) = -\nabla \cdot \vec{v}_f$$

The gradient ∇ in Cartesian coordinates is a vector with bases $\widehat{e}_x, \widehat{e}_y$, and $\widehat{e}_z$:

$$\nabla = \widehat{e}_x \frac{\partial}{\partial x} + \widehat{e}_y \frac{\partial}{\partial y} + \widehat{e}_z \frac{\partial}{\partial z}$$

For a one-dimensional plane wave, the equation of continuity in Cartesian coordinate can be simplified as:

$$\frac{\partial}{\partial t}\left(\frac{\rho - \rho_0}{\rho_0}\right) = -\left(\frac{\partial}{\partial x}\right)v_f$$

The equation of continuity can be derived from the law of conservation of mass. To simplify the derivation of the equation of continuity, we will use the one-dimensional plane wave in Cartesian coordinate as an example:

Mass flow rate: $Q_\rho = \left[\rho_o \cdot v_f \frac{mass}{unit\ area \cdot unit\ time}\right]$:
Mass passing through an unit area per unit time

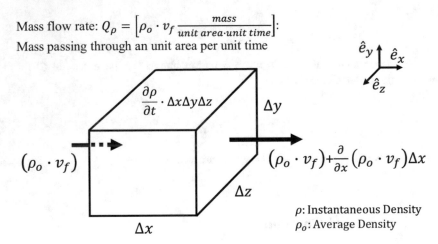

ρ: Instantaneous Density
ρ_o: Average Density

The above figure illustrates the mass coming in and out of a cubic unit. The mass flow rate, Q_ρ, is defined as the mass passing through a unit area per unit time. So, the amount of mass passing through a small area, say $\Delta y \Delta z$, is equal to $Q_\rho \Delta y \Delta z \Delta t$.

According to the law of conservation of mass, the increase of mass in the cubic unit is equal to the "mass flow in" minus "mass flow out", as shown in the figure below:

$$\frac{\partial \rho}{\partial t} \cdot \Delta t \cdot \Delta x \Delta y \Delta z \qquad\qquad \left(Q_\rho + \frac{\partial Q_\rho}{\partial x} \cdot \Delta x \right) \Delta y \Delta z \cdot \Delta t$$

Increase of mass	$=$	Mass in	$-$	Mass out

$$(Q_\rho) \cdot \Delta y \Delta z \cdot \Delta t$$

Mass flow rate

$$Q_\rho = \rho_o \cdot v_f \left[\frac{mass}{unit\ area \cdot unit\ time} \right]$$

Mass passing through
an unit area per unit time

ρ: Instantaneous Density
ρ_o: Average Density

The equation of continuity can be concluded as:

$$\frac{\partial \rho}{\partial t} \cdot \Delta t \cdot \Delta x \Delta y \Delta z = Q_\rho \cdot \Delta y \Delta z \cdot \Delta t - \left(Q_\rho + \frac{\partial Q_\rho}{\partial x} \Delta x \right) \cdot \Delta y \Delta z \cdot \Delta t$$

We will compare the above equation to the law of conservation of mass.

The left-hand side of the equation is the increase of mass in this cubic unit ($\frac{\partial \rho}{\partial t} \cdot \Delta t \cdot \Delta x \Delta y \Delta z$) after a finite time ($\Delta t$).

The right-hand side of the equation is the mass flow in ($Q_\rho \cdot \Delta y \Delta z \cdot \Delta t$) minus the mass flow out ($\left(Q_\rho + \frac{\partial Q_\rho}{\partial x} \Delta x \right) \cdot \Delta y \Delta z \cdot \Delta t$). The above equation can be simplified to:

$$\frac{\partial \rho}{\partial t} \cdot \Delta t \cdot \Delta x \Delta y \Delta z = - \left(\frac{\partial Q_\rho}{\partial x} \Delta x \right) \cdot \Delta y \Delta z \cdot \Delta t$$

and:

$$\frac{\partial \rho}{\partial t} = - \frac{\partial Q_\rho}{\partial x}$$

The mass flow Q_ρ is replaced with $\rho_o v_f$ to get:

$$\frac{\partial \rho}{\partial t} = - \frac{\partial}{\partial x} \left(\rho_o v_f \right)$$

This is the one-dimensional equation of continuity in Cartesian coordinates.

2.3   Equation of State

The third equation needed for deriving the acoustic wave equation is called the equation of state which is the most difficult equation to derive and comprehend among the three equations. This is because it involves both the kinetic theory of gases and the principle of conservation of energy. Despite this difficulty, the physics involved in the equation of state is both interesting and important for understanding acoustics.

The equation of state describes the relationship between air pressure change and mass density change as shown in the figure below:

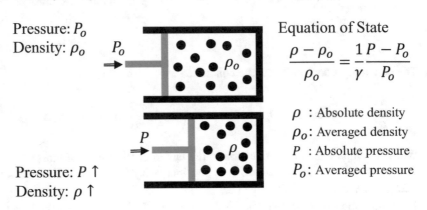

The equation of state can be expressed as:

$$\frac{P - P_o}{P_o} = \gamma \frac{\rho - \rho_o}{\rho_o}$$

where:

P_o is the average air pressure.
P is the instantaneous pressure.
ρ_o is the averaged air mass density.
ρ is the instantaneous mass density.
γ is the ratio of specific heats and is defined as:

$$\gamma \equiv \frac{3\alpha + 2}{3\alpha}$$

where α is a ratio between the total energy (rotational energy + translational energy) and the translational energy of molecules as:

$$\alpha \equiv \frac{\frac{1}{2}I_a\dot{\theta}_{ca}^2 + \frac{1}{2}m_a v_{ca}^2}{\frac{1}{2}m_a v_{ca}^2}$$

or:

$$\frac{1}{2}I_a\dot{\theta}_{ca}^2 = (\alpha - 1)\frac{1}{2}m_a v_{ca}^2$$

where m_a is the mass of a particle, I_{ca} is the mass moment of inertia, v_{ca} is the collision speed, and θ_{ca} is the spin speed of a particle.

The equation of state can be derived from (A) the law of conservation of energy and (B) pressure due to the kinetic theory of gases.

The following is a flowchart of the derivation of the equation of state:

Derivation of Equation of State

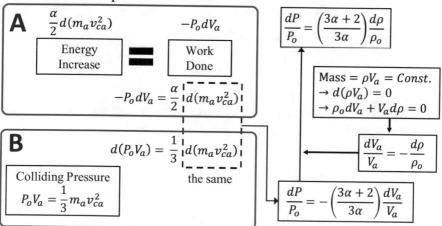

2.3.1   Energy Increase due to Work Done

The energy of a particle with mass m_a occupying a volume V_a is the summation of translational and rotational energy:

$$K_a = \frac{1}{2}m_a v_{ca}^2 + \frac{1}{2}I_a\dot{\theta}_{ca}^2$$

Assume the rotational kinetic energy is linearly related to the translational kinetic energy:

$$\frac{1}{2}I_a\dot{\theta}_{ca}^2 = (\alpha - 1)\frac{1}{2}m_a v_{ca}^2$$

where α is equal to or greater than 1. When the particle does not have rotational energy, $\alpha=1$; when the particle has rotational energy, $\alpha >1$:

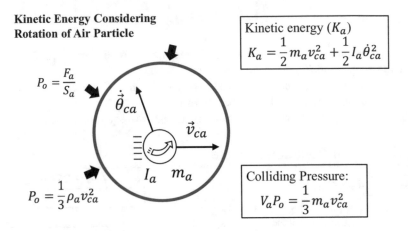

Kinetic Energy Considering Rotation of Air Particle

$$P_o = \frac{F_a}{S_a}$$

$$P_o = \frac{1}{3}\rho_a v_{ca}^2$$

Kinetic energy (K_a)
$$K_a = \frac{1}{2}m_a v_{ca}^2 + \frac{1}{2}I_a\dot{\theta}_{ca}^2$$

Colliding Pressure:
$$V_a P_o = \frac{1}{3}m_a v_{ca}^2$$

Using this linear relationship between rotational energy and translational energy, the total energy of a particle is given as follows:

$$K_a = \frac{1}{2}m_a v_{ca}^2 + \frac{1}{2}I_a\dot{\theta}_{ca}^2 = \frac{1}{2}m_a v_{ca}^2 + (\alpha - 1)\frac{1}{2}m_a v_{ca}^2 = \frac{\alpha}{2}m_a v_{ca}^2$$

According to the conservation of energy, in a closed system, the increase of energy is equal to the work done to the system:

$$\Delta K_a = \frac{\alpha}{2}d(m_a v_{ca}^2) \qquad W = -P_o dV_a$$

$$\boxed{\text{Energy Increase}} \ \mathbf{=} \ \boxed{\text{Work Done}}$$

The total energy K_a includes both translational energy and rotational energy and is equal to $\frac{\alpha}{2}m_a v_{ca}^2$, as shown before. And the work done by a constant external pressure P_o due to a system of decreased volume $-dV_a$ is $-P_o dV_a$; therefore:

$$\Delta K_a = W$$

$$\rightarrow \frac{\alpha}{2}d(m_a v_{ca}^2) = -P_o dV_a$$

2.3.2 *Pressure due to Colliding of Gases*

According to the kinetic theory, the pressure is directly related to the translational energy and is independent of the rotational energy. The pressure due to the particle collision is:

$$P_o = \frac{1}{3} \frac{m_a}{V_a} v_{ca}^2$$

The derivation of the collision pressure is based on the fundamentals of kinetic energy, as shown in the figure below:

Colliding Pressure (Kinetic theory of gases)

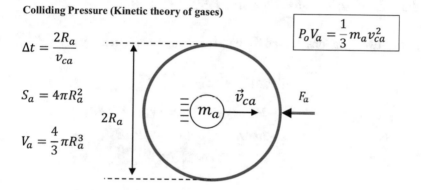

$$\Delta t = \frac{2R_a}{v_{ca}}$$

$$S_a = 4\pi R_a^2$$

$$V_a = \frac{4}{3}\pi R_a^3$$

$$P_o V_a = \frac{1}{3} m_a v_{ca}^2$$

According to Newton's law of motion, the force is equal to the change of momentum when the particle collides with the wall of the sphere as:

$$F_a = \frac{\Delta(m_a v_{ca})}{\Delta t} = \frac{2m_a v_{ca}}{\left(\frac{2R_a}{v_{ca}}\right)} = \frac{m_a v_{ca}^2}{R_a}$$

The pressure on the surface of the sphere is the force divided by the total surface area of the sphere as:

$$P_o = \frac{F_a}{S_a} = \frac{\left(\frac{m_a v_{ca}^2}{R_a}\right)}{4\pi R_a^2} = \frac{1}{3} \frac{m_a}{V_a} v_{ca}^2$$

Note that the pressure is directly related to the translational energy only (not include the rotational energy), and rewrite the relation as:

$$P_o V_a = \frac{1}{3} m_a v_{ca}^2$$

So, the change in translation energy will be:

$$d(P_o V_a) = d\left(\frac{1}{3} m_a v_{ca}^2\right)$$

2.3.3 Derivation of Equation of State

Equating two equations derived from the previous two sections by $d\left(m_a v_{ca}^2\right)$ yields:

$$\frac{-2}{\alpha} P_o dV_a = 3\, d(P_o V_a)$$

$$\rightarrow -2P_o \cdot dV_a = 3\alpha (dP \cdot V_a + P_o \cdot dV_a)$$

$$\rightarrow -3\alpha \cdot dP \cdot V_a = (3\alpha + 2)P_o \cdot dV_a$$

$$\rightarrow -\frac{dP}{P_o} = \left(\frac{3\alpha + 2}{3\alpha}\right)\frac{dV_a}{V_a}$$

Because the mass of the system remains constant, that is, $\rho_o V_a =$ constant, $\rho_o dV_a + d\rho V_a = 0$, or $\frac{dV_a}{V_a} = -\frac{d\rho}{\rho_o}$, the equation above changes to:

$$\frac{dP}{P_o} = \left(\frac{3\alpha + 2}{3\alpha}\right)\frac{d\rho}{\rho_o}$$

In the above equation, dP is the pressure difference between the absolute pressure P and the constant average pressure P_o ($P_o = 1\ atm$).

Similarly, in the above equation, $d\rho$ is the density difference between the absolute density ρ and the constant average density ρ_o.

Therefore, we can replace dP with $P - P_o$ and $d\rho$ with $\rho - \rho_o$ to get:

$$\frac{P - P_o}{P_o} = \left(\frac{3\alpha + 2}{3\alpha}\right)\frac{\rho - \rho_o}{\rho_o}$$

Finally, the ratio between the normalized pressure difference and the normalized density difference $\frac{(3\alpha + 2)}{3\alpha}$ is called the ratio of specific heats and is defined as:

$$\gamma \equiv \frac{3\alpha + 2}{3\alpha}$$

Replacing $\frac{(3\alpha+2)}{3\alpha}$ with the ratio of specific heats γ yields:

$$\frac{P - P_o}{P_o} = \gamma \frac{\rho - \rho_o}{\rho_o}$$

The ratio, γ, of specific heats is related to the ratio, α, between total energy and translation energy. The ratio, γ, for air can be measured ($\alpha = \frac{5}{3}$) and is approximately equal to:

$$\gamma = 1.4$$

Note that $\gamma = 1.4$ is approximately equivalent to $\alpha = \frac{5}{3}$ where α relates to the rotational energy and translational energy as mentioned earlier.

The ratio of specific heats, γ, is also referred to as the heat capacity ratio, the adiabatic index, adiabatic exponent, or Laplace's coefficient. The ratio of specific heats, γ, can be formulated as the heat capacity, Cp, at a constant pressure condition divided by the heat capacity, Cv, at a constant volume condition. Further discussion is beyond the scope of this course.

2.4   Derivation of Acoustic Wave Equation

The acoustic wave equation is based on the three physics principles, as introduced in the previous section.

The sound pressure can be related to the air density using the first principle:

I. Euler's force equation: $\frac{\partial}{\partial x}p = -\rho_o \frac{\partial v_f}{\partial t}$

The air density can be related to the flow velocity using the second principle:

II. The equation of continuity: $\frac{\partial}{\partial t}\left(\frac{\rho - \rho_o}{\rho_o}\right) = -\left(\frac{\partial}{\partial x}\right)v_f$

Finally, the air density can be related back to the sound pressure using the third principle:

III. The equation of state: $\frac{P - P_o}{P_o} = \gamma \frac{\rho - \rho_o}{\rho_o}$

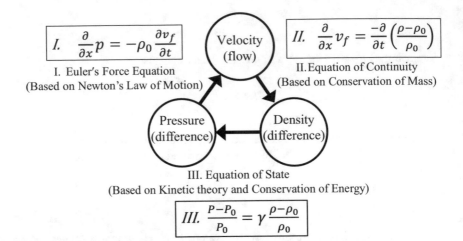

I. Euler's Force Equation
(Based on Newton's Law of Motion)

II. Equation of Continuity
(Based on Conservation of Mass)

III. Equation of State
(Based on Kinetic theory and Conservation of Energy)

From Euler's force equation, we have the relationship between pressure difference and flow velocity.

The first step is to replace the flow velocity with the air mass density so that the relationship between pressure difference and mass density can be formed. This is accomplished by combining (I) Euler's force equation and (II) the equation of continuity and eliminating the flow velocity, as shown in the figure below. Now we have a direct relationship between pressure difference and the air mass density.

The second step is to replace the air mass density with the pressure difference so that the acoustic wave equation has only one variable – the pressure difference $P - P_0$. This is accomplished by combining (III) equation of state and the relationship between the pressure difference and the air mass density from the previous step.

The third step is to replace the pressure difference with sound pressure, defined as $p = P - P_o$. This yields:

$$\frac{\partial^2}{\partial x^2}p = \frac{1}{\left(\frac{\gamma P_0}{\rho_0}\right)}\frac{\partial^2}{\partial t^2}p$$

This is the one-dimensional acoustic wave equation in Cartesian coordinates:

$$I. \quad \frac{\partial}{\partial x}p = -\rho_0 \frac{dv_f}{dt}$$

Derivation of the Acoustic Wave Equation

$$\frac{\partial^2}{\partial x^2}p = \frac{1}{\left(\frac{\gamma P_o}{\rho_o}\right)}\frac{\partial^2}{\partial t^2}p$$

$$II. \quad \frac{\partial v_f}{\partial x} = -\frac{\partial}{\partial t}\left(\frac{\rho - \rho_o}{\rho_o}\right)$$

$$III. \quad \frac{\rho - \rho_o}{\rho_o} = \frac{1}{\gamma}\frac{P - P_o}{P_o}$$

$$p = P - P_o$$

$$\frac{\partial^2}{\partial x^2}p = -\rho_0 \frac{d^2 v_f}{dxdt} = \rho_0 \frac{\partial^2}{\partial t^2}\left(\frac{\rho - \rho_o}{\rho_o}\right) = \frac{\rho_o}{\gamma}\frac{\partial^2}{\partial t^2}\left(\frac{P - P_o}{P_o}\right)$$

By introducing the speed of sound c as:

$$c^2 = \frac{\gamma P_o}{\rho_o}$$

Finally, we can summarize that the acoustic wave can be formatted in the general wave equation as:

$$\frac{\partial^2}{\partial x^2}p = \frac{1}{c^2}\frac{\partial^2}{\partial t^2}p$$

where c is the speed of the sound:

$$c = \sqrt{\frac{\gamma P_o}{\rho_o}}$$

where γ is the ratio of specific heats and the approximate value of γ for air is:

$$\gamma \equiv \frac{3\alpha + 2}{3\alpha} = 1.4$$

where α is a ratio between the total energy (rotational energy + translational energy) and the translational energy of molecules:

$$\alpha \equiv \frac{\frac{1}{2}I_a\dot{\theta}_{ca}^2 + \frac{1}{2}m_a v_{ca}^2}{\frac{1}{2}m_a v_{ca}^2}$$

The approximated value of α for air molecules is:

$$\alpha = \frac{5}{3}$$

2.5   Formulas for the Speed of Sound

The speed of the sound can be formulated using pressure, density, temperature, or bulk modulus. The following sections show three formulas for calculating the speed of sound.

2.5.1   Formula Using Pressure

In the previous section, the speed of the sound c was formulated using pressure and density as:

$$c = \sqrt{\frac{\gamma P_o}{\rho_o}}$$

where γ is the ratio of specific heats and the approximate value of γ for air is:

$\gamma = 1.4$ (for air at any temperature)

Example 2.1 Air at 1 atm, 15 C has a density:

$$\rho_o = 1.225 \left[\frac{Kg}{m^3}\right]$$

Calculate the speed of sound in air at 1 atm, 15 C based on the pressure, density, and ratio of specific heats. (Hint: Air has the ratio of specific heats $\gamma = 1.4$.)

Example 2.1 Solution The pressure at 1 atm is:

$$P_o = 1.01 \cdot 10^5 \ [Pa] \ (1 \ atm)$$

The speed of sound c at 15 C for air can be calculated based on the pressure P_o, density ρ_o, and ratio of specific heats γ as:

$$c = \sqrt{\frac{\gamma P_o}{\rho_o}} = \sqrt{\frac{1.4 \times 1.01 \cdot 10^5}{1.225}} \left[\frac{m}{s}\right] \cong 340 \left[\frac{m}{s}\right]$$

where:

$$\rho_o = 1.225 \left[\frac{Kg}{m^3}\right]$$

$$\gamma = 1.4$$

Note that in the formula for the speed of sound above, the speed c is a function of pressure P_o and density ρ_o. However, changing the pressure does not change the speed of sound because the effect from density ρ_o cancels out the effect from pressure P_o.

For example, at higher altitudes, the decrease of density is linearly proportional to the decrease of air pressure, and the speed of sound here will be the same as at sea level.

As a result, the temperature is the most significant factor in the speed of the sound and will be shown in the next section.

2.5.2 Formula Using Bulk Modulus

The speed of sound is related to the stiffness of the media. The coefficient of stiffness is also called bulk modulus and is defined as the ratio of pressure increase to volume decrease as:

$$B \equiv \frac{-dP}{\left(\frac{dV}{V_o}\right)}$$

Because the mass of the system remains constant, that is, $\rho_o V_a = \text{constant}$, $\rho_o dV_a + d\rho V_a = 0$, or $\frac{dV_a}{V_a} = -\frac{d\rho}{\rho_o}$, the equation of state can be rewritten using volume instead of density as:

$$\frac{P - P_o}{P_o} = \gamma \frac{\rho - \rho_o}{\rho_o}$$

$$\rightarrow \frac{dP}{P_o} = \gamma \frac{d\rho}{\rho_o}$$

$$\rightarrow \frac{dP}{P_o} = -\gamma \frac{dV_a}{V_a}$$

Substituting the equation above in the definition of bulk modulus yields:

$$B = \gamma P_o$$

The formula for the speed of sound (based on bulk modulus) can be obtained by substituting the bulk modulus into the formula for the speed of sound (based on pressure) as:

$$c = \sqrt{\frac{\gamma P_o}{\rho_o}}$$

$$\rightarrow c = \sqrt{\frac{B}{\rho_o}}$$

where B is the bulk modulus and ρ_o is the average density of the media.

The formula above for the speed of sound using bulk modulus is valid for both gases and liquids.

Example 2.2 Air at 1 atm, 15 C has a bulk modulus B and density ρ_o as:

$$B = 1.414 \times 10^5 \ [Pa]$$

$$\rho_o = 1.2 \left[\frac{Kg}{m^3}\right]$$

Calculate the speed of sound in air at 1 atm, 15 C based on the given bulk modulus B and density ρ_o.

Example 2.2 Solution The speed of sound c at 15 C for air can be calculated based on the given bulk modulus B and density ρ_o as:

$$c = \sqrt{\frac{B}{\rho_o}} = \sqrt{\frac{1.414 \times 10^5 \ [Pa]}{1.2 \left[\frac{Kg}{m^3}\right]}} \cong 340 \ \left[\frac{m}{s}\right]$$

where

$$B = 1.414 \times 10^5 \ [Pa]$$

$$\rho_o = 1.2 \left[\frac{Kg}{m^3}\right]$$

The speed of sound in both air and water can be calculated using the same formula in this section. The speeds of sound in air and water are tabulated below:

	Bulk modulus B [Pa]	Density ρ_o $\left[\frac{Kg}{m^3}\right]$	Speed of sound $\left[\frac{m}{s}\right]$
Air	1.414×10^5	1.2	340
Water	2.152×10^9	1000	1467

2.5.3   Formula Using Temperature

The ideal gas law states that:

$$PV = nN_A k_B T = nRT$$

where

n: number of moles
N_A: number of particles per mole (Avogadro constant)
k_B: Boltzmann's constant
R: ideal gas constant ($R = N_A k_B$)
T: absolute temperature ($K = 273.15 + C$)

The formula for the speed of sound (based on temperature) can be obtained by substituting the ideal gas law into the formula for the speed of sound (based on pressure) as:

$$c = \sqrt{\frac{\gamma P_o}{\rho_o}}$$

$$\rightarrow c = \sqrt{\gamma \frac{P}{\frac{Mn}{V}}} = \sqrt{\gamma \frac{PV}{Mn}} = \sqrt{\gamma \frac{nN_A k_B T}{Mn}} = \sqrt{\gamma \frac{N_A k_B T}{M}} = \sqrt{\gamma \frac{RT}{M}}$$

where

γ: ratio of specific heats $= 1.4$ (for air, N_2, and O_2 at any temperature)
M: molar mass of the gas (mass per moles)
R: ideal gas constant, $R = 8.314 \left[\frac{J}{mol\ K} = \frac{Kg \frac{m}{s^2} m}{mol\ K} \right]$
T: absolute temperature ($K = 273.15 + C$)

Example 2.3 The average molar mass for air at 15 C is 0.02895 $\left[\frac{Kg}{mol} \right]$. Calculate the speed of sound at 15 C based on the ratio of specific heats γ, molar mass M, ideal gas constant R, and temperature T.

Example 2.3 Solution The speed of sound in air at 15 C can be calculated based on the ratio of specific heats , molar mass M, ideal gas constant R, and temperature T as:

$$c = \sqrt{\gamma \frac{RT}{M}} = \sqrt{1.4 \frac{8.314 \left[\frac{Kg \frac{m}{s^2} m}{mol\ K} \right] \times 288.15[K]}{0.02895 \left[\frac{Kg}{mol} \right].}} \cong 340 \left[\frac{m}{s} \right]$$

where:

$$\gamma = 1.4$$

$$R = 8.314 \left[\frac{J}{mol\ K} = \frac{Kg\frac{m}{s^2}m}{mol\ K} \right]$$

$$T = 15C = 288.15K$$

$$M = 0.02895 \left[\frac{Kg}{mol} \right]$$

The speed of sound in air, pure oxygen, and pure nitrogen can be calculated using the same formula in this section. The speeds of sound in air, pure oxygen, and pure nitrogen are tabulated below:

	Molar mass (M) $[\frac{g}{mol}]$	Temperature(T) [C]	Speed of sound $[\frac{m}{s}]$
Air	28.95	15	340
O_2 (21% of air)	31.999	15	324
N_2 (78% of air)	28.013	15	346

Note that the ideal gas constant R is the same for all gases. The ratios of specific heats for N_2, O_2, and air are very close and are about 1.4. The ratios of specific heats could be different for different gases. For example, the ratio of specific heats for He is about 1.67.

2.5.4 Formula Using Colliding Speed

The pressure due to one particle collision was derived in Sect. 2.3.2 as:

$$P_o V_a = \frac{1}{3} m_a v_{ca}^2$$

Based on the equation above, the pressure for n moles of the particle is:

$$PV = \frac{1}{3} nM v_{ca}^2$$

Substituting the equation above into the formulas for the speed of sound based on pressure gives:

$$c = \sqrt{\frac{\gamma P_o}{\rho_o}}$$

$$\rightarrow c = \sqrt{\gamma \frac{P}{\frac{Mn}{V}}} = \sqrt{\gamma \frac{PV}{Mn}} = \sqrt{\gamma \frac{\frac{1}{3}nMv_{ca}^2}{Mn}} = \sqrt{\frac{1}{3}\gamma v_{ca}^2}$$

where:

v_{ca} is the RMS colliding speed of particles.

For air at 15 C, the RMS colliding speed of particles can be calculated backward from the formula above as:

$$v_{ca} = \sqrt{\frac{3c^2}{\gamma}} = \sqrt{\frac{3 \times 340^2}{1.4}} = 498 \left[\frac{m}{s}\right]$$

Note that the RMS colliding speed of air particles is faster than the speed of sound and is related to the ratio, γ, of specific heats. Also, the ratio, γ, of specific heats is related to the ratio, α, as:

$$\gamma \equiv \frac{3\alpha + 2}{3\alpha}$$

The corresponding value of α of air is:

$$\alpha = \frac{5}{3}$$

where α is a ratio between the total energy (rotational energy + translational energy) and the translational energy of molecules as:

$$\frac{1}{2}I_a\dot{\theta}_{ca}^2 + \frac{1}{2}m_a v_{ca}^2 = \alpha \frac{1}{2}m_a v_{ca}^2$$

2.6 Homework Exercises

Exercise 2.1 The bulk modulus B and the density ρ_o of certain seawater are:

$$B = 2.307 \times 10^9 \ [Pa]$$

$$\rho_o = 1025 \left[\frac{Kg}{m^3}\right]$$

Determine the speed of sound in this certain seawater.

(Answer) 1500 $\left[\frac{m}{s}\right]$

Exercise 2.2 Sound can propagate through gaseous *He* which has a molar mass:

$$M = 4.003 \left[\frac{g}{mol} \right]$$

Determine the speed of sound in *He* at 15 *C*.

Hint: The ideal gas constant is the same for all gases. The ratio of specific heats for He is 1.67.

(Answer) 1000 $\left[\frac{m}{s} \right]$

Exercise 2.3 The average molar mass of the molecules in the air is given as:

$$M = 28.95 \left[\frac{g}{mol} \right]$$

What is the speed of sound at 30 *C*?

(Answer) 349.12 $\left[\frac{m}{s} \right]$

Chapter 3
Solutions of Acoustic Wave Equation

In this chapter, you will learn how to construct both traveling waves and standing waves by solving the acoustic wave equation derived in the previous chapter. We will limit the scope of this chapter to only one-dimensional plane waves. Spherical waves will be constructed by solving the acoustic wave equation derived in spherical coordinate systems in Chap. 5.

We will use three building blocks to construct both traveling waves and standing waves. Each building block represents one type of *basic* solution under a certain requirement. The foundation for these three building blocks is the acoustic wave equation. They are summarized as follows:

The Foundation: The acoustic wave equation
Building Block 1: Basic complex solutions (BCSs)
Requirement: $\frac{\omega}{k} = \pm c$
Building Block 2: Basic traveling waves (BTW)
Requirement: complex conjugate pairs
Building Block 3: Basic standing waves (BSWs)
Requirement: same amplitude and opposite directions

© The Author(s), under exclusive license to Springer Nature Switzerland AG 2021
H. Lin et al., *Lecture Notes on Acoustics and Noise Control*,
https://doi.org/10.1007/978-3-030-88213-6_3

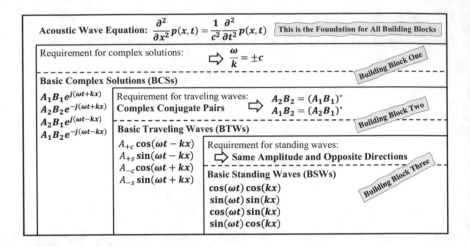

In Building Block 1, basic complex solutions (BCSs) are constructed based on the acoustic wave equation under the requirement $\frac{\omega}{k} = \pm c$. BCSs are the most direct solutions of the acoustic wave equation. But because BCSs are complex numbers, they have no physical attributes by themselves.

In Building Block 2, basic traveling waves (BTWs) are constructed using the BCSs under the requirement that the BCSs must be a complex conjugate pair. BTWs are the addition of a complex conjugate pair. Unlike BCSs, BTWs are real numbers and have physical attributes of traveling waves. Any traveling wave can be constructed by using the four BTWs.

In Building Block 3, basic standing waves (BSWs) are constructed using the BTWs under the requirement that two traveling waves are traveling in the opposite direction and have exactly the same amplitude. Any standing wave can be constructed by using the four BSWs.

These "basic" solutions and waves will be used as building blocks for acoustic analysis throughout this course. Basic solutions and waves allow for a more in-depth analysis of acoustics and noise control.

Based on the traveling waves, this chapter will demonstrate how to construct standing waves by adding two traveling waves that are moving in the opposite direction and having the same amplitude:

$$
\begin{aligned}
p_s(x, t) &= p_-(x, t) + p_+(x, t) \\
&= A \cos\left(\omega t + kx + \phi_-\right) + A \cos\left(\omega t - kx + \phi_+\right) \\
&= 2A \cos\left[\omega t + \left(\frac{\phi_- + \phi_+}{2}\right)\right] \cos\left[kx + \left(\frac{\phi_- - \phi_+}{2}\right)\right]
\end{aligned}
$$

3.1   Review of Partial Differential Equations

3.1.1   Complex Solutions of a Partial Differential Equation

The acoustic wave equation is a second-order partial differential equation. The acoustic wave equation for a one-dimensional plane wave in Cartesian coordinates is:

$$\frac{\partial^2}{\partial x^2} p(x,t) = \frac{1}{c^2} \frac{\partial^2}{\partial t^2} p(x,t)$$

where p is sound pressure, x is a dimension, t is time, and c is the propagation speed of sound.

It can be directly shown (see Example 3.1) that when $\frac{\omega}{k} = \pm c$, the following four basic complex solutions:

$$p(x,t) = A e^{\,j(\omega t + kx + \phi)}$$

$$p(x,t) = A e^{\,-j(\omega t + kx + \phi)}$$

$$p(x,t) = A e^{\,j(\omega t - kx + \phi)}$$

$$p(x,t) = A e^{\,-j(\omega t - kx + \phi)}$$

satisfy the one-dimensional acoustic wave equation of $p(x,t)$:

$$\frac{\partial^2}{\partial x^2} p(x,t) = \frac{1}{c^2} \frac{\partial^2}{\partial t^2} p(x,t)$$

where A is an amplitude (real number) and ϕ is a phase shift. The detailed derivation will be shown in Sect. 3.2. The following example is proof of the solutions:

Example 3.1   A complex exponential function $p(x,t)$ is given as:

$$p(x,t) = A e^{\,j(\omega t + kx + \phi)}$$

where A is the amplitude (real number) and ϕ is the phase (angle). Show that when $\frac{\omega}{k} = \pm c$, the above complex exponential function $p(x,t)$ satisfies the one-dimensional acoustic wave equation:

$$\frac{\partial^2}{\partial x^2} p(x,t) = \frac{1}{c^2} \frac{\partial^2}{\partial t^2} p(x,t)$$

Example 3.1 Solution Step 1: The left-hand side of the given acoustic wave equation is $\frac{\partial^2}{\partial x^2}p(x,t)$.

Calculate $\frac{\partial^2}{\partial x^2}p$ by substituting the given function $p(x,t)$ into $\frac{\partial^2}{\partial x^2}p(x,t)$ to arrive at:

$$\frac{\partial^2}{\partial x^2}p = \frac{\partial}{\partial x}\left\{\frac{\partial}{\partial x}\left[Ae^{j(\omega t+kx+\phi)}\right]\right\}$$

$$= \frac{\partial}{\partial x}\left\{jk\left[Ae^{j(\omega t+kx+\phi)}\right]\right\}$$

$$= -k^2\left[Ae^{j(\omega t+kx+\phi)}\right]$$

Step 2: The right-hand side of the given acoustic wave equation is $\frac{\partial^2}{\partial t^2}p(x,t)$. Calculate $\frac{\partial^2}{\partial t^2}p$ by substituting the given function $p(x,t)$ into $\frac{\partial^2}{\partial t^2}p(x,t)$ to get:

$$\frac{\partial^2}{\partial t^2}p = \frac{\partial}{\partial t}\left\{\frac{\partial}{\partial t}\left[Ae^{j(\omega t+kx+\phi)}\right]\right\}$$

$$= \frac{\partial}{\partial t}\left[j\omega Ae^{j(\omega t+kx+\phi)}\right]$$

$$= -\omega^2\left[Ae^{j(\omega t+kx+\phi)}\right]$$

Step 3: Substituting the calculated $\frac{\partial^2}{\partial x^2}p$ and $\frac{\partial^2}{\partial t^2}p$ into the acoustic wave equation will give:

$$\frac{\partial^2}{\partial x^2}p = \frac{1}{c^2}\frac{\partial^2}{\partial t^2}p$$

$$\rightarrow \quad -k^2\left[Ae^{j(\omega t+kx+\phi)}\right] = -\frac{1}{c^2}\omega^2\left[Ae^{j(\omega t+kx+\phi)}\right]$$

$$\rightarrow \quad \frac{k^2}{\omega^2} = \frac{1}{c^2}$$

$$\rightarrow \quad \frac{\omega}{k} = \pm c$$

It shows that when $\frac{\omega}{k} = \pm c$, the complex exponential function $p(x,t)$ satisfies the one-dimensional acoustic wave equation.

3.1.2 *Trigonometric Solutions of a Partial Differential Equation*

Similar to the complex solutions of the partial differential equation covered in the previous section, it can be directly shown that when $\frac{\omega}{k} = \pm c$, the following four basic complex solutions:

$$p(x,t) = A_{+c} \cos\left(\omega t - kx + \phi_+\right)$$

$$p(x,t) = A_{+s} \sin\left(\omega t - kx + \phi_+\right)$$

$$p(x,t) = A_{-c} \cos\left(\omega t + kx + \phi_-\right)$$

$$p(x,t) = A_{-s} \sin\left(\omega t + kx + \phi_-\right)$$

satisfy the one-dimensional acoustic wave equation of $p(x,t)$:

$$\frac{\partial^2}{\partial x^2} p(x,t) = \frac{1}{c^2}\frac{\partial^2}{\partial t^2} p(x,t)$$

where A is an amplitude (real number) and ϕ is a phase shift. The detailed derivation will be shown in Sect. 3.2.

Example 3.2 A backward traveling wave $p_-(x,t)$ (Form 1:R_{IP}) is given as:

$$p_-(x,t) = A_{-s} \sin\left(\omega t + kx\right)$$

where A_{-s} is an amplitude. Show that if $\frac{\omega}{k} = \pm c$, the above trigonometric function $p(x,t)$ satisfies the one-dimensional acoustic wave equation:

$$\frac{\partial^2}{\partial x^2} p(x,t) = \frac{1}{c^2}\frac{\partial^2}{\partial t^2} p(x,t)$$

Example 3.2 Solution Step 1: The left-hand side of the given acoustic wave equation is:

$$\frac{\partial^2}{\partial x^2} p(x,t)$$

Calculate $\frac{\partial^2}{\partial x^2} p$ by substituting the given function $p(x,t)$ into $\frac{\partial^2}{\partial x^2} p(x,t)$ to get:

$$\frac{\partial^2}{\partial x^2}p = \frac{\partial}{\partial x}\left\{\frac{\partial}{\partial x}[A_{-s}\sin(\omega t + kx)]\right\} = A_{-s}\frac{\partial}{\partial x}\left\{\frac{\partial}{\partial x}[\sin(\omega t + kx)]\right\}$$

$$= A_{-s}\frac{\partial}{\partial x}\{k[\cos(\omega t + kx)]\} = -A_{-s}k^2\sin(\omega t + kx)$$

Step 2: The right-hand side of the given acoustic wave equation is $\frac{\partial^2}{\partial t^2}p(x,t)$. Calculate $\frac{\partial^2}{\partial t^2}p$ by substituting the given function $p(x,t)$ into $\frac{\partial^2}{\partial t^2}p(x,t)$ to get:

$$\frac{\partial^2}{\partial t^2}p = \frac{\partial}{\partial t}\left\{\frac{\partial}{\partial t}[A_{-s}\sin(\omega t + kx)]\right\} = A_{-s}\frac{\partial}{\partial t}\left\{\frac{\partial}{\partial t}[\sin(\omega t + kx)]\right\}$$

$$= A_{-s}\frac{\partial}{\partial t}[\omega\cos(\omega t + kx)] = -A_{-s}\omega^2\sin(\omega t + kx)$$

Step 3: Substituting the calculated $\frac{\partial^2}{\partial x^2}p$ and $\frac{\partial^2}{\partial t^2}p$ into the acoustic wave equation yields:

$$\frac{\partial^2}{\partial x^2}p = \frac{1}{c^2}\frac{\partial^2}{\partial t^2}p$$

$$\rightarrow \quad -A_{-s}k^2\sin(\omega t + kx) = -\frac{1}{c^2}A_{-s}\omega^2\sin(\omega t + kx)$$

$$\rightarrow \quad \frac{k^2}{\omega^2} = \frac{1}{c^2}$$

$$\rightarrow \quad \frac{\omega}{k} = \pm c$$

It shows that if $\frac{\omega}{k} = \pm c$, the trigonometric function $p(x,t)$ satisfies the one-dimensional acoustic wave equation.

3.2 Four Basic Complex Solutions

The general sound wave equation in Cartesian coordinates is:

$$\left(\frac{\partial^2}{\partial x^2} + \frac{\partial^2}{\partial y^2} + \frac{\partial^2}{\partial z^2}\right)p(x,y,z,t) = \frac{1}{c^2}\frac{\partial^2}{\partial t^2}p(x,y,z,t)$$

The above general sound wave equation is only valid in Cartesian coordinates. It is not valid in cylindrical coordinates and spherical coordinates.

In reality, a point sound source radiates spherical waves and is usually formulated in spherical coordinates.

To prepare for transferring the acoustic wave equation to spherical coordinates, the acoustic wave equation is formulated in vector format, which is independent of the choice of coordinates and can be transferred to other coordinate systems. The acoustic wave equation in vector form is:

$$\nabla^2 p(x,t) = \frac{1}{c^2} \frac{\partial^2}{\partial t^2} p(x,t)$$

where the symbol ∇ is the gradient operator and ∇^2 ("del squared") is called a Laplacian operator. The physical meaning of the gradient ∇ is the same in any coordinate system. However, the actual formulation of the gradient ∇ depends on the choice of a coordinate system.

The gradient ∇ in Cartesian coordinates is a vector with bases $\widehat{e}_x, \widehat{e}_y$, and $\widehat{e}_z$:

$$\nabla = \widehat{e}_x \frac{\partial}{\partial x} + \widehat{e}_y \frac{\partial}{\partial y} + \widehat{e}_z \frac{\partial}{\partial z}$$

And the Laplacian operator ∇^2 in Cartesian coordinates is:

$$\nabla^2 = \frac{\partial^2}{\partial x^2} + \frac{\partial^2}{\partial y^2} + \frac{\partial^2}{\partial z^2}$$

If a coordinate system is chosen so that the direction of propagation coincides with the x-axis, then the spatial derivatives of the pressure with respect to the y-axis and z-axis are zero. For a one-dimensional plane acoustic wave in the x-direction, the sound pressure $p(x,t)$ is independent of y and z. Therefore:

$$\frac{\partial p}{\partial y} = 0$$

$$\frac{\partial p}{\partial z} = 0$$

Substituting the above two conditions into the three-dimensional acoustic wave equation will cast the one-dimensional acoustic wave equation as follows:

$$\frac{\partial^2}{\partial x^2} p(x,t) = \frac{1}{c^2} \frac{\partial^2}{\partial t^2} p(x,t)$$

Or:

$$\frac{\partial^2 p}{\partial x^2} = \frac{1}{c^2}\frac{\partial^2 p}{\partial t^2} \tag{3.1}$$

The differential equation is separable if a solution can be cast in the following form:

$$p(x,t) = X(x)T(t) \tag{3.2}$$

and it satisfies the acoustic wave equation. If this assumed solution is substituted into Eq.(3.1), the following equation with two independent functions $X(x)$ and $T(t)$ is derived:

$$\frac{\partial^2 X(x)}{\partial x^2}T(t) = \frac{1}{c^2}X(x)\frac{\partial^2 T(t)}{\partial t^2}$$

or, indicating derivatives of the functions $X(x)$ and $T(t)$ by a prime (') and rearranging the terms equivalently:

$$\frac{X''(x)}{X(x)} = \frac{1}{c^2}\frac{T''(t)}{T(t)}$$

It is important to realize here that $X(x)$ and $T(t)$ are independent functions and that the only way that the two sides of the above expression can be equal is if they are equal to a constant. Now, variable k is introduced to relate two sides of the above equation as:

$$\frac{X''(x)}{X(x)} = \frac{1}{c^2}\frac{T''(t)}{T(t)} = -k^2$$

This final form of the equation above provides the two separate ordinary differential equations (ODEs) given below:

$$X''(x) + k^2 X(x) = 0 \tag{3.3}$$

$$T''(t) + \omega^2 T(t) = 0 \tag{3.4}$$

where:

$$\omega^2 = k^2 c^2$$

Since both the wavenumber (k) and the speed of sound (c) are positive, the angular velocity (ω) is defined as:

$$\omega = \pm kc$$

Eq.(3.3) and Eq.(3.4) are ordinary differential equations (ODEs) and can be solved by several different techniques. The following are two approaches that solve the ODEs above.

The first approach is to assume the following two real trigonometric solutions:

$$T(x) = A_t \cos(\omega t + \theta_t)$$
$$X(x) = A_x \cos(kx + \theta_x)$$

This approach can give us the standing wave solutions by simply solving the acoustic wave equation. This approach will be used for finding standing waves in resonant cavities (Chap. 7).

However, the first approach cannot provide in-depth knowledge of the relationship between traveling waves and standing waves. For this reason, the second approach is introduced to gain a better understanding of acoustic waves. Note that two approaches will give us the same standing wave solutions.

The second approach is to assume a complex exponential solution as:

$$X(x) = A e^{rx} \tag{3.5}$$
$$T(t) = B e^{st} \tag{3.6}$$

where A, B, r, and s are complex numbers. Substituting the above two equations into Eq. (3.3) and Eq. (3.4) yields:

$$r = \pm jk \tag{3.7}$$
$$s = \pm j\omega \tag{3.8}$$

Therefore, the general solutions of the separated acoustic wave equations are:

$$X(x) = A_1 e^{jkx} + A_2 e^{-jkx} \tag{3.9}$$
$$T(t) = B_1 e^{j\omega t} + B_2 e^{-j\omega t} \tag{3.10}$$

where A_1, A_2, B_1, and B_2 are complex numbers.

Substituting Eq. (3.9) and Eq. (3.10) into the acoustic wave equation Eq. (3.2) will yield:

$$\begin{aligned} p(x,t) &= X(x)T(t) \\ &= \left(A_1 e^{jkx} + A_2 e^{-jkx}\right)\left(B_1 e^{j\omega t} + B_2 e^{-j\omega t}\right) \end{aligned} \tag{3.11}$$

and can be expanded as:

$$p(x, t) = A_1 B_1 e^{j(\omega t + kx)} + A_1 B_2 e^{-j(\omega t - kx)}$$
$$+ A_2 B_1 e^{j(\omega t - kx)} + A_2 B_2 e^{-j(\omega t + kx)} \tag{3.12}$$

where A_1, A_2, B_1, and B_2 are complex numbers.

The four terms in the above equation can be considered as:

Four Basic Complex Solutions (BCS)

	Complex Function	Complex Conjugate Pair
Forward Wave	$A_2 B_1 e^{j(\omega t - kx)}$	$A_1 B_2 e^{-j(\omega t - kx)}$
Backward Wave	$A_1 B_1 e^{j(\omega t + kx)}$	$A_2 B_2 e^{-j(\omega t + kx)}$

Because each term of the above equation is a complex solution of the acoustic wave equation and every term is an independent solution, the four basic complex solutions when in complex conjugate pairs can be used to construct any traveling wave.

In addition, every BCS is a complex solution that satisfies the one-dimensional acoustic wave equation Eq.(3.1). This can be proved directly by substituting the BCS into the acoustic wave equation, as shown in Example 2.3 in the previous chapter.

Although the complex solution above $p(x, t)$ is a mathematical solution to the acoustic wave equation, a complex function $p(x, t)$ cannot represent a physical acoustic pressure. For the complex solution $p(x, t)$ to represent an acoustic pressure, it must be a real number resulting from the addition of complex conjugate pairs.

The constant complex coefficients of the four BCSs can be expressed with complex exponentials of explicit phases (C_{EP}) as follows:

$$A_1 = A_1 e^{j\alpha_1} \tag{3.13}$$

$$A_2 = A_2 e^{j\alpha_2} \tag{3.14}$$

$$B_1 = B_1 e^{j\beta_1} \tag{3.15}$$

$$B_2 = B_2 e^{j\beta_2} \tag{3.16}$$

where A_1, A_2, B_1, and B_2 are magnitudes and α_1, α_2, β_1, and β_1 are phases.

Substituting Eq. (3.13) and Eq. (3.16) into Eq.(3.12) yields:

$$p(x, t) = A_1 B_1 e^{j(\omega t + kx + \alpha_1 + \beta_1)} + A_1 B_2 e^{-j(\omega t - kx - \alpha_1 - \beta_2)}$$
$$+ A_2 B_1 e^{j(\omega t - kx + \alpha_2 + \beta_1)} + A_2 B_2 e^{-j(\omega t + kx - \alpha_2 - \beta_2)} \tag{3.17}$$

where A_1, A_2, B_1, B_2, α_1, α_2, β_1, and β_1 are real unknown constants that can be determined by the boundary and initial conditions.

3.3 Four Basic Traveling Waves

Because sound pressure $p(x, t)$ is a real physical quantity, the complex solution $p(x, t)$ must be real after applying boundary and initial conditions that are also real numbers. Therefore, the real solution of sound pressure $p(x, t)$ must be constructed by two complex conjugate pairs as a result of the following constraints:

$$A_2 B_2 = (A_1 B_1)^*$$
$$A_1 B_2 = (A_2 B_1)^*$$

or:

$$A_1 B_1 = A_2 B_2 \equiv A_-$$
$$A_1 B_2 = A_2 B_1 \equiv A_+$$
$$\alpha_1 + \beta_1 = -\alpha_2 - \beta_2 \equiv \theta_-$$
$$\alpha_2 + \beta_1 = -\alpha_1 - \beta_2 \equiv \theta_+$$

where we define four new real variables A_-, A_+, θ_-, and θ_+ with a subscript "-" to indicate backward waves and a subscript "+" to indicate forward waves.

Based on the above four constraints and four new real variables, the general solution $p(x, t)$ is a real number because of the addition of the following two complex conjugate pairs:

$$p(x, t) = \left[A_- e^{j(\omega t + kx + \theta_-)} + A_- e^{-j(\omega t + kx + \theta_-)} \right] \\ + \left[A_+ e^{j(\omega t - kx + \theta_+)} + A_+ e^{-j(\omega t - kx + \theta_+)} \right] \quad \text{(Form 3 : } C_{EP})$$

Note that there are four real unknown variables A_-, A_+, θ_-, and θ_+ that can be determined by two boundary conditions and two initial conditions. The first two terms above represent a backward wave $p_-(x, t)$, and the last two terms represent a forward wave $p_+(x, t)$.

The general wave solution $p(x, t)$ is the addition of a backward wave $p_-(x, t)$ and a forward wave $p_+(x, t)$:

$$p(x, t) = p_-(x, t) + p_+(x, t) \\ = A_- \cos(\omega t + kx + \phi_-) + A_+ \cos(\omega t - kx + \phi_+) \quad \text{(Form 2 : } R_{EP})$$

where the backward wave $p_-(x, t)$ and the forward wave $p_+(x, t)$ can be expressed in Form 1 and Form 2 as:

$$p_\pm(x, t) = A_\pm \cos(\omega t \mp kx + \phi_\pm) \quad \text{(Form 2 : } R_{EP}) \\ = A_{\pm c} \cos(\omega t \mp kx) - A_{\pm s} \sin(\omega t \mp kx) \quad \text{(Form 1 : } R_{IP})$$

Note that when $\frac{\omega}{k} = \pm c$, it can be shown that every term in the general solutions in both Form 1 and Form 2 satisfies the one-dimensional acoustic wave equation below:

$$\frac{\partial^2}{\partial x^2} p(x,t) = \frac{1}{c^2} \frac{\partial^2}{\partial t^2} p(x,t)$$

In (**Form 1:R_{IP}**), the sound pressure is formulated with four unknown variables $A_{-c}, A_{+c}, A_{-s},$ and A_{+s} as:

$$
\begin{aligned}
p(x,t) &= p_+(x,t) + p_-(x,t) \\
&= A_{+c} \cos(\omega t - kx) - A_{+s} \sin(\omega t - kx) \\
&\quad + A_{-c} \cos(\omega t + kx) - A_{-s} \sin(\omega t + kx)
\end{aligned}
\qquad \text{(Form 1 : } R_{IP})
$$

The four basic traveling waves (BTW) are defined as:

Four Basic Traveling Waves (BTW) in Form 1

	Cosine Function	Sine Function
Forward Wave	$A_{+c} \cos(\omega t - kx)$	$A_{+s} \sin(\omega t - kx)$
Backward Wave	$A_{-c} \cos(\omega t + kx)$	$A_{-s} \sin(\omega t + kx)$

The sound pressure in four equivalent forms is shown below for easier reference:

$$
\begin{aligned}
p(x,t) &= A_{+c} \cos(\omega t - kx) - A_{+s} \sin(\omega t - kx) \\
&\quad + A_{-c} \cos(\omega t + kx) - A_{-s} \sin(\omega t + kx) \qquad &\text{(Form 1:} R_{IP}) \\
p(x,t) &= A_+ \cos(\omega t - kx + \phi_+) + A_- \cos(\omega t + kx + \phi_-) \qquad &\text{(Form 2:} R_{EP}) \\
p(x,t) &= \tfrac{1}{2} A_+ e^{j(\omega t - kx + \phi_+)} + \tfrac{1}{2} A_+ e^{-j(\omega t - kx + \phi_+)} \\
&\quad + \tfrac{1}{2} A_- e^{j(\omega t + kx + \phi_-)} + \tfrac{1}{2} A_- e^{-j(\omega t + kx + \phi_-)} \qquad &\text{(Form 3:} C_{EP}) \\
p(x,t) &= \tfrac{1}{2}\left[(A_{+c} + jA_{+s}) e^{j(\omega t - kx)} + (A_{+c} - jA_{+s}) e^{-j(\omega t - kx)} \right] \\
&\quad + \tfrac{1}{2}\left[(A_{-c} + jA_{-s}) e^{j(\omega t + kx)} + (A_{-c} - jA_{-s}) e^{-j(\omega t + kx)} \right] \qquad &\text{(Form 4:} C_{IP})
\end{aligned}
$$

As shown above, the BTWs can also be expressed in four equivalent forms introduced in Chap. 1.

3.4 Four Basic Standing Waves

A standing wave is the addition of two traveling waves with the same amplitude, traveling in the opposite direction. Therefore, it is required that:

$$A_+ = A_- = A$$

With the above standing wave condition, a standing wave can be depicted as:

$$p_s(x,t) = p_+(x,t) + p_-(x,t)$$
$$= A \cos(\omega t - kx + \phi_+) + A \cos(\omega t + kx + \phi_-)$$

where A, ϕ_+, and ϕ_- are real unknown constants that can be determined by given boundary and initial conditions.

The above equation can also be expressed as:

$$p_s(x,t) = 2A \cos\left[\omega t + \left(\tfrac{\phi_- + \phi_+}{2}\right)\right] \cos\left[kx + \left(\tfrac{\phi_- - \phi_+}{2}\right)\right] \qquad \text{(Form 2 : } R_{EP})$$

The standing wave equation above can be simplified using two new variables, ϕ_t and ϕ_x, to become:

$$p_s(x,t) = 2A \cos[\omega t + \phi_t] \cos[kx + \phi_x] \qquad \text{(Form 2 : } R_{EP})$$

where the two new variables, ϕ_t and ϕ_x, are defined as:

$$\phi_t \equiv \left(\frac{\phi_- + \phi_+}{2}\right)$$

$$\phi_x \equiv \left(\frac{\phi_- - \phi_+}{2}\right)$$

Expand the general standing wave $p_s(x,t)$ above to get (Form 1 : R_{IP}) from (Form 2 : R_{EP}) as follows:

$$
\begin{aligned}
p_s(x,t) \;=\;& 2A \cos[\omega t + \phi_t] \cos[kx + \phi_x] \qquad &\text{(Form 2 : } R_{EP}) \\
=\;& 2A[\cos(\omega t) \cos(\phi_t) - \\
& \sin(\omega t) \sin(\phi_t)][\cos(kx) \cos(\phi_x) - \\
& \sin(kx) \sin(\phi_x)] \\
=\;& 2A \cos(\phi_t) \cos(\phi_x) \cos(\omega t) \cos(kx) + \\
& 2A \sin(\phi_t) \sin(\phi_x) \sin(\omega t) \sin(kx) \\
& -2A \sin(\phi_t) \cos(\phi_x) \sin(\omega t) \cos(kx) \\
& - 2A \cos(\phi_t) \sin(\phi_x) \cos(\omega t) \sin(kx) \\
=\;& A_{cc} \cos(\omega t) \cos(kx) + A_{ss} \sin(\omega t) \qquad &\text{(Form1 : } R_{IP}) \\
& \sin(kx) + A_{sc} \sin(\omega t) \cos(kx) + A_{cs} \cos(\omega t) \sin(kx)
\end{aligned}
$$

where:

$$A_{cc} = 2A \cos(\phi_t) \cos(\phi_x); \qquad A_{ss} = 2A \sin(\phi_t) \sin(\phi_x)$$
$$A_{sc} = -2A \sin(\phi_t) \cos(\phi_x); \qquad A_{cs} = -2A \cos(\phi_t) \sin(\phi_x)$$

where each term of the above equation is a basic standing wave (BSW):

Four basic standing waves (BSWs):

$$p_s(x, t) = \cos(\omega t)\cos(kx)$$
$$p_s(x, t) = \sin(\omega t)\sin(kx)$$
$$p_s(x, t) = \sin(\omega t)\cos(kx)$$
$$p_s(x, t) = \cos(\omega t)\sin(kx)$$

BSWs can be expressed by the multiplication of a spatial function and a temporal function. In addition, each BSW can be expressed by the addition of a forward traveling and a backward traveling with the same amplitude as:

$$p_s(x, t) = \cos(\omega t)\cos(kx) = \frac{1}{2}(\cos(\omega t + kx) + \cos(\omega t - kx))$$

$$p_s(x, t) = \sin(\omega t)\sin(kx) = \frac{1}{2}(\cos(\omega t - kx) - \cos(\omega t + kx))$$

$$p_s(x, t) = \sin(\omega t)\cos(kx) = \frac{1}{2}(\sin(\omega t + kx) + \sin(\omega t - kx))$$

$$p_s(x, t) = \cos(\omega t)\sin(kx) = \frac{1}{2}(\sin(\omega t + kx) - \sin(\omega t - kx))$$

The proofs of the above four equations are in the problems.

When the arguments of sine and cosine functions include phases ϕ_t, ϕ_x, ϕ_+, and ϕ_-, since cosine and sine only differ by a phase of $\pi/2$, all sine functions can be changed to cosine functions, and the use of only cosine functions is enough to represent the standing waves.

Example 3.3 Prove that the following relationship between a standing wave and the combination of a forward wave and a backward wave is true:

$$\cos(\omega t - kx + \phi_+) + \cos(\omega t + kx + \phi_-)$$
$$= 2\cos\left[\omega t + \left(\frac{\phi_- + \phi_+}{2}\right)\right]\cos\left[kx + \left(\frac{\phi_- - \phi_+}{2}\right)\right]$$

Example 3.3 Solution Compare the given equation:

$$\cos(\omega t - kx + \phi_+) + \cos(\omega t + kx + \phi_-)$$
$$= 2\cos\left[\omega t + \left(\frac{\phi_- + \phi_+}{2}\right)\right]\cos\left[kx + \left(\frac{\phi_- - \phi_+}{2}\right)\right]$$

to the following trigonometric property:

$$\cos\left(a+b\right)+\cos\left(a-b\right)=2\cos\left(a\right)\cos\left(b\right)$$

Comparing the right-hand sides of the two equations above yields:

$$2\cos\left(a\right)\cos\left(b\right)=2\cos\left[\omega t+\left(\frac{\phi_-+\phi_+}{2}\right)\right]\cos\left[kx+\left(\frac{\phi_--\phi_+}{2}\right)\right]$$

Let:

$$a=\omega t+\left(\frac{\phi_-+\phi_+}{2}\right);$$

$$b=kx+\left(\frac{\phi_--\phi_+}{2}\right)$$

Based on the definition of a and b, the addition and the subtraction of a and b result in:

$$a+b=\left[\omega t+\left(\frac{\phi_-+\phi_+}{2}\right)\right]+\left[kx+\left(\frac{\phi_--\phi_+}{2}\right)\right]=\omega t+kx+\phi_-$$

$$a-b=\left[\omega t+\left(\frac{\phi_-+\phi_+}{2}\right)\right]-\left[kx+\left(\frac{\phi_--\phi_+}{2}\right)\right]=\omega t-kx+\phi_+$$

Finally, we can conclude from the trigonometric property that:

$$\cos\left(a+b\right)+\cos\left(a-b\right)=2\cos\left(a\right)\cos\left(b\right)$$

will give:

$$\rightarrow\cos\left(\omega t-kx+\phi_+\right)+\cos\left(\omega t+kx+\phi_-\right)$$

$$=2\cos\left[\omega t+\left(\frac{\phi_-+\phi_+}{2}\right)\right]\cos\left[kx+\left(\frac{\phi_--\phi_+}{2}\right)\right]$$

Note that it is harder and more complicated to get a and b by equating the left-hand sides of the equations.

3.5   Conversion Between Traveling and Standing Waves

In this section, conversions between traveling and standing waves are demonstrated in two examples (Examples 3.4 and 3.5) using mathematical formulations of sound waves.

Traveling waves $p_{\pm}(x,t)$ can be categorized into four basic traveling waves (BTW) expressed in (Form 2 : R_{IP}) as:

$$p_+(x,t) = A_{+c} \cos(\omega t - kx)$$

$$p_+(x,t) = A_{+s} \sin(\omega t - kx)$$

$$p_-(x,t) = A_{-c} \cos(\omega t + kx)$$

$$p_-(x,t) = A_{-s} \sin(\omega t + kx)$$

or:

Four Basic Traveling Waves (BTW) in Form 1

	Cosine Function	Sine Function
Forward Wave	$A_{+c} \cos(\omega t - kx)$	$A_{+s} \sin(\omega t - kx)$
Backward Wave	$A_{-c} \cos(\omega t + kx)$	$A_{-s} \sin(\omega t + kx)$

When ωt and kx have *different signs*, such as $(\omega t - kx)$, it is a forward traveling wave. When ωt and kx have the *same signs*, such as $(\omega t + kx)$, it is a backward traveling wave.

Because the above four BTWs are independent functions, any traveling wave can be constructed from these four BTWs. Because these four BTWs are simple, yet can fully represent any traveling wave, these four BTWs will be used in acoustic analysis throughout this course.

A standing wave can be constructed with two traveling waves that have the same amplitude and are traveling in the opposite direction.

For example, choose a forward traveling wave $p_+(x,t)$ and a backward traveling wave $p_-(x,t)$ as follows:

$$p_+(x,t) = A_{+c} \cos(\omega t - kx)$$

$$p_-(x,t) = A_{-c} \cos(\omega t + kx)$$

Assume these two traveling waves have the same amplitude as:

$$A_{+c} = A_{-c} \equiv A$$

It can be shown that the addition of these two traveling waves is a standing wave $p_s(x,t)$ as:

$$p_s(x,t) = p_+(x,t) + p_-(x,t)$$
$$= A \cos(\omega t - kx) + A \cos(\omega t + kx)$$
$$= A \left[\cos(\omega t) \cos(kx) + \sin(\omega t) \sin(kx) \right]$$
$$+ A \left[\cos(\omega t) \cos(kx) - \sin(\omega t) \sin(kx) \right]$$
$$= 2A \cos(\omega t) \cos(kx)$$

Note that the two traveling waves must have the same amplitude and travel in opposite directions to construct a standing wave.

Example 3.4 The following is a harmonic function $p(x,t)$ of one-dimensional sound pressure in Cartesian coordinates:

$$p(x,t) = \mathrm{Re}\left[P \cos(kx) e^{j(\omega t)} \right]$$

where P is a real number.

(a) Show that the harmonic function $p(x,t)$ can be constructed by multiplying two trigonometric cosine functions:

$$p(x,t) = P \cos(kx) \cos(\omega t)$$

(b) Show that the harmonic function $p(x,t)$ can be constructed by adding two trigonometric cosine functions:

$$p(x,t) = \frac{P}{2} \left[\cos(\omega t + kx) + \cos(\omega t - kx) \right]$$

Note that each term in the above function has no phase and can be considered as either Form 1 or Form 2.

Example 3.4 Solution

(a) The given harmonic function $p(x,t)$ is:

$$p(x,t) = \mathrm{Re}\left[P \cos(kx) e^{j(\omega t)} \right]$$

Since both P and $\cos(kx)$ are real numbers, we can move these real parts to outside of the "Re" function as:

$$p(x,t) = P \cos(kx) \cdot \mathrm{Re}\left[e^{j(\omega t)} \right]$$

Remove the "Re" function by the addition of the corresponding complex conjugate pair and dividing the whole thing by two results in:

$$p(x,t) = P\cos(kx)\frac{1}{2}\left[e^{j(\omega t)} + e^{-j(\omega t)}\right]$$

Change the complex conjugate function pair to a trigonometric cosine function according to Euler's formula:

$$p(x,t) = P\,\cos(\omega t)\cos(kx)$$

(b) From the above equation, the multiplication of two trigonometric cosine functions can be expressed as the addition of two trigonometric cosine functions (see Sect. 1.8):

$$p(x,t) = P\,\cos(\omega t)\cos(kx) = \frac{P}{2}\left[\cos(\omega t + kx) + \cos(\omega t - kx)\right]$$

Example 3.5 Use the following combination of a forward wave $p_+(x,t)$ and a backward wave $p_-(x,t)$ to answer every question of this problem:

$$p_-(x,t) = \cos(\omega t + kx); \qquad p_+(x,t) = -\sin(\omega t - kx + \pi)$$

a) Calculate the standing wave (real number) produced by the forward wave and the backward wave.
b) Find the locations of the peaks and valleys of the standing wave in terms of wavelength. Given $k = \frac{2\pi}{\lambda}$.

Problem 3.5 Solution
a) Because:

$$2\cos(a)\cos(b) = \cos(a+b) + \cos(a-b)$$

Let:
$$a = \omega t + \left(\frac{\phi_- + \phi_+}{2}\right); \quad b = kx + \left(\frac{\phi_- - \phi_+}{2}\right)$$
Substituting a and b into the first equation gives:

$$2\cos\left[\omega t + \left(\frac{\phi_- + \phi_+}{2}\right)\right]\cos\left[kx + \left(\frac{\phi_- - \phi_+}{2}\right)\right]$$

$$= \cos\left(\omega t - kx + \phi_+\right) + \cos\left(\omega t + kx + \phi_-\right)$$

For comparing the given forward and backward waves to the above formula, convert the sine function of the given forward wave $p_+(x, t)$ to a cosine function as:

$$p_-(x, t) = \cos\left(\omega t + kx\right)$$

$$p_+(x, t) = -\sin\left(\omega t - kx + \pi\right) = -\cos\left(\frac{\pi}{2} - \omega t + kx - \pi\right)$$

$$= -\cos\left(\omega t - kx + \frac{\pi}{2}\right) = \cos\left(\omega t - kx + \frac{3\pi}{2}\right)$$

Comparing the converted forward and backward waves to the formulas and letting:

$$\phi_+ = \frac{3\pi}{2}, \qquad \phi_- = 0$$

$$\rightarrow \frac{\phi_- + \phi_+}{2} = \frac{3\pi}{4} \quad and \quad \frac{\phi_- - \phi_+}{2} = \frac{-3\pi}{4}$$

give:

$$p_+(x, t) + p_-(x, t) = \cos\left(\omega t - kx + \frac{3\pi}{2}\right) + \cos\left(\omega t + kx\right)$$

$$= 2\cos\left(\omega t + \frac{3\pi}{4}\right)\cos\left(kx - \frac{3\pi}{4}\right)$$

or:

$$= -2\sin\left(\omega t + \frac{\pi}{4}\right)\sin\left(kx - \frac{\pi}{4}\right)$$

b) The locations of the peaks and valleys in terms of the wavelength are:

Peak at $kx = \frac{3\pi}{4} \rightarrow \frac{2\pi}{\lambda}x = \frac{3\pi}{4} \rightarrow x = \frac{3\lambda}{8}$

Valley at $kx = \frac{7\pi}{4} \rightarrow \frac{2\pi}{\lambda}x = \frac{7\pi}{4} \rightarrow x = \frac{7\lambda}{8}$

3.6   Wavenumber, Angular Frequency, and Wave Speed

The backward wave $p_-(x, t)$ and the forward wave $p_+(x, t)$ can be expressed in Form 2 (Chap. 1) as:

$$p_\pm(x, t) = A_\pm \cos\left(\omega t \mp kx + \phi_\pm\right) \qquad\qquad (\text{Form 2}: R_{EP})$$

Because the harmonic wave repeats after every 2π, the above cosine function will be the same when the increase of angle, $\Delta\theta$, is an integer of 2π as:

$$\cos\left(\omega t \mp kx + \phi_\pm\right) = \cos\left(\omega t \mp kx + \phi_\pm + \Delta\theta\right)$$
$$= \cos\left(\omega t \mp kx + \phi_\pm + 2\pi \cdot i\right)$$

where:

$\Delta\theta = 2\pi \cdot i$.
i is an integer.

Based on this property, we can define wavelength λ and period T from the following relations:

$$\Delta\theta = \omega T = 2\pi \rightarrow \omega = \frac{2\pi}{T}$$

$$\Delta\theta = k\lambda = 2\pi \rightarrow k = \frac{2\pi}{\lambda}$$

Based on the two equations above, we can conclude the relationship between k and ω as:

$$k\lambda = \omega T \rightarrow \omega = \frac{\lambda}{T}k = ck$$

where c is the speed of the wave and is equal to the wavelength (λ) divided by the period (T) as:

$$c \equiv \frac{\lambda}{T}$$

In real applications, angular frequency ω is a given property. Also, the speed of sound in the air is a known value, which just slightly varies with meteorological conditions. Hence, based on the known speed of sound and given a specific angular frequency, we can calculate the wavenumber and the wavelength of the acoustic wave as:

$$k = \frac{\omega}{c}$$

and

$$\lambda = \frac{2\pi}{k} = 2\pi \frac{c}{\omega} = 2\pi \frac{c}{2\pi f} = \frac{c}{f}$$

where the circular frequency f (unit in Hz) is related to the angular frequency ω (unit in rad/s) as:

$$\omega = 2\pi f$$

or

$$\omega \left[\frac{rad}{time}\right] = 2\pi [rad] \, f \left[\frac{1}{time}\right]$$

By replacing ω with ck, the backward wave $p_-(x, t)$ and the forward wave $p_+(x, t)$ can be expressed in Form 2 as:

$$\begin{aligned} p_\pm(x, t) \quad &= A_\pm \cos (\omega t \mp kx + \phi_\pm) \quad &\text{(Form 2 : } R_{EP}) \\ &= A_\pm \cos [k(ct \mp x) + \phi_\pm] \quad &\text{(Form 2 : } R_{EP}) \end{aligned}$$

or:

$$\begin{aligned} p_+(x, t) &= A_+ \cos [k(ct - x) + \phi_+] \quad &\text{(Form 2 : } R_{EP}) \\ p_-(x, t) &= A_- \cos [k(ct + x) + \phi_-] \quad &\text{(Form 2 : } R_{EP}) \end{aligned}$$

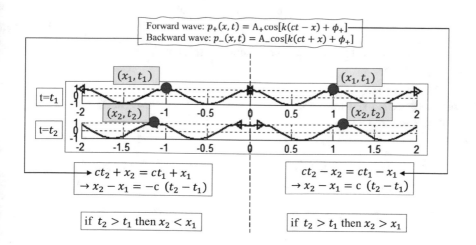

To determine the propagation direction of a wave, we can trace one point (a constant value of pressure p) in the wave from (x_1, t_1) to (x_2, t_2).

For the forward wave p_+ above, where the signs of ω and k are different, we have:

$$ct_2 - x_2 = ct_1 - x_1$$
$$\rightarrow c(t_2 - t_1) = x_2 - x_1$$

Therefore, if $t_2 > t_1$, then $x_2 > x_1$, and the wave is a forward wave traveling from left (x_1) to right (x_2).

For the backward wave p_- above, where the signs of ω and k are the same, we have:

$$ct_2 + x_2 = ct_1 + x_1$$
$$\rightarrow c(t_2 - t_1) = -(x_2 - x_1)$$

Therefore, if $t_2 > t_1$, then $x_1 > x_2$, and the wave is a backward wave traveling from right (x_1) to left (x_2).

In this course, the solutions of the acoustic wave equation are represented as harmonic waves. Using Fourier analysis, any traveling waves can be represented by the summation of a series of harmonic waves.

However, another common approach for analyzing the wave equation is by using the function in moving wave form, $f(x + ct)$ or $f(x - ct)$. Any moving wave function $f(x + ct)$ or $f(x - ct)$ is a solution of the acoustic wave equation. This can be easily proved by substituting the function to the wave equation and using the following relations:

$$\frac{\partial}{\partial x} f(x \pm ct) = f'(x \pm ct), \quad \text{and} \quad \frac{\partial^2}{\partial x^2} f(x \pm ct) = f''(x \pm ct)$$

$$\frac{\partial}{\partial t} f(x \pm ct) = \pm c f'(x \pm ct), \quad \text{and} \quad \frac{\partial^2}{\partial t^2} f(x \pm ct) = c^2 f''(x \pm ct)$$

All BTWs are the special cases (i.e., harmonic waves) of these moving wave functions. At a fixed location, say $x = 0$, any arbitrary wave function, $p(t)$, can be decomposed into a series of harmonic waves:

$$p(t) = A_o + 2 \sum_{k=1}^{N/2-1} [A_k \cos(\omega_k t) - B_k \sin(\omega_k t)] + A_{N/2}$$

where:

$$A_k + jB_k = \frac{1}{N} \sum_{i=0}^{N-1} p_i \cdot \left(e^{-j(2\pi/N)} \right)^{k \cdot i}$$

This is the discrete Fourier transform formula we will be using to calculate the weighted sound pressure level spectrum in Sect. 9.5.

3.7 Visualization of Acoustic Waves

3.7.1 Plotting Traveling Wave

The solutions of the acoustic wave equation expressed in both complex exponential functions and real trigonometric functions are derived in the previous chapter, and only Form 2 : R_{EP} is shown below for reference:

$$p(x, t) = p_-(x, t) + p_+(x, t)$$
$$= A_- \cos (\omega t + kx + \phi_-) + A_+ \cos (\omega t - kx + \phi_+) \quad \text{(Form 2 : } R_{EP})$$

where $\cos(\omega t + kx + \phi_-)$ is the *backward wave* since ωt and kx have the *same signs* and $\cos (\omega t - kx + \phi_+)$ is the *forward wave* since ωt and kx have *different signs*.

For simplicity, let both phase shift angles ϕ_- and ϕ_+ be zero and both amplitudes A_- and A_+ be one, and we thus arrive at:

$$p(x, t) = p_-(x, t) + p_+(x, t)$$
$$= A_- \cos (\omega t + kx) + A_+ \cos (\omega t - kx) \quad \text{(Form 2 : } R_{EP})$$

For plotting a backward wave, assume a positive increment of time $\Delta t = \frac{T}{8}$ and a negative decrement in space (left-hand side of the origin) $\Delta x = \frac{-\lambda}{8}$. The backward wave becomes:

$$p_-(x, t) = \cos (\omega t + kx) = \cos \left[\omega \left(t + \frac{T}{8} \right) + k \left(x + \frac{-\lambda}{8} \right) \right]$$

For plotting a forward wave, assume a positive increment of time $\Delta t = \frac{T}{8}$ and a positive increment in space (right-hand side of the origin) $\Delta x = \frac{\lambda}{8}$. The forward wave then becomes:

$$p_+(x, t) = \cos (\omega t - kx) = \cos \left[\omega \left(t + \frac{T}{8} \right) - k \left(x + \frac{+\lambda}{8} \right) \right]$$

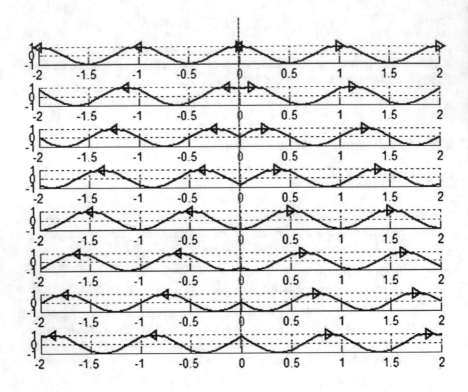

3.7.2   Plotting Standing Wave

This section will demonstrate how standing waves work. The pressure function of a standing wave is separated into two pressure functions of a backward wave and a forward wave:

$$p(x,t) = 2\cos(\omega t)\cos(kx) = p_-(x,t) + p_+(x,t)$$

For a source at $x = 0$, the pressure functions for the backward wave and the forward wave are:

Backward wave:

$$p_-(x,t) = \cos(\omega t + kx)$$

Forward wave:

$$p_+(x,t) = \cos(\omega t - kx)$$

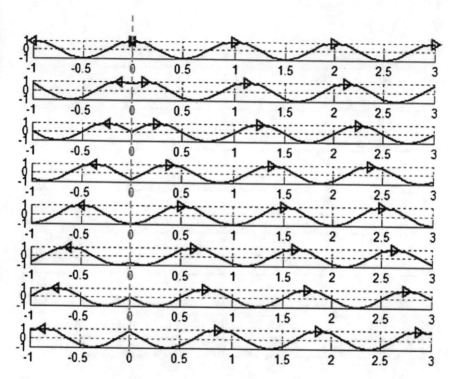

For a source at $x = 2\lambda$, the pressure functions for the backward wave and the forward wave are as follows:

$$p_- = \cos(\omega t + k(x - 2\lambda)) = \cos(\omega t + kx - 2k\lambda)$$
$$= \cos(\omega t + kx - 4\pi) = \cos(\omega t + kx)$$
$$p_+ = \cos(\omega t - k(x - 2\lambda)) = \cos(\omega t - kx + 2k\lambda)$$
$$= \cos(\omega t - kx + 4\pi) = \cos(\omega t - kx)$$

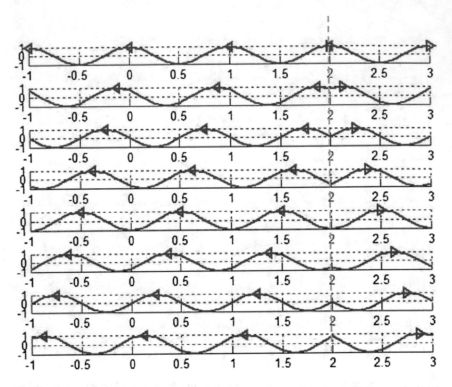

Combining the waves induced by the sources at $x = 0$ and $x = 2\lambda$, we arrive at the standing wave between $x = (0,\ 2\lambda)$:

$$p = p_- + p_+ = \cos(\omega t + kx) + \cos(\omega t - kx) = 2\cos(\omega t)\cos(kx)$$

In addition, the wave at the left of $x = 0$ yields:

$$p = p_- + p_- = 2\cos(\omega t + kx)$$

Moreover, the wave at the right of $x = 2\lambda$ thusly yields:

$$p = p_+ + p_+ = 2\cos(\omega t - kx)$$

The following figure shows the motion of a standing wave (shown in blue curves) as the result of the addition of a forward traveling wave (shown in black curves) and a backward traveling wave (shown in red curved) between $x = (0,\ 2\lambda)$:

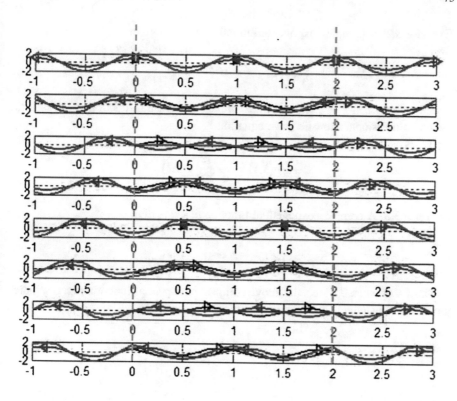

3.8 Homework Exercises

Exercise 3.1 A complex exponential function $p(x, t)$ is given as:

$$p(x, t) = e^{j\left(\omega t + kx - \frac{\pi}{2}\right)} \qquad [C_{EP}]$$

Show that when $\frac{\omega}{k} = \pm c$, the above complex exponential function $p(x, t)$ satisfies the one-dimensional acoustic wave equation:

$$\frac{\partial^2}{\partial x^2} p(x, t) = \frac{1}{c^2} \frac{\partial^2}{\partial t^2} p(x, t)$$

Exercise 3.2 A complex exponential function $p(x, t)$ is given as:

$$p(x,t) = e^{j\left(\omega t - kx - \frac{\pi}{2}\right)} \qquad [C_{EP}]$$

Show that when $\frac{\omega}{k} = \pm c$, the above complex exponential function $p(x, t)$ satisfies the one-dimensional acoustic wave equation:

$$\frac{\partial^2}{\partial x^2} p(x,t) = \frac{1}{c^2} \frac{\partial^2}{\partial t^2} p(x,t)$$

Exercise 3.3 A complex exponential function $p(x, t)$ is give as:

$$p(x,t) = e^{-j\left(\omega t - kx - \frac{\pi}{2}\right)} \qquad [C_{EP}]$$

Show that when $\frac{\omega}{k} = \pm c$, the above complex exponential function $p(x, t)$ satisfies the one-dimensional acoustic wave equation:

$$\frac{\partial^2}{\partial x^2} p(x,t) = \frac{1}{c^2} \frac{\partial^2}{\partial t^2} p(x,t)$$

Exercise 3.4 A backward traveling wave $p_-(x, t)$ with a phase shift (Form 2:R_{EP}) is given as:

$$p_-(x,t) = A_- \cos\left(\omega t + kx + \phi_-\right)$$

where A_- is an amplitude. Show that if $\frac{\omega}{k} = \pm c$, the above trigonometric cosine function $p(x, t)$ satisfies the one-dimensional acoustic wave equation:

$$\frac{\partial^2}{\partial x^2} p(x,t) = \frac{1}{c^2} \frac{\partial^2}{\partial t^2} p(x,t)$$

Exercise 3.5 Prove that the following relationship between a standing wave and the combination of a forward wave and a backward wave is true:

$$\sin\left(\omega t\right) \sin\left(kx\right) = \frac{-1}{2}\left(\cos\left(\omega t + kx\right) - \cos\left(\omega t - kx\right)\right)$$

Exercise 3.6 Prove that the following relationship between a standing wave and the combination of a forward wave and a backward wave is true:

$$\sin\left(\omega t\right)\cos\left(kx\right) = \frac{1}{2}\left(\sin\left(\omega t + kx\right) + \sin\left(\omega t - kx\right)\right)$$

Exercise 3.7 Find the relations between phases θ_t, θ_x, ϕ_+, and ϕ_- that satisfy the following relationship between a standing wave and the combination of a forward wave and a backward wave:

$$\cos\left(\omega t + \theta_t\right)\cos\left(kx + \theta_x\right) = \cos\left(\omega t + kx + \phi_-\right) + \cos\left(\omega t - kx + \phi_+\right) \quad (3.2)$$

(Answers): (You must show all of your work for full credit!)

$$\phi_- = \theta_t + \theta_x$$
$$\phi_+ = \theta_t - \theta_x$$

Exercise 3.8 Find the relations between phases θ_t, θ_x, ϕ_+, and ϕ_- that satisfy the following relationship between a standing wave and the combination of a forward wave and a backward wave:

$$2\sin\left(\omega t + \theta_t\right)\cos\left(kx + \theta_x\right) = \sin\left(\omega t + kx + \phi_-\right) + \sin\left(\omega t - kx + \phi_+\right)$$

(Answers): (You must show all of your work for full credit!)

$$\phi_- = \theta_t + \theta_x$$
$$\phi_+ = \theta_t - \theta_x$$

Exercise 3.9 The following is a harmonic function $p(x, t)$ of one-dimensional sound pressure in Cartesian coordinates:

$$p(x,t) = Re\left[P\sin(kx)e^{j(\omega t)}\right]$$

where P is a real number.

a) Show that the harmonic function $p(x, t)$ can be constructed by multiplying two trigonometric cosine functions as:

$$p(x,t) = P\cos(\omega t)\sin\left(kx\right)$$

b) Show that the harmonic function $p(x, t)$ can be constructed by adding two trigonometric cosine functions as:

$$p(x, t) = \frac{P}{2} \left[\cos \left(kx + \omega t - \frac{\pi}{2} \right) + \cos \left(kx - \omega t - \frac{\pi}{2} \right) \right]$$

Note that each term in the above function has a phase of $\frac{\pi}{2}$ and is considered as Form 2.

Exercise 3.10 The following is a harmonic function $p(x, t)$ of one-dimensional sound pressure in Cartesian coordinates:

$$p(x, t) = \text{Re} \left[jP \cos (kx) e^{j(\omega t)} \right]$$

where P is a real number.

(a) Show that the harmonic function $p(x, t)$ can be constructed by multiplying two trigonometric cosine functions:

$$p(x, t) = -P \sin (\omega t) \cos (kx)$$

(b) Show that the harmonic function $p(x, t)$ can be constructed by adding two trigonometric sine functions:

$$p(x, t) = \frac{-P}{2} [\sin (\omega t + kx) + \sin (\omega t - kx)]$$

Note that each term in the above function has no phase and can be considered as Form 1.

Exercise 3.11 The following is a harmonic function $p(x, t)$ of one-dimensional sound pressure in Cartesian coordinates:

$$p(x, t) = \text{Re} \left[A \sin \left(kx - \frac{\pi}{2} \right) e^{j(\omega t)} \right]$$

where A is a real number.

(a) Show that the harmonic function $p(x, t)$ can be constructed by multiplying two trigonometric cosine functions:

$$p(x, t) = -A \cos (kx) \cos (\omega t)$$

(b) Show that the harmonic function $p(x, t)$ can be constructed by adding two trigonometric cosine functions:

$$p(x,t) = \frac{-A}{2}[\cos{(\omega t + kx)} + \cos{(\omega t - kx)}]$$

Exercise 3.12 Use the following combination of a forward wave $p_+(x,t)$ and a backward wave $p_-(x,t)$ to answer every question of this problem:

$$p_-(x,t) = \sin\left(\omega t + kx - \frac{\pi}{2}\right)$$

$$p_+(x,t) = -\sin{(\omega t - kx)}$$

a) Calculate the standing wave (real number) produced by the forward wave and the backward.
b) Find the locations of the peaks and valleys in terms of wavelength.

(Answers):

a) $2\cos\left(\omega t + \frac{\pi}{4}\right)\cos\left(kx - \frac{5\pi}{4}\right)$
a) Peak: $x = \frac{5\lambda}{8}$; Valley: $x = \frac{\lambda}{8}$

Chapter 4
Acoustic Intensity and Specific Acoustic Impedance

In the previous chapter, we constructed the sound pressures $p(x, t)$ of both traveling waves and standing waves based on the acoustic wave equation. In this chapter, you will learn how to calculate flow velocity $u(x, t)$, acoustic intensity, and specific acoustic impedance for given sound pressure. These quantities are commonly used in the analysis of acoustics and noise control. This chapter is organized as follows:

First, you will learn how to calculate the flow velocity $u(x, t)$ for a given sound pressure $p(x, t)$ using Euler's force equation that was derived in Chap. 2. This process can be reversed to calculate the sound pressure $p(x, t)$ for a given flow velocity $u(x, t)$ using Euler's force equation as shown in the figure below:

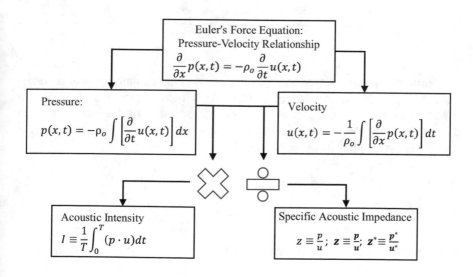

© The Author(s), under exclusive license to Springer Nature Switzerland AG 2021
H. Lin et al., *Lecture Notes on Acoustics and Noise Control*,
https://doi.org/10.1007/978-3-030-88213-6_4

Second, you will learn how to calculate acoustic intensity I from sound pressure $p(x,t)$ and its corresponding flow velocity $u(x,t)$. The acoustic intensity I is simply the average of the sound pressure $p(x,t)$ multiplied by the flow velocity $u(x,t)$ over the period T as shown in the figure above. The formula of acoustic intensity I for both traveling waves and standing waves will be derived.

Third, you will learn how to calculate specific acoustic impedance z as real numbers. The real number specific acoustic impedance z is simply the sound pressure $p(x,t)$ divided by the flow velocity $u(x,t)$ as shown in the figure above. The formulas of specific acoustic impedance z for both traveling waves and standing waves will be derived. Note that the specific acoustic impedance can be formulated in either the real number format or the complex number format. Formulations of the real number specific acoustic impedance z for traveling waves and standing waves are shown in the table below and will be developed in Sect. 4.4.

	$p(x,t)$	$u(x,t)$	p_{RMS}	I	z
Traveling Wave	$p_\pm = A_{\pm c}\cos(\omega t \mp kx)$ Form 1: R_{IP}		$p^2_{\pm RMS} = \dfrac{A^2_{\pm c}}{2}$		
	$p_\pm = A_{\pm s}\sin(\omega t \mp kx)$ Form 1: R_{IP}	$u_\pm = \pm\dfrac{1}{\rho_o c}p_\pm(x,t)$	$p^2_{\pm RMS} = \dfrac{A^2_{\pm s}}{2}$	$I_\pm = \pm\dfrac{1}{\rho_o c}p^2_{\pm RMS}$	$z_\pm = \pm\rho_o c$
	$p_\pm = A_\pm\cos(\omega t \mp kx + \theta_\pm)$ Form 2: R_{EP}		$p^2_{\pm RMS} = \dfrac{A^2_\pm}{2}$		
Standing Wave	$p_s = A\cos(\omega t)\cos(kx)$	$u_s = \dfrac{A}{\rho_o c}\sin(\omega t)\sin(kx)$	$p^2_{sRMS} = \dfrac{A^2}{2}\cos^2(kx)$		$z_s = \rho_o c\,\dfrac{\cos(\omega t)\cos(kx)}{\sin(\omega t)\sin(kx)}$
	$p_s = A\sin(\omega t)\sin(kx)$	$u_s = \dfrac{A}{\rho_o c}\cos(\omega t)\cos(kx)$	$p^2_{sRMS} = \dfrac{A^2}{2}\sin^2(kx)$	$I_s = 0$	$z_s = \rho_o c\,\dfrac{\sin(\omega t)\sin(kx)}{\cos(\omega t)\cos(kx)}$
	$p_s = A\cos(\omega t)\sin(kx)$	$u_s = \dfrac{-A}{\rho_o c}\sin(\omega t)\cos(kx)$	$p^2_{sRMS} = \dfrac{A^2}{2}\sin^2(kx)$		$z_s = -\rho_o c\,\dfrac{\cos(\omega t)\sin(kx)}{\sin(\omega t)\cos(kx)}$
	$p_s = A\sin(\omega t)\cos(kx)$	$u_s = \dfrac{-A}{\rho_o c}\cos(\omega t)\sin(kx)$	$p^2_{sRMS} = \dfrac{A^2}{2}\cos^2(kx)$		$z_s = -\rho_o c\,\dfrac{\sin(\omega t)\cos(kx)}{\cos(\omega t)\sin(kx)}$

Fourth, the acoustic impedance z expressed as complex numbers will be introduced in Sect. 4.5. This is an advanced topic but is important for analyzing acoustic waves in pipes. The formulas of complex number acoustic impedance will be discussed and derived. An application of complex number acoustic impedance for filter designs is applied in Chaps. 12 and 13.

4.1 Pressure-Velocity Relationship

The relationship between $p(x,t)$ and $u(x,t)$, based on Euler's force equation, is:

$$\frac{\partial}{\partial x}p(x,t) = -\rho_o\frac{\partial}{\partial t}u(x,t)$$

For a given pressure $p(x, t)$, the flow velocity $u(x, t)$ can be formulated as:

$$u(x, t) = -\frac{1}{\rho_o} \int \left[\frac{\partial}{\partial x} p(x, t) \right] dt$$

Formulas of pressure-velocity relationships for traveling waves $u_\pm(x, t)$ and standing waves $u_s(x, t)$ will be developed based on Euler's force equation shown above. The following is a summary of formulas of pressure-velocity relationships that will be derived in this section:

$$u_\pm(x, t) = \pm \frac{1}{\rho_o c} p_\pm(x, t) \qquad \text{(BTW)}$$

$$u_s(x, t) = \frac{1}{\rho_o c} \sin(\omega t) \sin(kx) \text{ if } p_s(x, t) = \cos(\omega t) \cos(kx) \qquad \text{(BTW)}$$

$$u_s(x, t) = \frac{1}{\rho_o c} \cos(\omega t) \cos(kx) \text{ if } p_s(x, t) = \sin(\omega t) \sin(kx) \qquad \text{(BTW)}$$

$$u_s(x, t) = -\frac{1}{\rho_o c} \sin(\omega t) \cos(kx) \text{ if } p_s(x, t) = \cos(\omega t) \sin(kx) \qquad \text{(BTW)}$$

$$u_s(x, t) = -\frac{1}{\rho_o c} \cos(\omega t) \sin(kx) \text{ if } p_s(x, t) = \sin(\omega t) \cos(kx) \qquad \text{(BTW)}$$

Note that the flow velocity $u_\pm(x, t)$ above was referred to as v_f in Chap. 2. The $u_\perp(x, t)$ is the macroscopic flow velocity v_f and should not be confused with the microscopic colliding velocity v_c. Also, (BTW) indicates basic traveling waves, and BSW indicates basic standing waves as discussed in Chap. 3.

4.1.1 Pressure-Velocity Relationships for BTW

Even though there are no linear relationships between pressure and velocity for general complex waves, there are linear relationships between pressure and velocity for *forward and backward traveling waves*. The formulations are shown below.

Forward Traveling Waves

$$p_+(x, t) = \cos(\omega t - kx + \phi)$$

Or:

$$\rightarrow u_+(x, t) = +\frac{1}{\rho_o c} p_+(x, t)$$

$$\rightarrow p_+(x, t) = \rho_o c \, u_+(x, t)$$

Sound Pressure   Molecular Flow Velocity

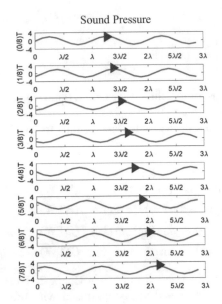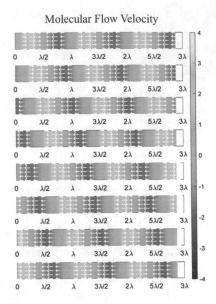

Backward Traveling Waves

$$p_-(x,t) = \cos(\omega t + kx + \phi)$$

Or:

$$\rightarrow u_-(x,t) = -\frac{1}{\rho_o c} p_-(x,t)$$

$$\rightarrow p_-(x,t) = -\rho_o c\, u_-(x,t)$$

The following is an example of how to prove the formulations above.
The pressure of a backward traveling wave is:

$$p_-(x,t) = \cos(\omega t + kx + \phi)\left[\text{or} = \frac{1}{2}\left(e^{j(\omega t+kx+\phi)} + e^{-j(\omega t+kx+\phi)}\right)\right]$$

According to Euler's force equation, the velocity can be calculated by first taking the derivative of the pressure with respect to x and then taking its integration with respect to time t as:

$$\frac{\partial}{\partial x} p_-(x,t) = -k\,\sin(\omega t + kx + \phi)$$

$$\left[or = \frac{jk}{2} \left(e^{j(\omega t + kx + \phi)} - e^{-j(\omega t + kx + \phi)} \right) \right]$$

$$\rightarrow u_-(x,t) = -\frac{1}{\rho_o} \int \left[\frac{\partial}{\partial x} p_-(x,t) \right] dt$$

$$= -\frac{1}{\rho_o c} \cos(\omega t + kx + \phi)$$

$$\left[or = -\frac{1}{\rho_o c} \frac{1}{2} \left(e^{j(\omega t + kx + \phi)} + e^{-j(\omega t + kx + \phi)} \right) \right]$$

$$= -\frac{1}{\rho_o c} p_-(x,t)$$

So, for backward traveling waves, the pressure and velocity relationship is:

$$u_-(x,t) = -\frac{1}{\rho_o c} p_-(x,t)$$

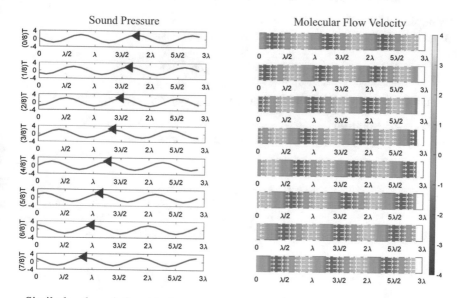

Similarly, the relationship between pressure and velocity for forward traveling waves can be derived as:

$$u_+(x,t) = \frac{1}{\rho_o c} p_+(x,t)$$

Therefore, we get the relationship between pressure and velocity as:

$$\rightarrow u_\pm(x,t) = \pm\frac{1}{\rho_o c} p_\pm(x,t) \qquad\qquad [BTW]$$

4.1.2   Pressure-Velocity Relationships for BSW

The pressure-velocity relationships for standing waves are different from the pressure-velocity relationships for the previous traveling waves.

The relationships between pressure and velocity for basic standing waves are:

$$p_s(x,t) = \cos(\omega t)\cos(kx) \rightarrow u(x,t) = \frac{1}{\rho_o c}\sin(\omega t)\sin(kx)$$

$$p_s(x,t) = \sin(\omega t)\sin(kx) \rightarrow u(x,t) = \frac{1}{\rho_o c}\cos(\omega t)\cos(kx)$$

$$p_s(x,t) = \sin(\omega t)\cos(kx) \rightarrow u(x,t) = -\frac{1}{\rho_o c}\cos(\omega t)\sin(kx)$$

$$p_s(x,t) = \cos(\omega t)\sin(kx) \rightarrow u(x,t) = -\frac{1}{\rho_o c}\sin(\omega t)\cos(kx)$$

The following is an example to validate the statement above.

Example 4.1
Given a basic standing wave with two cosine functions:

$$p_s(x,t) = \cos(\omega t)\cos(kx)$$

Find the corresponding particle velocity using Euler's force equation.

Example 4.1 Solution
The flow velocity can be calculated using Euler's force equation:

$$u_s(x,t) = \frac{-1}{\rho_o}\int\left\{\frac{\partial}{\partial x}p(x,t)\right\}dt$$

$$= \frac{-1}{\rho_o}\int\left\{\frac{\partial}{\partial x}[\cos(\omega t)\cos(kx)]\right\}dt$$

$$= \frac{-1}{\rho_o}\int\left\{\cos(\omega t)\frac{\partial}{\partial x}[\cos(kx)]\right\}dt$$

$$= \frac{-1}{\rho_o}\int\{-k\cos(\omega t)\sin(kx)\}dt$$

$$= \frac{1}{\rho_o}k\cdot\sin(kx)\int\cos(\omega t)dt$$

$$= \frac{1}{\rho_o} \frac{k}{\omega} \sin(kx) \sin(\omega t)$$

$$= \frac{1}{\rho_o c} \sin(kx) \sin(\omega t)$$

From the results above, we can conclude that:

$$u_s(x, t) \neq \frac{1}{\rho_o c} p(x, t)$$

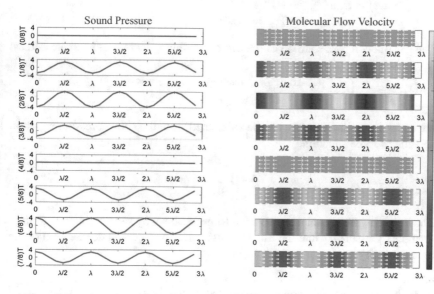

The following is another example of validating these pressure-velocity relationships.

Example 4.2

Given a basic standing wave with cosine and sine functions:

$$p_s(x, t) = \cos(\omega t) \sin(kx)$$

Find the corresponding particle velocity using Euler's force equation.

Example 4.2 Solution

The flow velocity can be calculated using Euler's force equation:

$$u(x, t) = -\frac{1}{\rho_o} \int \left[\frac{\partial}{\partial x} p(x, t) \right] dt = -\frac{k}{\rho_o \omega} \sin(\omega t) \cos(kx)$$

$$= -\frac{1}{\rho_o c} \sin(\omega t) \cos(kx) \neq -\frac{1}{\rho_o c} p(x, t)$$

4.1.3 Pressure-Velocity Relationships in Complex Function Form

The general complex solution of sound pressure for acoustic waves is:

$$p(x,t) = A_1 B_1 e^{j(\omega t + kx)} + A_1 B_2 e^{-j(\omega t - kx)}$$
$$+A_2 B_1 e^{j(\omega t - kx)} + A_2 B_2 e^{-j(\omega t + kx)} \qquad \text{(Form 4 : } C_{IP})$$

or:

$$p(x,t) = P_+ e^{j(\omega t - kx)} + P_+^* e^{-j(\omega t - kx)}$$
$$+P_- e^{j(\omega t + kx)} + P_-^* e^{-j(\omega t + kx)} \qquad \text{(Form 4 : } C_{IP})$$

The general complex solution of the corresponding velocity can be expressed as:

$$u(x,t) = \frac{1}{\rho_o c}\left(P_+ e^{j(\omega t - kx)} + P_+^* e^{-j(\omega t - kx)} - P_- e^{j(\omega t + kx)} - P_-^* e^{-j(\omega t + kx)}\right)$$

where each term of the velocity can be calculated from the above pressure equation using Euler's force equation:

$$u_\pm(x,t) = \pm\frac{1}{\rho_o c}p_\pm(x,t)$$

The following two examples prove the formula above by substituting one term of the complex pressure $p_\pm$ into the Euler's force equation to calculate one term of the complex velocity $u_\pm$.

Example 4.3
Assume that a pressure function is constructed with only the first term of the general solution of sound pressure as shown at the beginning of this section and that the other coefficients (complex numbers) are all zero. In this case:

$$p_-(x,t) = e^{j(\omega t + kx)}$$

Use Euler's force equation to get the flow velocity $u_-(x,t)$ from the given pressure $p_-(x,t)$.

Example 4.3 Solution

$$\frac{\partial}{\partial x}p_-(x,t) = jk\, e^{j(\omega t + kx)}$$

$$u_-(x,t) = -\frac{1}{\rho_o}\int\left[\frac{\partial}{\partial x}p_-(x,t)\right]dt = -\frac{1}{\rho_o}\int\left[jk\, e^{j(\omega t + kx)}\right]dt$$

$$= -\frac{k}{\rho_o \omega}e^{j(\omega t + kx)} = -\frac{1}{\rho_o c}e^{j(\omega t + kx)} = -\frac{1}{\rho_o c}p_-(x,t)$$

Example 4.4

Use Euler's force equation to get the flow velocity $u_+(x, t)$ from the given pressure as shown below:

$$p_+(x, t) = e^{j(\omega t - kx)}$$

Example 4.4 Solution

$$\frac{\partial}{\partial x} p_+(x, t) = -jk \, e^{j(\omega t - kx)}$$

$$u_+(x, t) = -\frac{1}{\rho_o} \int \left[\frac{\partial}{\partial x} p_-(x, t) \right] dt = -\frac{1}{\rho_o} \int \left[-jk \, e^{j(\omega t - kx)} \right] dt$$

$$= \frac{k}{\rho_o \omega} e^{j(\omega t - kx)} = \frac{1}{\rho_o c} e^{j(\omega t - kx)} = \frac{1}{\rho_o c} p_+(x, t)$$

Based on the two examples above, we can conclude that:

$$u_\pm(x, t) = \pm \frac{1}{\rho_o c} p_\pm(x, t)$$

Remark:

Note that for general complex waves, there are no linear relationships between pressure and velocity:

$$u(x, t) \neq \frac{1}{\rho_o c} p(x, t) \neq -\frac{1}{\rho_o c} p(x, t)$$

4.2 RMS Pressure

The root-mean-square (RMS) pressure is defined as:

$$p_{RMS}^2 \equiv \frac{1}{T} \int_0^T p^2 \, dt$$

where T is the period of the pressure wave.

Formulas of RMS pressure for traveling waves $p_\pm(x, t)$ and standing waves $p_s(x, t)$ will be developed based on the definition of RMS pressure shown above. The following is a summary of formulas of RMS pressure that will be derived in this section:

$$p^2_{\pm RMS} \equiv \frac{A^2}{2} \qquad\qquad\qquad\qquad\qquad\qquad\qquad\text{[BTW]}$$

$$p^2_{sRMS} = \frac{A^2}{2}\cos^2(kx) \text{ if } p_s(x,t) = A\cos(\omega t)\cos(kx) \quad\text{[BSW]}$$

$$p^2_{sRMS} = \frac{A^2}{2}\sin^2(kx) \text{ if } p_s(x,t) = A\sin(\omega t)\sin(kx) \quad\text{[BSW]}$$

$$p^2_{sRMS} = \frac{A^2}{2}\sin^2(kx) \text{ if } p_s(x,t) = A\cos(\omega t)\sin(kx) \quad\text{[BSW]}$$

$$p^2_{sRMS} = \frac{A^2}{2}\cos^2(kx) \text{ if } p_s(x,t) = A\sin(\omega t)\cos(kx) \quad\text{[BSW]}$$

4.2.1 RMS Pressure of BTW

The RMS pressure of traveling waves is defined as follows:

$$p^2_{+RMS} \equiv \frac{A^2}{2}$$

$$p^2_{-RMS} \equiv \frac{A^2}{2}$$

where A is the amplitude of the traveling waves of the four basic traveling waves:

$$p_-(x,t) = A \cdot \cos(\omega t + kx)$$
$$p_+(x,t) = A \cdot \cos(\omega t - kx)$$
$$p_-(x,t) = A \cdot \sin(\omega t + kx)$$
$$p_+(x,t) = A \cdot \sin(\omega t - kx)$$

The following is an example of validating these relationships:

For forward traveling wave $p_+(x,t) = \cos(\omega t - kx)$, the RMS pressure is by definition:

$$p^2_{RMS} \equiv \frac{1}{T}\int_0^T p^2 dt = \frac{1}{T}\int_0^T (A\cdot\cos(\omega t - kx))^2 dt$$

For general traveling waves, RMS pressure is independent of the choice of location. Therefore, we can calculate the square of RMS pressure at $x = 0$. Therefore:

$$p^2_{RMS} = A^2 \frac{1}{T}\int_0^T \cos^2(\omega t) dt$$

Because $\omega = \frac{2\pi}{T}$, replace ω with $\frac{2\pi}{T}$ to get:

$$p_{RMS}^2 = A^2 \frac{1}{T} \int_0^T \cos^2\left(\frac{2\pi}{T}t\right) dt$$

This means that the square of RMS pressure is an integral over a period 2π. Note that the geometry relationship is used to yield:

$$p_{RMS}^2 = A^2 \frac{1}{T} \int_0^T \cos^2\left(\frac{2\pi}{T}t\right) dt = A^2 \frac{1}{2} = \frac{A^2}{2}$$

4.2.2   RMS Pressure of BSW

The square of the RMS pressure of standing waves can be illustrated as:

$$p_s(x,t) = A \cos(\omega t) \cos(kx) \rightarrow p_{sRMS}^2 = \frac{A^2}{2} \cos^2(kx)$$

$$p_s(x,t) = A \sin(\omega t) \sin(kx) \rightarrow p_{sRMS}^2 = \frac{A^2}{2} \sin^2(kx)$$

$$p_s(x,t) = A \cos(\omega t) \sin(kx) \rightarrow p_{sRMS}^2 = \frac{A^2}{2} \sin^2(kx)$$

$$p_s(x,t) = A \sin(\omega t) \cos(kx) \rightarrow p_{sRMS}^2 = \frac{A^2}{2} \cos^2(kx)$$

The following is an example to validate one of the above relationships between pressure and RMS pressure:

The square of the RMS pressure is defined as:

$$P_{RMS}^2 = \frac{1}{T} \int_0^T p^2 \, dt = \frac{1}{T} \int_0^T (A \cos(\omega t) \cos(kx))^2 \, dt$$

Bringing the time-independent function $\cos(kx)$ outside of the integral yields:

$$P_{RMS}^2 = A^2 \cos^2(kx) \frac{1}{T} \int_0^T \cos^2(\omega t) dt = \frac{A^2}{2} \cos^2(kx)$$

$$P_{RMS} = \left| \frac{A}{\sqrt{2}} \cos(kx) \right|$$

Example 4.5 (Traveling Wave with Known Velocity)

Use the following harmonic motion as velocity $u(t)$ to answer every question of this problem:

$$u(t) = V_o \cos\left(\omega t + \frac{\pi}{2}\right); \quad V_o \text{ is a real number.}$$

A large, rigid wall is vibrating with a velocity $u(t)$ perpendicular to its planar surface. The motion of the wall creates plane waves that propagate into the air. Considering that the flow velocity of the plane wave on the surface of the wall ($x = 0$) would be equal to the velocity of the wall:

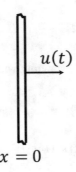

a) Obtain the expressions for the flow velocity at any point in space (not just on the wall surface) of the plane wave radiating from the wall in terms of the wall vibration velocity.

b) Obtain the expressions for the sound pressure at any point in space (not just on the wall surface) of the plane wave radiating from the wall in terms of the wall vibration velocity.

c) What is the root-mean-square (RMS) pressure at any point in space if the wall velocity $V_o = 0.01$ m/s.

Example 4.5 Solution

a) The general solution of the forward plane wave with an amplitude of A:

$$u(x, t) = A \cos(\omega t - kx + \theta)$$

Find the unknown amplitude A and phase delay θ by the given boundary conditions at $x = 0$:

$$u(x = 0, t) = V_o \cos\left(\omega t + \frac{\pi}{2}\right)$$

Substituting the boundary condition into general solution yields:

$$V_o \cos \left(\omega t + \frac{\pi}{2} \right) = A \cos \left(\omega t + \theta \right)$$

Hence:

$$\theta = \frac{\pi}{2}$$ (and)

$$A = V_o,$$

Therefore, the general expression of the particle velocity at any point to the right of the wall is:

$$u(x, t) = V_o \cos \left(\omega t - kx + \frac{\pi}{2} \right)$$

b) The acoustic pressure can be calculated from the flow velocity using Euler's force equation.

For a forward traveling wave, the acoustic pressure is related to the flow velocity as follows:

$$p_+(x, t) = \rho_o c\, u(x, t)$$

$$= \rho_o c V_o \cos \left(\omega t - kx + \frac{\pi}{2} \right)$$

c) Substituting the following known variables:

$$V_o = 0.01 \frac{m}{s}$$

$$\rho_o c = 415 \ [rayles] \ or \ \left[Pa \frac{s}{m} \right]$$

into pressure function $p(x, t)$ above yields:

$$p_+(x, t) = 4.15 \cos \left(\omega t - kx + \frac{\pi}{2} \right) \ [Pa]$$

For traveling waves, forward or backward, the *RMS* pressure is independent of space.
Hence:

$$P_{RMS}^2 = \frac{1}{2}(4.15)^2 \ [Pa^2]$$

$$\rightarrow \quad P_{RMS} = \sqrt{\frac{(4.15)^2}{2}} \ [Pa]$$

4.3   Acoustic Intensity

The instantaneous intensity $I_i(t)$ of a sound wave is defined as the instantaneous rate of work done by one element of the fluid on an adjacent element per unit area.

Hence:

$$I_i(t) = \frac{w}{S} = \frac{F}{S}u = p(t)u(t)$$

where:

S is surface area.

F is force.

w is power, which is defined as $\frac{\text{work done}}{\text{time}} = \frac{\text{force}*\text{distance}}{\text{time}} = \text{force} * \text{velocity}$.

The acoustic intensity I is simply the time average of $I_i(t)$. Hence:

$$I \equiv \frac{1}{T}\int_0^T I_i(t)dt$$

$$\rightarrow I \equiv \frac{1}{T}\int_0^T (p \cdot u)dt$$

Formulas of acoustic intensity I for traveling waves $u_{\pm}(x,t)$ and standing waves $u_s(x,t)$ will be developed based on the definition of acoustic intensity shown above. The following two formulas of pressure-velocity relationships will be derived in this section:

$$I_{\pm} = \frac{\pm 1}{\rho_o c}p_{\pm RMS}^2 \quad \text{(BTW)}$$
$$I_s = 0 \qquad\qquad \text{(BSW)}$$

4.3.1   Acoustic Intensity of BTW

The acoustic intensity $I_{\pm}$ of traveling waves is:

$$I_{\pm} = \pm\frac{1}{\rho_o c}p_{\pm RMS}^2$$

where the RMS pressure of traveling waves is:

$$p_{\pm RMS}^2 \equiv \frac{A^2}{2}$$

where A is the amplitude of the traveling waves of the four basic traveling waves:

$$p_-(x, t) = A \cdot \cos(\omega t + kx)$$
$$p_+(x, t) = A \cdot \cos(\omega t - kx)$$
$$p_-(x, t) = A \cdot \sin(\omega t + kx)$$
$$p_+(x, t) = A \cdot \sin(\omega t - kx)$$

To validate the acoustic intensity formula for the forward traveling waves, the following forward traveling waves are shown as an example below:

$$p_+(x, t) = A \cdot \cos(\omega t - kx)$$

Based on Euler's force equation, the flow velocity is given by:

$$p_+(x, t) = A \cdot \cos(\omega t - kx) \rightarrow u_+(x, t) = +\frac{1}{\rho_o c} p_+(x, t)$$

Hence, the time average of instantaneous intensity can be written as:

$$I_+ = \frac{1}{T} \int_0^T p_+ u_+ dt = \frac{1}{\rho_o c} \frac{1}{T} \int_0^T p_+^2 dt = \frac{1}{\rho_o c} p_{+RMS}^2$$

To validate the acoustic intensity formula for backward traveling waves, the following backward traveling waves are shown as an example below:

$$p_-(x, t) = A \cdot \cos(\omega t + kx)$$

Based on Euler's force equation, the flow velocity is:

$$p_-(x, t) = A \cdot \cos(\omega t + kx) \rightarrow u_-(x, t) = -\frac{1}{\rho_o c} p_-(x, t)$$

Hence, the time average of instantaneous intensity can be written as:

$$I_- = \frac{1}{T} \int_0^T p_- u_- dt = \frac{-1}{\rho_o c} \frac{1}{T} \int_0^T p_-^2 dt = \frac{-1}{\rho_o c} p_{-RMS}^2$$

4.3.2 Acoustic Intensity of BSW

The acoustic intensity of a standing wave is:

$$I_s = 0$$

The following is an example to validate the above statement.

Using the following basic standing wave as an example yields:

$$p_s(x, t) = A cos(\omega t) cos (kx)$$

The flow velocity of a standing wave by Euler's force equation is:

$$u_s(x, t) = \frac{A}{\rho_o c} \sin (kx) \sin (\omega t)$$

The acoustic intensity can be calculated by using the pressure and the velocity:

$$
\begin{aligned}
I_s &= \frac{1}{T} \int_0^T p \cdot u \, dt \\
&= \frac{1}{\rho_o c} \frac{1}{T} \int_0^T A \cos (\omega t) \cos (kx) A \sin (\omega t) \sin (kx) \, dt \\
&= \frac{1}{\rho_o c} \frac{A^2}{T} \cos (kx) \sin (kx) \int_0^T \cos (\omega t) \sin (\omega t) \, dt \\
&= \frac{1}{\rho_o c} A^2 \frac{1}{4} \sin (2kx) \frac{1}{T} \int_0^T \sin \left(\frac{4\pi}{T} t\right) dt = 0
\end{aligned}
$$

4.4 Specific Acoustic Impedance Expressed as Real Numbers

The real number specific acoustic impedance (real impedance) z is defined as the ratio of acoustic pressure to associated flow velocity as follows:

$$z \equiv \frac{p}{u}$$

The usage of these formulas of real impedance for standing waves is limited because they are time functions. However, because the real impedance is easier to understand and formulate, we will focus on real impedance in this section. The specific acoustic impedance expressed as complex numbers (complex impedance) will be introduced in the next section (Sect. 4.5).

Formulas of real impedance for traveling waves $p_\pm(x, t)$ and standing waves $p_s(x, t)$ will be developed based on the definition of real impedance shown above. The following is a summary of formulas of real impedance that will be derived in this section:

$$z_{\pm} = \pm \rho_o c \qquad \qquad \text{(BTW)}$$

$$z_s = \rho_o c \frac{\cos(\omega t)}{\sin(\omega t)} \frac{\cos(kx)}{\sin(kx)} \text{ if } p_s(x,t) = A\cos(\omega t)\cos(kx) \quad \text{(BSW)}$$

$$z_s = \rho_o c \frac{\sin(\omega t)}{\cos(\omega t)} \frac{\sin(kx)}{\cos(kx)} \text{ if } p_s(x,t) = A\sin(\omega t)\sin(kx) \quad \text{(BSW)}$$

$$z_s = -\rho_o c \frac{\cos(\omega t)}{\sin(\omega t)} \frac{\sin(kx)}{\cos(kx)} \text{ if } p_s(x,t) = A\cos(\omega t)\sin(kx) \quad \text{(BSW)}$$

$$z_s = -\rho_o c \frac{\sin(\omega t)}{\cos(\omega t)} \frac{\cos(kx)}{\sin(kx)} \text{ if } p_s(x,t) = A\sin(\omega t)\cos(kx) \quad \text{(BSW)}$$

4.4.1 *Specific Acoustic Impedance of BTW*

Specific acoustic impedance of traveling waves is denoted by:

$$z_{\pm} = \pm \rho_o c$$

For forward and backward traveling waves:

$$p_-(x,t) = A \cdot \cos(\omega t + kx)$$
$$p_+(x,t) = A \cdot \cos(\omega t - kx)$$
$$p_-(x,t) = A \cdot \sin(\omega t + kx)$$
$$p_+(x,t) = A \cdot \sin(\omega t - kx)$$

Let's use the following backward traveling wave as an example:

$$p_-(x,t) = A \cdot \cos(\omega t + kx)$$

The corresponding flow velocity by Euler's force equation is:

$$u_-(x,t) = -\frac{1}{\rho_o c} p_-(x,t)$$

Hence, specific acoustic impedance is by definition denoted as:

$$z_- \equiv \frac{p_-}{u_-} = \frac{p_-(x,t)}{\left[-\frac{1}{\rho_o c} p_-(x,t)\right]} = -\rho_o c$$

Similarly, for a forward traveling wave:

$$z_+ \equiv \frac{p_+}{u_+} = \frac{p_+(x,t)}{\left[\frac{1}{\rho_o c} p_+(x,t)\right]} = +\rho_o c$$

With a given specific acoustic impedance, the flow velocity can be calculated from pressure and vice versa as follows:

$$p_-(x,t) = A \cdot \cos\left(\omega t + kx\right) \rightarrow u_-(x,t) = -\frac{1}{\rho_o c} p_-(x,t) \leftrightarrow p_-(x,t)$$

$$= -\rho_o c u_-(x,t)$$

The unit of specific acoustic impedance is:

$$\frac{kg}{m^3}\frac{m}{s} = \frac{kg}{m^2}\frac{m}{s^2}\frac{s}{m} = Pa\frac{s}{m} = rayl$$

where *rayl* is a unit created in honor of John William Strutt, Baron Rayleigh.

The product has greater acoustic significance for the medium than ρ_o and c along and is also called the characteristic impedance of the medium. The value of characteristic impedance for air is 415 *rayl*.

4.4.2   Specific Acoustic Impedance of BSW

The specific acoustic impedance of standing waves is designated as:

$$p_s(x,t) = A \cos\left(\omega t\right) \cos\left(kx\right) \rightarrow \qquad z_s = \rho_o c \frac{\cos\left(\omega t\right)}{\sin\left(\omega t\right)} \frac{\cos\left(kx\right)}{\sin\left(kx\right)}$$

$$p_s(x,t) = A \sin\left(\omega t\right) \sin\left(kx\right) \rightarrow \qquad z_s = \rho_o c \frac{\sin\left(\omega t\right)}{\cos\left(\omega t\right)} \frac{\sin\left(kx\right)}{\cos\left(kx\right)}$$

$$p_s(x,t) = A \cos\left(\omega t\right) \sin\left(kx\right) \rightarrow \qquad z_s = -\rho_o c \frac{\cos\left(\omega t\right)}{\sin\left(\omega t\right)} \frac{\sin\left(kx\right)}{\cos\left(kx\right)}$$

$$p_s(x,t) = A \sin\left(\omega t\right) \cos\left(kx\right) \rightarrow \qquad z_s = -\rho_o c \frac{\sin\left(\omega t\right)}{\cos\left(\omega t\right)} \frac{\cos\left(kx\right)}{\sin\left(kx\right)}$$

The following is an example to validate one of the above relationships between pressure and specific acoustic impedance.

Let's use the following standing wave as an example.

$$p_s(x,t) = A \cos\left(\omega t\right) \cos\left(kx\right)$$

The flow velocity of a standing wave, by Euler's force equation, is expressed as:

$$u_s(x,t) = \frac{1}{\rho_o c} \sin\left(\omega t\right) \sin\left(kx\right)$$

Specific acoustic impedance is written as:

$$z \equiv \frac{p}{u} = \frac{A \cos(\omega t) \cos(kx)}{\frac{1}{\rho_o c} A \sin(\omega t) \sin(kx)} = \rho_o c \frac{\cos(\omega t)}{\sin(\omega t)} \frac{\cos(kx)}{\sin(kx)}$$

Note that the specific acoustic impedance defined in real p and u is *a time function*, $\cos(\omega t)$ and $\sin(\omega t)$, even at a fixed location x. Therefore, *the uses of the real acoustic impedance are limited*:

$$p_s(x, t) = A \cos(\omega t) \cos(kx) = p_+ + p_- = \frac{1}{2}[A \cos(\omega t - kx) + A \cos(\omega t + kx)]$$

$$= \frac{1}{4}\left[\left(Ae^{j(\omega t - kx)} + Ae^{-j(\omega t - kx)}\right) + \left(Ae^{j(\omega t + kx)} + Ae^{-j(\omega t + kx)}\right)\right]$$

$$= \frac{1}{4}\left[\left(Ae^{j(\omega t - kx)} + Ae^{j(\omega t + kx)}\right) + \left(Ae^{-j(\omega t - kx)} + Ae^{-j(\omega t + kx)}\right)\right]$$

$$= \frac{1}{4}\left[\left(Ae^{-jkx} + Ae^{jkx}\right)e^{j\omega t} + \left(Ae^{jkx} + Ae^{-jkx}\right)e^{-j\omega t}\right] = \frac{1}{2}(p + p^*)$$

The flow velocity of a standing wave, by Euler's force equation, is expressed as:

$$u_s(x, t) = \frac{1}{4\rho_o c}\left[\left(Ae^{-jkx} - Ae^{jkx}\right)e^{j\omega t} + \left(Ae^{jkx} - Ae^{-jkx}\right)e^{-j\omega t}\right] = \frac{1}{2}(u + u^*)$$

Complex specific acoustic impedance is by definition:

$$\rightarrow z = \frac{p}{u} = \rho_0 c \frac{e^{-jkx} + e^{jkx}}{e^{-jkx} - e^{jkx}} = j\rho_0 c \frac{\cos(kx)}{\sin(kx)} = j\rho_0 c \cot(kx)$$

$$z^* = \frac{p^*}{u^*} = \rho_0 c \frac{e^{jkx} + e^{-jkx}}{e^{jkx} - e^{-jkx}} = -j\rho_0 c \frac{\cos(kx)}{\sin(kx)} = -j\rho_0 c \cot(kx)$$

Note that complex specific acoustic impedance is zero at $\cos(kx) = 0$ and is infinite at $\sin(kx) = 0$. (Let $A_{+c} = A_{-c} = A/2$; $A_{+s} = A_{-s} = 0$).

Example 4.6 (Traveling Wave with Known Pressure)
Use the following backward traveling wave to answer every question of this problem:

$$p_-(x, t) = P_0 \sin\left(\omega t + kx - \frac{\pi}{2}\right)$$

a) Obtain the expressions for the flow velocity at any point in space.
b) What is the RMS pressure of the wave at any point in space?
c) What is the RMS flow velocity of the wave at any point in space?

d) What is the acoustic intensity of the wave at any point in space?
e) What is the specific acoustic impedance of the wave at any point in space?

Example 4.6 Solution
A backward traveling wave can be represented as:

$$p_-(x,t) = P_0 \sin\left(\omega t + kx - \frac{\pi}{2}\right)$$

a) For a backward traveling wave, the velocity by Euler's force equation is given by:

$$u_-(x,t) = \frac{-1}{\rho_o c} p_-(x,t) = \frac{-P_0}{\rho_o c} \sin\left(\omega t + kx - \frac{\pi}{2}\right)$$

b) For a traveling wave, forward or backward, the *RMS* pressure P_{-RMS} is given by:
   $P_{-RMS}^2 = \frac{1}{2} P_o^2$, where P_o is the amplitude of the pressure $p_-(x,t)$

   Hence:

$$P_{-RMS} = \frac{1}{\sqrt{2}} |P_o|$$

c) The RMS velocity U_{-RMS} is similar to the RMS pressure, given by:

$U_{-RMS}^2 = \frac{1}{2} \left(\frac{-P_0}{\rho_o c}\right)^2$, where $\frac{-P_0}{\rho_o c}$ is the amplitude of velocity $u_-(x,t)$

   Thus:

$$U_{-RMS} = \frac{1}{\sqrt{2}\rho_o c} |P_o|$$

d) The intensity of backward traveling waves is:

$$I_- = \frac{-1}{\rho_o c} p_{-RMS}^2$$

where p_{-RMS}^2 was calculated in part (b) as:

$$P_{-RMS} = \frac{1}{\sqrt{2}} |P_o|$$

Hence:

$$I_- = \frac{-1}{\rho_o c} p^2_{-RMS} = \frac{-1}{2} \frac{P_o^2}{\rho_0 c}$$

e) The specific acoustic impedance is by definition given by:

$$z_- = \frac{p_-}{u_-} = \frac{p_-}{\left(\frac{-p_-}{\rho_0 c}\right)} = -\rho_0 c$$

Example 4.7 (Standing Wave Pattern)

Use the combination of the following waves to answer every question of this problem:

$$p_-(x,t) = X_o \cos(\omega t + kx); \qquad p_+(x,t) = X_o \cos(\omega t - kx)$$

a) Calculate the standing wave (real number) produced by the forward traveling wave and the backward traveling wave.
b) What are the wave amplitude and the wavelength of the standing wave?
c) Sketch the resulting wave pattern, and indicate the location of peaks and valleys in terms of wavelength.
d) Calculate the root-mean-square (RMS) pressure at $x = \frac{\pi}{k}$.
e) Calculate the acoustic intensity of the standing wave at any point in space.
f) Calculate the specific acoustic impedance of the standing wave at any point in space.

Example 4.7 Solution

A standing wave pressure is constructed by the following two traveling waves with opposite traveling directions:

$$p_-(x,t) = X_o \cos(\omega t + kx)$$
$$p_+(x,t) = X_o \cos(\omega t - kx)$$

a) A standing wave pressure can be obtained by:

$$P_s = p_- + p_+ = X_o \cos(\omega t + kx) + X_o \cos(\omega t - kx) = 2X_o \cos(\omega t) \cos(kx)$$

b) The wave amplitude is $2X_o$, and the wavelength is $\lambda = \frac{2\pi}{k}$:
c) The wave pattern when $t = 0$ yields $\cos(\omega t) = 1$:

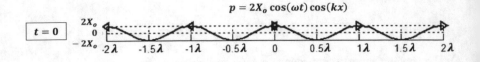

$$p = 2X_o \cos(\omega t)\cos(kx)$$

d) The *RMS* pressure of a standing wave is:

$$p_s(x,t) = (2X_o)\cos(\omega t)\cos(kx) \rightarrow p^2_{sRMS} = \frac{(2X_o)^2}{2}\cos^2(kx)$$

Hence, when:

$$x = \frac{\pi}{k} \quad \rightarrow \quad \cos(kx) = \cos(\pi) = -1$$

and:

$$p^2_{RMS} = \frac{(2X_o)^2}{2}$$

$$P_{RMS} = \sqrt{2}|X_o|$$

e) The acoustic intensity of a standing wave is given by (see note for explanation):

$$I_s = \frac{1}{T}\int_0^T pu\,dt = 0$$

f) Specific acoustic impedance is defined as:

$$z \equiv \frac{p}{u}; \quad z \equiv \frac{p}{u}; z^* \equiv \frac{p^*}{u^*}$$

Based on the formula for specific acoustic impedance of a standing wave:

$$p_s(x,t) = A\cos(\omega t)\cos(kx) \rightarrow z = \rho_o c \frac{\cos(\omega t)}{\sin(\omega t)}\frac{\cos(kx)}{\sin(kx)}$$

$$p_s(x,t) = \frac{1}{2}(p + p^*)$$

$$= \frac{1}{2}\left[\left(Ae^{-jkx} + Ae^{jkx}\right)e^{j\omega t} + \left(Ae^{jkx} + Ae^{-jkx}\right)e^{-j\omega t}\right]$$

$$u_s(x, t) = \frac{1}{2}(u + u^*)$$

$$= \frac{1}{2\rho_0 c}\left[\left(Ae^{-jkx} - Ae^{jkx}\right)e^{j\omega t} + \left(Ae^{jkx} - Ae^{-jkx}\right)e^{-j\omega t}\right]$$

$$\rightarrow z = \frac{p}{u} = \rho_0 c\,\frac{e^{-jkx} + e^{jkx}}{e^{-jkx} - e^{jkx}} = j\rho_0 c\,\frac{\cos(kx)}{\sin(kx)} = j\rho_0 c\cot(kx)$$

$$\rightarrow z^* = \frac{p^*}{u^*} = \rho_0 c\,\frac{e^{jkx} + e^{-jkx}}{e^{jkx} - e^{-jkx}} = -j\rho_0 c\,\frac{\cos(kx)}{\sin(kx)} = -j\rho_0 c\cot(kx)$$

4.5 Specific Acoustic Impedance Expressed as Complex Numbers

Acoustic impedance z expressed as complex numbers (real impedance) will be introduced in this section. This is an advanced topic but is important for analyzing acoustic waves in pipes for filter designs (Chaps. 12 and 13).

4.5.1 Issues with Real Impedance

The specific acoustic impedance defined in real p and u is a time function even at a fixed location. Therefore, the uses of the real acoustic impedance are limited as described below.

For analyzing the sound pressures in connected pipes, equilibrium equations of pressure and flow (see Chap. 11 for details) use functions of space kr. Using any function of time ωt in an equilibrium equation is impossible for analyzing the sound pressures in connected pipes. For this reason, only specific impedances defined by complex numbers of pressure and velocity are used for the analysis in filter design (Chaps. 11 and 12).

The specific acoustic impedance can be location-dependent, but they need to be time-independent for filter designs using pipes. Therefore, the time-dependent terms such as $\cos(\omega t)$ or $\sin(\omega t)$ must be eliminated in the specific acoustic impedance.

4.5.2 Definition of Complex Impedance

A universal technique to obtain a specific acoustic impedance that is *not* time-dependent is to choose terms that are associated with either $e^{j(\omega t)}$ or $e^{-j(\omega t)}$ such as:

$$z \equiv \frac{p}{u} = \frac{P_+e^{\ j(\omega t-kx)}+P_-e^{\ j(\omega t+kx)}}{\frac{1}{\rho_o c}(P_+e^{\ j(\omega t-kx)}-P_-e^{\ j(\omega t+kx)})} \qquad \text{associated with } e^{j(\omega t)} \text{ only}$$

$$z^* \equiv \frac{p^*}{u^*} = \frac{P_+^*e^{-j(\omega t-kx)}+P_-^*\ e^{-j(\omega t+kx)}}{\frac{1}{\rho_o c}\left(P_+^*e^{-j(\omega t-kx)}-P_-^*\ e^{-j(\omega t+kx)}\right)} \qquad \text{associated with } e^{-j(\omega t)} \text{ only}$$

Note that the complex specific acoustic impedances z and z^* are a complex conjugate pair. z is defined by the pressure p and velocity u which are both only associated with $e^{j(\omega t)}$ (the first and the third terms in the equation below). Its complex conjugate part z^* is defined by the pressure p^* and velocity u^* which are both only associated with $e^{-j(\omega t)}$ (the second and the fourth terms in the equation below). The complex pressure p and velocity u were defined in Sect. 4.1.3 and are relisted below for easier reference:

$$p(x,t) = P_+e^{\ j(\omega t-kx)} + P_+^*e^{-j(\omega t-kx)} + P_-e^{\ j(\omega t+kx)} + P_-^*\ e^{-j(\omega t+kx)}$$

$$u(x,t) = \frac{1}{\rho_o c}\left(P_+e^{\ j(\omega t-kx)} + P_+^*e^{-j(\omega t-kx)} - P_-e^{\ j(\omega t+kx)} - P_-^*\ e^{-j(\omega t+kx)}\right)$$

Remarks:
It can be easily observed that after canceling the common terms $e^{j\omega t}$ in z and $e^{-j\omega t}$ in z^*, the complex specific acoustic impedances z and z^* defined above are time-independent as shown below:

$$z \equiv \frac{p}{u} = \frac{P_+e^{\ j(\omega t-kx)} + P_-e^{\ j(\omega t+kx)}}{\frac{1}{\rho_o c}\left(P_+e^{\ j(\omega t-kx)} - P_-e^{\ j(\omega t+kx)}\right)} = \frac{P_+e^{-jkx} + P_-e^{jkx}}{\frac{1}{\rho_o c}\left(P_+e^{-jkx} - P_-e^{jkx}\right)}$$

$$z^* \equiv \frac{p^*}{u^*} = \frac{P_+^*e^{-j(\omega t-kx)} + P_-^*\ e^{-j(\omega t+kx)}}{\frac{1}{\rho_o c}\left(P_+^*e^{-j(\omega t-kx)} - P_-^*\ e^{-j(\omega t+kx)}\right)} = \frac{P_+^*e^{jkx} + P_-^*\ e^{-jkx}}{\frac{1}{\rho_o c}\left(P_+^*e^{jkx} - P_-^*\ e^{-jkx}\right)}$$

The definition above for the complex specific acoustic impedance is valid for any combination of a forward and backward traveling wave.

A specific acoustic impedance in a complex form is defined in complex p and u for each part of a conjugate complex pair. Then the complex specific acoustic impedance will depend on the location only as shown in the following tables:

Complex specific acoustic impedance for plane waves associated with $e^{j\omega t}$

$(e^{j\omega t})$	$p(x,t),\ u(x,t)$	$z(x) = p/u$
$p_+ + p_-$ $u_+ + u_-$	$p : [(A_{+c} + jA_{+s})e^{-jkx} + (A_{-c} + jA_{-s})e^{jkx}]e^{j\omega t}$ $u : \frac{1}{\rho_o c}[(A_{+c} + jA_{+s})e^{-jkx} - (A_{-c} + jA_{-s})e^{jkx}]e^{j\omega t}$	$\rho_o c\ \frac{(A_{+c}+jA_{+s})e^{-jkx}+(A_{-c}+jA_{-s})e^{jkx}}{(A_{+c}+jA_{+s})e^{-jkx}-(A_{-c}+jA_{-s})e^{jkx}}$
p_+ u_+	$p : [(A_{+c} + jA_{+s})e^{-jkx}e^{j\omega t}$ $u : (A_{+c} + jA_{+s})e^{-jkx}e^{j\omega t}/\rho_o c$	$\rho_o c$
p_- u_-	$p : [(A_{-c} + jA_{-s})e^{jkx}e^{j\omega t}$ $u : -(A_{-c} + jA_{-s})e^{jkx}e^{j\omega t}/\rho_o c$	$-\rho_o c$

Complex specific acoustic impedance for plane waves associated with $e^{-j\omega t}$

$(e^{-j\omega t})$	$p^*(x, t), u^*(x, t)$	$z^*(x) = p^*/u^*$
$p_+^* + p_-^*$	$p^* : [(A_{+c} - jA_{+s})e^{jkx} + (A_{-c} - jA_{-s})e^{-jkx}]e^{-j\omega t}$	$\rho_o c \frac{(A_{+c}-jA_{+s})e^{jkx}+(A_{-c}-jA_{-s})e^{-jkx}}{(A_{+c}-jA_{+s})e^{jkx}-(A_{-c}-jA_{-s})e^{-jkx}}$
$u_+^* + u_-^*$	$u^* : \frac{1}{\rho_o c}[(A_{+c} - jA_{+s})e^{jkx} - (A_{-c} - jA_{-s})e^{-jkx}]e^{-j\omega t}$	

Terms associated with $e^{-j\omega t}$ are the conjugates of the terms associated with $e^{j\omega t}$.
Usually, the analysis of sound transmission in pipes only needs to compute for one part associated with $e^{j\omega t}$. The results of p* and u* associated with $e^{-j\omega t}$ are just the conjugates of the resulting **p** and **u** using the complex **z** associated with $e^{j\omega t}$.

The final real solutions p and u will be the summation of the pair of the complex solution.

4.6 Computer Program

Plotting Traveling and Standing Waves
A standing wave, $p_s(t, x)$, is the addition of a forward wave, $p_+(t, x)$, and a backward wave, $p_-(t, x)$, as shown below:

$$
\begin{aligned}
p_+(t, x) &= A_c \cos (\omega t - kx) - A_s \sin (\omega t - kx) \quad \textbf{(Form 1}:R_{IP}) \\
&= A \cos (\omega t - kx + \phi) \quad \textbf{(Form 2}:R_{EP}) \\
p_-(t, x) &= A_c \cos (\omega t + kx) - A_s \sin (\omega t + kx) \quad \textbf{(Form 1}:R_{IP}) \\
&= A \cos (\omega t + kx + \phi) \quad \textbf{(Form 2}:R_{EP}) \\
p_s(t, x) &= p_+ + p_- \\
&= 2A \cos (\omega t + \phi) \cos (kx) \quad \text{See Exercise 3.4}
\end{aligned}
$$

Use MATLAB to plot the pressure and velocity of a forward traveling wave (Part A), a backward traveling wave (Part B), and a standing wave (Part C) of three wavelengths using the following parameters:

$$
A_c = 0 \ [m]; A_s = -2 \ [m]; \omega = 2\pi \left[\frac{rad}{s}\right]; k = 2\pi \left[\frac{rad}{m}\right]
$$

Plot eight subplots to capture the wave motion with a time increment of T divided by eight. For calculating the velocity, assume $\rho_o c = 1.s$

Part A:
- Study and understand the provided MATLAB code.
- Complete the function "getForwardWave " which calculates the pressure and velocity from A, w, k, Time, and XDir.
- Run the code to plot the following eight subplots representing the forward wave motion during a period T:

$$
p_+(t, x) = A \cos (\omega t - kx + \phi)
$$

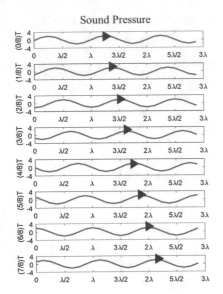

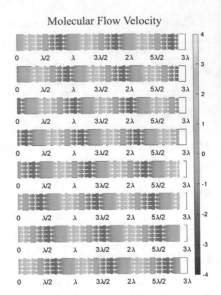

Part B:

- Complete the function "getBackwardWave" which calculates the pressure and velocity from A, w, k, Time, and XDir.
- Run the code to plot the following eight subplots representing the backward wave motion during a period T:

$$p_-(t, x) = A \cos (\omega t + kx + \phi)$$

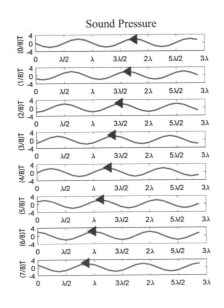

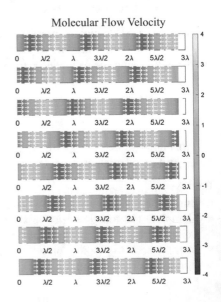

Part C:

- Complete the function "getStandingWave " which calculates the pressure and velocity from A, w, k, Time, and XDir.

- Run the code to plot the following eight subplots representing the standing wave motion during a period T:

$$p_s(t,x) = p_+ + p_-$$
$$= 2A\cos(\omega t + \phi)\cos(kx)$$

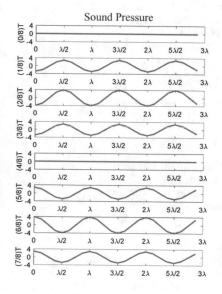

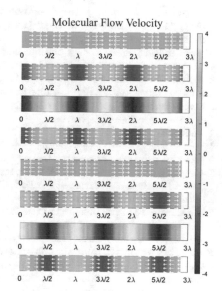

MATLAB Code:

```
function ANC_PRJ021_TravelingAndStandingWaves
clear all % removes all variables
close all % deletes all figures
%-------------------------------------------------------%
% Section 1: Define the Variables and Parameters
%-------------------------------------------------------%
% A standing wave,ps(t,x),is the addition of a forward wave, p+(t,x),
% and a backward wave, p-(t,x) as shown below
% p+(t,x)=Ac cos(wt-kx)+As sin(wt-kx)                        (Form 1)
%      =A cos(wt-kx+phi)                                     (Form 2)
% p-(t,x)=Ac cos(wt+kx)+As sin(wt+kx)                        (Form 1)
%      =A cos(wt+kx+phi)                                     (Form 2)
% ps(t,x)=P+(t,x)+p-(t,x)
```

```
%       =A cos(wt-kx+phi)+A cos(wt+kx+phi)                          (Form 2)
%       =2A cos(wt+phi)*cos(kx) --- see the Homework Eroblem 3.4
% note that the p+ and p- have the same amplitude A
% define the parameters of the wave function
iCase=1;
if iCase == 1 % phase phi = 0
Ac=0;% the coefficient of the cosine function in Form 1
As=2;% the coefficient of the sine funciton in Form 1
elseif iCase == 2
Ac=3^(1/2);  % the coefficient of the cosine function in Form 1
As=-1;     % the coefficient of the sine funciton in Form 1
end
w=2*pi/1;   % the angular frequency [rad/sec] in Form 1 and Form 2
k=2*pi/1;   % the wave number [rad/m] in Form 1 and Form 2
% define the number of times to take a snapshot
nTimePerPeriod=8;     % the number of data per period (T)
% plot only one period since it repeats every period
% define the parameters for plotting [space related]
nDataPerWaveLength=8;   % the number of data per period (T)
nWaveLength=3;       % the number of periods for plotting
%---------------------------------------------------------%
% Section 2: Setup
%---------------------------------------------------------%
% get the period (T) and a time array
T =2*pi/w;                    % the period of the wave [sec]
TimeIncrement=T/nTimePerPeriod;     % [sec]
nTime=nTimePerPeriod;              % the number of time data
TimeArray=(0:1:nTime-1)'*TimeIncrement; % [sec]
% get the data array of the location in the X-direction
L=2*pi/k; % [m] , note that k=2*pi/L
XDataIncrement=L/nDataPerWaveLength;  % [m]
nXData=nDataPerWaveLength*nWaveLength; % the number of space data
XDataArray=(0:1:nXData-1)'*XDataIncrement; % [m]
% get the data array of the location in the Y-direction
% (Y-direction is a dummy direction and does not exist)
YDataIncrementDummy=XDataIncrement*0.15;  % [m] a dummy value
nYDataDummy=5; % a dummy number(to adjust the arrow size)
YDataArrayDummy=(0:1:nYDataDummy-1)'*YDataIncrementDummy; % [m]
% get A and phi in Form 2 from From Ac and As in Form 1
[A,phi]=get_A_and_phi(Ac, As); % the same as PROJECT 1.9.1
%---------------------------------------------------------%
% Section 3: Calculation
%---------------------------------------------------------%
% Index of Cases
% iCaseWaveDirection = 1; % Forward Traveling Wave
% iCaseWaveDirection = 2; % Backward Traveling Wave
% iCaseWaveDirection = 3; % Standing Wave
for iCaseWaveDirection=1:1:3
  % calculate the cosine function
  for kTime=1:1:nTimePerPeriod
    for kXData=1:1:nXData
      % extract the scalars from the arrays
      % scalars are used temporally in the for-loop
```

```
     TimeTemp=TimeArray(kTime,1);
     XDataTemp=XDataArray(kXData,1);
     if iCaseWaveDirection == 1 % forward wave
       % get the pressure from A and phi                          (Form 2)
       [pressTemp,velocTemp]=...
         getForwardWave(A,w,k,TimeTemp,XDataTemp,phi);
       % put the pressure and velocity in the matrices
       pressColMatrix(1,kXData)=pressTemp; % pressure
       veloXColMatrix(1,kXData)=velocTemp; % velocity
       veloYColMatrix(1,kXData)=0;       % dummy value
       % adjust the arrow size in the figures
       arrow_scale_factor=1/20; % adjust the arrows size
     elseif iCaseWaveDirection == 2 % backward wave
       [pressTemp,velocTemp]=...
         getBackwardWave(A,w,k,TimeTemp,XDataTemp,phi);
       % put the pressure and velocity in the matrices
       pressColMatrix(1,kXData)=pressTemp; % pressure
       veloXColMatrix(1,kXData)=velocTemp; % velocity
       veloYColMatrix(1,kXData)=0;       % dummy value
       % adjust the arrow size in figure
       arrow_scale_factor=1/20;
     elseif iCaseWaveDirection == 3 % standing wave
       [pressTemp,velocTemp]=...
         getStandingWave(A,w,k,TimeTemp,XDataTemp,phi);
       pressColMatrix(1,kXData)=pressTemp; % pressure
       veloXColMatrix(1,kXData)=velocTemp; % velocity
       veloYColMatrix(1,kXData)=0;       % dummy value
       % adjust the arrow size in the figures
       arrow_scale_factor=1/40;
     end
   end
   %-------------------------------------------------------------%
   % Section 4: Plotting
   %-------------------------------------------------------------%
   % the meshed data of the pressure for the surface plots
   onesRowMatrix=ones(nYDataDummy,1);
   pressGrid=onesRowMatrix*pressColMatrix;
   veloXGrid=onesRowMatrix*veloXColMatrix;
   veloYGrid=onesRowMatrix*veloYColMatrix;
   % figures setup
   figure(210+iCaseWaveDirection);
   set(gcf,'Position',[100 100 1000 600])
   subplot(8,2,kTime*2-1)
   plot(XDataArray,pressColMatrix,'lineWidth',2)
   xlim([0 3])
   xticks([0 0.5 1 1.5 2 2.5 3])
   xticklabels({'0','\lambda/2','\lambda','3\lambda/2',...
     '2\lambda','5\lambda/2','3\lambda'})
   ylim([-4 4])
   yticks([-4 0 4])
   str=strcat('(',num2str(kTime-1),'/8)T');
   ylabel(str)
   % plot the pressure colormap
```

```
    subplot(8,2,kTime*2-0)
    pcolor(XDataArray,YDataArrayDummy,pressGrid); shading interp;
    xlim([0 3])
    xticks([0 0.5 1 1.5 2 2.5 3])
    xticklabels({'0','\lambda/2','\lambda','3\lambda/2',...
      '2\lambda','5\lambda/2','3\lambda'})
    set(gca,'YTick',[])
    cb=colorbar('location','eastoutside');
    set(cb,'position',[.92 .125 .01 .795])
    caxis([-4 4])
    hold on
    % plot the velocity vectors
    subplot(8,2,kTime*2-0)
    h=quiver(XDataArray,YDataArrayDummy,...
      veloXGrid.*arrow_scale_factor,...
      veloYGrid.*arrow_scale_factor);
    xlim([0 3])
    ylim([0 max(max(YDataArrayDummy))])
    set(gca,'YTick',[])
    set(h,'AutoScale','off')
    set(h,'LineWidth',1)
    set(h,'color','w')
    hold off
  end
  % save the figure
  if iCaseWaveDirection == 1 % forward wave
  saveas(gcf,strcat('Figure_ANC_PRJ021_ForwardWaves'),'emf')
  elseif iCaseWaveDirection == 2 % backward wave
  saveas(gcf,strcat('Figure_ANC_PRJ021_BackwardWaves'),'emf')
  elseif iCaseWaveDirection == 3 % standing wave
  saveas(gcf,strcat('Figure_ANC_PRJ021_StandingWaves'),'emf')
  end
end
end
%----------------------------------------------------------------%
% Section 5: Functions
%----------------------------------------------------------------%
function [pressure,velocity]=getForwardWave(A,w,k,Time,XDir,phi)
% p+(t,x)=Ac cos(wt-kx)+As sin(wt-kx)                        (Form 1)
%       =A cos(wt-kx+phi)                                    (Form 2)
% modify the following two lines for the pressure and velocity
% assume loc=0
pressure=A*cos(w*Time-k*XDir+phi);  % pressure of a forward wave
velocity=pressure;               % velocity of a forward wave
end
function [pressure,velocity]=getBackwardWave(A,w,k,Time,XDir,phi)
% p-(t,x)=Ac cos(wt+kx)+As sin(wt+kx)                        (Form 1)
%       =A cos(wt+kx+phi)                                    (Form 2)
% modify the following two lines for the pressure and velocity
% assume loc=0
pressure=0;  % pressure of a backward wave
velocity=0;  % velocity of a backward wave
end
```

```
function [pressure,velocity]=getStandingWave(A,w,k,Time,XDir,phi)
% ps(t,x)=P+(t,x)+p-(t,x)
%      =A cos(wt-kx+phi)+A cos(wt+kx+phi)              (Form 2)
%      =2A cos(wt+phi)*cos(kx) --- see the homework Problem 3.4
% modify the following two lines for the pressure and velocity
% assume loc=0
pressure=0; % pressure of standing wave
velocity=0; % velocity of standing wave
end
% from Project 9.1.1
function [A, phi]=get_A_and_phi(Ac, As)
% p(t) = Ac cos(wt) + As sin(wt)                       (Form 1)
%      = A cos(wt + phi)                               (Form 2)
A=(Ac^2+As^2)^(1/2);
phi=atan2(As,Ac);
end
```

4.7 Homework Exercises

Exercise 4.1 (Traveling Wave with Known Velocity)

Use one of the harmonic motions shown below (chosen by the instructor) as the velocity $u(t)$ to answer every question of this exercise (a–c):

i. $u(t) = U_0 \sin\left(\omega t - \frac{\pi}{2}\right)$; U_0 is a real number.

ii. $u(t) = Re\left(U_0 e^{j(\omega t - \pi)}\right)$; U_0 is a real number.

iii. $u(t) = U_0 \cos\left(\omega t + \frac{\pi}{2}\right)$; U_0 is a real number.

iv. $u(t) = Re\left(U_0 e^{j\left(\omega t + \frac{\pi}{2}\right)}\right)$; U_0 is a real number.

A large, rigid wall is vibrating with a velocity $u(t)$ perpendicular to its planar surface. The motion of the wall creates plane waves that propagate into the air. Consider the particle velocity of the plane wave on the surface of the wall ($x = 0$) to be equal to the velocity of the wall:

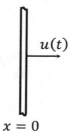

$x = 0$

a) Obtain the expressions for the flow velocity at any point in space (not just on the wall surface) of the plane wave radiating from the wall in terms of the wall vibration velocity.

b) Obtain the expressions for the sound pressure at any point in space (not just on the wall surface) of the plane wave radiating from the wall in terms of the wall vibration velocity.

c) What is the root-mean-square (RMS) pressure at any point in space if the wall velocity $U_0 = 0.01$ *m/s*.

Use 415 rayls for the characteristic impedance of air.
(Answers):

i, ii

(a) $U_o \sin \left(\omega t - kx - \frac{\pi}{2} \right)$; (b) $\rho_o c\, U_o \sin \left(\omega t - kx - \frac{\pi}{2} \right)$; (c) $\sqrt{\frac{(4.15)^2}{2}}$[Pa]

iii, iv

(a) $U_o \cos \left(\omega t - kx + \frac{\pi}{2} \right)$; (b) $\rho_o c\, U_o \cos \left(\omega t - kx + \frac{\pi}{2} \right)$; (c) $\sqrt{\frac{(4.15)^2}{2}}$[Pa]

Exercise 4.2 (Traveling Wave with Known Velocity)
Use the following harmonic motion $u(t)$ as velocity to answer every question of this exercise:

$$u(t) = Re \left(0.02\, e^{\,j\left(\omega t + \frac{\pi}{2}\right)} \right) \left[\frac{m}{s} \right]$$

A large, rigid wall is vibrating with a velocity $u(t)$ perpendicular to its planar surface. The motion of the wall creates *plane waves* that propagate into the air. Considering the particle velocity of the plane wave on the surface of the wall $(x = 0)$ would be equal to the velocity of the wall:

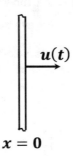

$u(t)$

$x = 0$

a) Obtain the expressions for the flow velocity at any point in space (not just on the wall surface) of the plane wave radiating from the wall in terms of the wall vibration velocity.

b) Obtain the expressions for the sound pressure at any point in space (not just on the wall surface) of the plane wave radiating from the wall in terms of the wall vibration velocity. Use 415 rayls for the characteristic impedance of air.

c) What is the root-mean-square (RMS) pressure at any point in space?

Use 415 rayls for the characteristic impedance $(\rho_o c)$ of air.

(Answers):   (a) $0.02 \cos\left(\omega t - kx + \frac{\pi}{2}\right)\left[\frac{m}{s}\right]$;   (b) $8.3\cos\left(\omega t - kx + \frac{\pi}{2}\right)[\text{Pa}]$;
(c) $\sqrt{\frac{(8.3)^2}{2}}[\text{Pa}]$

Exercise 4.3 (Traveling Wave with Known Pressure)

Use one of the traveling waves shown below (chosen by the instructor) to answer every question of this exercise (a–e):

$$p_-(x,t) = P_0 \cos\left(\omega t + kx + \frac{\pi}{2}\right)$$

$$p_+(x,t) = P_0 \cos\left(\omega t - kx + \pi\right)$$

$$p_+(x,t) = P_0 \sin\left(\omega t - kx - \pi\right)$$

a. Obtain the expressions for the particle velocity at any point in space.
b. What is the RMS pressure of the wave at any point in space?
c. What is the RMS flow velocity of the wave at any point in space?
d. What is the acoustic intensity of the wave at any point in space?
e. What is the specific acoustic impedance of the wave at any point in space?

(Answers):

i. (a) $\frac{-1}{\rho_o c} P_0 \cos\left(\omega t + kx + \frac{\pi}{2}\right)$; (b) $\frac{1}{\sqrt{2}}|P_o|$; (c) $\frac{1}{\sqrt{2}\rho_o c}|P_o|$; (d) $\frac{-1}{2}\frac{P_o^2}{\rho_o c}$; (e) $-\rho_o c$ [ray]

ii. (a) $\frac{P_0}{\rho_o c} \cos\left(\omega t - kx + \pi\right)$; (b) $\frac{1}{\sqrt{2}}|P_o|$; (c) $\frac{1}{\sqrt{2}\rho_o c}|P_o|$; (d) $\frac{1}{2}\frac{P_o^2}{\rho_o c}$; (e) $\rho_o c$ [ray]

iii. (a) $\frac{P_0}{\rho_o c} \sin\left(\omega t - kx - \pi\right)$; (b) $\frac{1}{\sqrt{2}}|P_o|$; (c) $\frac{1}{\sqrt{2}\rho_o c}|P_o|$; (d) $\frac{1}{2}\frac{P_o^2}{\rho_o c}$; (e) $\rho_o c$ [ray]

Exercise 4.4 (Standing Wave Pattern)

Use one of the combinations of a forward traveling wave $p_+(x,t)$ and a backward traveling wave $p_-(x,t)$ shown (chosen by the instructor) to answer every question of this exercise (a–f):

$$p_+(x,t) = A \cos\left(\omega t - kx\right) = \frac{1}{2}\left(Ae^{j(\omega t - kx)} + Ae^{-j(\omega t - kx)}\right)$$

$$p_-(x,t) = A \cos\left(\omega t + kx\right) = \frac{1}{2}\left(Ae^{j(\omega t + kx)} + Ae^{-j(\omega t + kx)}\right)$$

$$p_+(x,t) = -A \cos\left(\omega t - kx\right) = \frac{1}{2}\left(-Ae^{j(\omega t - kx)} - Ae^{-j(\omega t - kx)}\right)$$

$$p_-(x,t) = A \cos\left(\omega t + kx\right) = \frac{1}{2}\left(Ae^{j(\omega t + kx)} + Ae^{-j(\omega t + kx)}\right)$$

$$p_+(x,t) = A \sin\left(\omega t - kx\right) = \frac{1}{2}\left(-jAe^{j(\omega t - kx)} + jAe^{-j(\omega t - kx)}\right)$$

$$p_-(x,t) = A \sin(\omega t + kx) = \frac{1}{2}\left(-jAe^{j(\omega t+kx)} + jAe^{-j(\omega t+kx)}\right)$$

$$p_+(x,t) = -A \sin(\omega t - kx) = \frac{1}{2}\left(jAe^{j(\omega t-kx)} - jAe^{-j(\omega t-kx)}\right)$$

$$p_-(x,t) = A \sin(\omega t + kx) = \frac{1}{2}\left(-jAe^{j(\omega t+kx)} + jAe^{-j(\omega t+kx)}\right)$$

a) Calculate the standing wave (real number) produced by the forward and backward traveling waves.
b) What are the wave amplitude and the wavelength of the standing wave?
c) Sketch the resulting wave pattern and indicate the location of peaks and valleys in terms of wavelength.
d) Calculate the root-mean-square (RMS) pressure at $x = \frac{\pi}{k}$.
e) Calculate the acoustic intensity of the standing wave at any point in space.
f) Calculate the specific acoustic impedance of the standing wave at any point in space.

(Answers):

i. (a) $-2A \cos(\omega t) \cos(kx)$; (b) 2A, $\frac{2\pi}{k}$; (c) N/A; (d) $\sqrt{2}|A|$; (e) 0;

(f) $z = \rho_o c \frac{\cos(\omega t)}{\sin(\omega t)} \frac{\cos(kx)}{\sin(kx)}$; $z = j\rho_o c \cot(kx)$; $z^* = -j\rho_o c \cot(kx)$

ii. (a) $-2A \sin(\omega t) \sin(kx)$; (b) 2A, $\frac{2\pi}{k}$; (c) N/A; (d) 0; (e) 0;

(f) $z = \rho_o c \frac{\sin(\omega t)}{\cos(\omega t)} \frac{\sin(kx)}{\cos(kx)}$; $z = -j\rho_o c \tan(kx)$; $z^* = j\rho_o c \tan(kx)$

iii. (a) $2A \sin(\omega t) \cos(kx)$; (b) 2A, $\frac{2\pi}{k}$; (c) N/A; (d) $\sqrt{2}|A|$; (e) 0;

(f) $z = -\rho_o c \frac{\sin(\omega t)}{\cos(\omega t)} \frac{\cos(kx)}{\sin(kx)}$; $z = j\rho_o c \cot(kx)$; $z^* = -j\rho_o c \cot(kx)$

iv. (a) $2A \cos(\omega t) \sin(kx)$; (b) 2A, $\frac{2\pi}{k}$; (c) N/A; (d) 0; (e) 0;

(f) $z = -\rho_o c \frac{\cos(\omega t)}{\sin(\omega t)} \frac{\sin(kx)}{\cos(kx)}$; $z = -j\rho_o c \tan(kx)$; $z^* = j\rho_o c \tan(kx)$

4.8   References

4.8.1   Derivatives of Trigonometric and Complex Exponential Functions

Derivatives of trigonometric functions and complex exponential functions:

$$\frac{d}{dx}[\cos[f(x)]] = -\sin[f(x)]\frac{df(x)}{dx} = -\sin[f(x)]f'(x)$$

$$\frac{d}{dx}[\sin[f(x)]] = +\cos[f(x)]\frac{df(x)}{dx} = +\cos[f(x)]f'(x)$$

$$\frac{d}{dx}\left[e^{f(x)}\right] = e^{f(x)}\frac{df(x)}{dx} = e^{f(x)}f'(x)$$

Reversed:

$$\int \cos[f(x)]f'(x)dx = +\sin[f(x)]$$

$$\int \sin[f(x)]f'(x)dx = -\cos[f(x)]$$

$$\int e^{f(x)}f'(x)dx = e^{f(x)}$$

4.8.2   Trigonometric Integrals

Some trigonometric integrals are:

$$\int_0^{2\pi} \cos(\theta)d\theta = \int_0^{2\pi} \sin(\theta)d\theta = 0$$

$$\int_0^{2\pi} \cos(\theta)\sin(\theta)d\theta = 0$$

$$\frac{1}{2\pi}\int_0^{2\pi} \cos^2(\theta)d\theta = \frac{1}{2\pi}\int_0^{2\pi}\left(\frac{1+\cos(2\theta)}{2}\right)d\theta$$

$$= \frac{1}{2\pi}\frac{1}{2}\int_0^{2\pi} 1d\theta + \frac{1}{2\pi}\frac{1}{2}\int_0^{2\pi} \cos(2\theta)d\theta = \frac{1}{2}$$

$$\frac{1}{2\pi}\int_0^{2\pi}\sin^2(\theta)d\theta = \frac{1}{2\pi}\int_0^{2\pi}\left(\frac{1-\cos(2\theta)}{2}\right)d\theta$$

$$= \frac{1}{2\pi}\frac{1}{2}\int_0^{2\pi}1d\theta - \frac{1}{2\pi}\frac{1}{2}\int_0^{2\pi}\cos(2\theta)d\theta = \frac{1}{2}$$

Change 2π to period T:

$$\frac{1}{T}\int_0^T\cos(\omega t)dt = \frac{1}{T}\int_0^T\sin(\omega t)dt = \frac{1}{T}\int_0^T\cos\left(\frac{2\pi}{T}t\right)dt = \frac{1}{T}\int_0^T\sin\left(\frac{2\pi}{T}t\right)dt = 0$$

$$\frac{1}{T}\int_0^T\cos(\omega t)\sin(\omega t)dt = 0$$

$$\frac{1}{T}\int_0^T\cos^2(\omega t)dt = \frac{1}{T}\int_0^T\left(\frac{1+\cos(2\omega t)}{2}\right)dt = \frac{1}{2}\frac{1}{T}\int_0^T1dt + \frac{1}{2}\frac{1}{T}\int_0^T\cos\left(\frac{4\pi}{T}t\right)dt = \frac{1}{2}$$

$$\frac{1}{T}\int_0^T\sin^2(\omega t)dt = \frac{1}{T}\int_0^T\left(\frac{1-\cos(2\omega t)}{2}\right)dt = \frac{1}{2}\frac{1}{T}\int_0^T1dt - \frac{1}{2}\frac{1}{T}\int_0^T\cos\left(\frac{4\pi}{T}t\right)dt = \frac{1}{2}$$

Chapter 5
Solutions of Spherical Wave Equation

In the previous chapter, formulas for sound pressure, flow velocity, acoustic intensity, and specific acoustic impedance of plane waves were formulated in Cartesian coordinates. In this chapter, formulas for these properties will be developed in spherical coordinates.

The formulations of these properties in a spherical coordinate system are more useful than in a Cartesian coordinate system in terms of applications and numerical calculations because any vibrating surface can be treated as a point source. A point source radiates sound in radial directions and can be easily formulated in a spherical coordinate system.

The acoustic wave solutions in Cartesian coordinates were derived in the previous chapter. The acoustic wave solutions in spherical coordinates will be derived in this chapter. The following is a summary table of the acoustic wave solutions in both Cartesian and spherical coordinate systems:

General forms	Cartesian coordinates	Spherical coordinates
Position vector: $\vec{r}$	$\vec{r} = x\widehat{e}_x + y\widehat{e}_y + z\widehat{e}_z$	$\vec{r} = r\widehat{e}_r + \theta\widehat{e}_\theta + \phi\widehat{e}_\phi$
Gradient operator: ∇	$\nabla = \frac{\partial}{\partial x}\widehat{e}_x + \frac{\partial}{\partial y}\widehat{e}_y + \frac{\partial}{\partial z}\widehat{e}_z$	$\nabla = \frac{\partial}{\partial r}\widehat{e}_r + \frac{1}{r}\frac{\partial}{\partial \theta}\widehat{e}_\theta + \frac{1}{r^2 \sin(\theta)}\frac{\partial}{\partial \varphi}\widehat{e}_\varphi$
The Laplacian operator: ∇^2	$\nabla^2 = \frac{\partial^2}{\partial x^2}\ (1 - Dim)$	$\nabla^2 = \frac{\partial^2}{\partial r^2} + \frac{2}{r}\frac{\partial}{\partial r}\ (1 - Dim)$
Euler's force equation: $\nabla p = -\rho_0 \frac{\partial}{\partial t}\vec{u}$	$u(x,t) = -\frac{1}{\rho_o} \int \left[\frac{\partial}{\partial x}p(x,t)\right] dt$	$u(r,t) = -\frac{1}{\rho_o} \int \left[\frac{\partial}{\partial r}p(r,t)\right] dt$
Wave equations: $\nabla^2 p = \frac{1}{c^2}\frac{\partial^2}{\partial t^2}p$	$\frac{\partial^2}{\partial x^2}p(x,t) = \frac{1}{c^2}\frac{\partial^2}{\partial t^2}p(x,t)$	$\frac{\partial^2}{\partial r^2}[rp(r,t)] = \frac{1}{c^2}\frac{\partial^2}{\partial t^2}[rp(r,t)]$
$p_\pm$	$p_\pm(x, t) = A_\pm \cos(\omega t \mp kx + \theta_\pm)$ Form 2 : R_{EP}	$rp(r,t) = A \cos(\omega t - kr + \theta)$ Form 2 : R_{EP}

(continued)

General forms	Cartesian coordinates	Spherical coordinates
$u_\pm$	$u_\pm = \pm\frac{1}{\rho_o c} p_\pm(x,t)$	$r u(r,t) = \frac{A}{\rho_o c \cos\phi} \cos(\omega t - kr + \theta - \phi)$
p_{RMS}	$p^2_{\pm RMS} \equiv \frac{A^2}{2}$	$r^2 p^2_{RMS} \equiv \frac{A^2}{2}$
$I = pu$	$I_\pm = \pm\frac{1}{\rho_o c} p^2_{\pm RMS}$	$I(r) = \frac{1}{\rho_o c} p^2_{RMS}$
$z = \frac{p}{u}$	$z_\pm = \pm\rho_o c$	$z = \rho_o c \cos(\phi) \frac{\cos(\omega t - kr + \theta)}{\cos(\omega t - kr + \theta - \phi)}$
z	See table in Chap. 4	$z = \rho_o c \cos(\phi) e^{j\phi}$
w	$w \equiv \int_s I(r)\, ds$	$w \equiv \int_s I(r)\, ds = 4\pi a^2 I(a) = \frac{2\pi A^2}{\rho_o c}$

5.1 Spherical Coordinate System

Spherical coordinates consist of three coordinates: r, θ, and φ. These denote the direct distance to a point in space, the lateral angle, and the vertical angle to get there. These three coordinates describe a point in 3D space, just as do three Cartesian coordinates x, y, and z.

The geometry of the spherical coordinates is defined below:

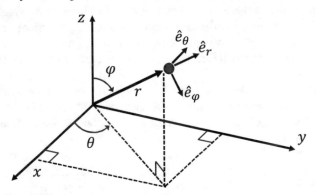

Definition of the coordinates in spherical coordinates

When a vector is expressed in vector format, this expression is independent of the choice of the coordinate systems. A vector can be formulated in the Cartesian coordinates and the spherical coordinates as follows:

$$\vec{v} = x\hat{e}_x + y\hat{e}_y + z\hat{e}_z \qquad \text{(Cartesian coord)}$$

$$\vec{v} = r\hat{e}_r + \theta\hat{e}_\theta + \phi\hat{e}_\varphi \qquad \text{(Spherical coord)}$$

Gradient operator ∇ is also independent of the choice of coordinate systems. The gradient operator can also be formulated in the Cartesian coordinates and the spherical coordinates as follows:

$$\nabla = \frac{\partial}{\partial x}\hat{e}_x + \frac{\partial}{\partial y}\hat{e}_y + \frac{\partial}{\partial z}\hat{e}_z \qquad \text{(Cartesian coord)}$$

$$\nabla = \frac{\partial}{\partial r}\hat{e}_r + \frac{1}{r}\frac{\partial}{\partial \theta}\hat{e}_\theta + \frac{1}{r^2 \sin(\theta)}\frac{\partial}{\partial \varphi}\hat{e}_\varphi \qquad \text{(Spherical coord)}$$

When the acoustic field is independent of θ and φ, the gradient operator simplifies to:

$$\nabla = \frac{\partial}{\partial x}\hat{e}_x \qquad \text{(Cartesian coord, 1D)}$$

$$\nabla = \frac{\partial}{\partial r}\hat{e}_r \qquad \text{(Spherical coord, 1D)}$$

The Laplacian operator ∇^2 can be formulated in both coordinate systems as:

$$\nabla^2 = \frac{\partial^2}{\partial x^2} + \frac{\partial^2}{\partial y^2} + \frac{\partial^2}{\partial z^2} \qquad \text{(Cartesian coord)}$$

$$\nabla^2 = \frac{\partial^2}{\partial r^2} + \frac{2}{r}\frac{\partial}{\partial r} + \frac{1}{r^2 \sin(\theta)}\frac{\partial}{\partial \theta}\left[\sin(\theta)\frac{\partial}{\partial \theta}\right] + \frac{1}{r^2 \sin(\theta)}\frac{\partial^2}{\partial \varphi^2} \qquad \text{(Spherical coord)}$$

When the acoustic field is independent of θ and φ, the Laplacian operator simplifies to:

$$\nabla^2 = \frac{\partial^2}{\partial x^2} \qquad \text{(Cartesian coord, 1D)}$$

$$\nabla^2 = \frac{\partial^2}{\partial r^2} + \frac{2}{r}\frac{\partial}{\partial r} \qquad \text{(Spherical coord, 1D)}$$

5.2 Wave Equation in Spherical Coordinate System

The acoustic wave equation can be formulated in the vector format as:

$$\nabla^2 p = \frac{1}{c^2}\frac{\partial^2}{\partial t^2}p \qquad \text{(Vector Format)}$$

where the Laplacian operator for Cartesian coordinate and spherical coordinate is shown in the previous section. Because the vector format is independent of the choice of coordinate system, the Laplacian operator, ∇^2, in the equation above can be formulated in both the spherical and Cartesian coordinate systems as shown in the previous section. The one-dimensional wave equation in the spherical coordinate system can be obtained by replacing the Laplacian operator in vector format in the wave equations as:

$$\left(\frac{\partial^2}{\partial r^2} + \frac{2}{r}\frac{\partial}{\partial r}\right)p(r,t) = \frac{1}{c^2}\frac{\partial^2}{\partial t^2}p(r,t) \qquad \text{(Spherical coord)}$$

The one-dimensional acoustic wave equation in the spherical coordinate system is a function of r and t. The wave equation above can be transferred to the form of the wave equation in the Cartesian coordinate system by defining a new function $\phi(r,t)$:

$$\psi(r,t) \equiv rp(r,t)$$

By replacing $p(r,t)$ in the above wave equation with $\psi(r,t)/r$, the wave equation becomes:

$$\frac{\partial^2}{\partial r^2}\psi(r,t) = \frac{1}{c^2}\frac{\partial^2}{\partial t^2}\psi(r,t) \qquad \text{(Spherical coord)}$$

By comparing the wave equations in the spherical coordinate above to the wave equation in the Cartesian coordinate as shown below:

$$\frac{\partial^2}{\partial x^2}p(x,t) = \frac{1}{c^2}\frac{\partial^2}{\partial t^2}p(x,t) \qquad \text{(Cartesian coord)}$$

We can see that the wave equation in the spherical coordinate is the same as the wave equation in the Cartesian coordinate. Therefore, the solutions, $p(x,t)$, of the wave equation in the Cartesian coordinate, derived in Chap. 2, can be used as the solutions, $\psi(r,t)$, of the wave equation in the spherical coordinate.

Example 5.1
Show that the following two equations are the same:

$$\frac{\partial^2}{\partial r^2}[rp(r,t)] = \frac{1}{c^2}\frac{\partial^2}{\partial t^2}[rp(r,t)]$$

$$\left(\frac{\partial^2}{\partial r^2} + \frac{2}{r}\frac{\partial}{\partial r}\right)p(r,t) = \frac{1}{c^2}\frac{\partial^2}{\partial t^2}p(r,t)$$

Example 5.1 Solution
The Left-Hand Side

The left-hand side of the first equation becomes:

$$\frac{\partial^2}{\partial r^2}[rp(r,t)] = \frac{\partial}{\partial r}\left\{\frac{\partial}{\partial r}[rp(r,t)]\right\}$$

$$= \frac{\partial}{\partial r}\left\{p(r,t) + r\frac{\partial}{\partial r}p(r,t)\right\}$$

$$= \frac{\partial}{\partial r}p(r,t) + \frac{\partial}{\partial r}p(r,t) + r\frac{\partial^2}{\partial r^2}p(r,t)$$

$$= 2\frac{\partial}{\partial r}p(r,t) + r\frac{\partial^2}{\partial r^2}p(r,t)$$

The Right-Hand Side

Since r and t are independent variables, the right-hand side of the equation becomes:

$$\frac{1}{c^2}\frac{\partial^2}{\partial t^2}[rp(r,t)] = r\frac{1}{c^2}\frac{\partial^2}{\partial t^2}p(r,t)$$

The Combined Equation

Since the left-hand side equals the right-hand side, the original wave equation $p(r,t)$ can be cast as:

$$\text{Left Hand Side} = \text{Right Hand Side}$$

$$\frac{\partial^2}{\partial r^2}[rp(r,t)] = \frac{1}{c^2}\frac{\partial^2}{\partial t^2}[rp(r,t)]$$

$$\rightarrow 2\frac{\partial}{\partial r}p(r,t) + r\frac{\partial^2}{\partial r^2}p(r,t) = r\frac{1}{c^2}\frac{\partial^2}{\partial t^2}p(r,t)$$

$$\rightarrow \frac{\partial^2}{\partial r^2}p(r,t) + \frac{2}{r}\frac{\partial}{\partial r}p(r,t) = \frac{1}{c^2}\frac{\partial^2}{\partial t^2}p(r,t)$$

$$\rightarrow \left(\frac{\partial^2}{\partial r^2} + \frac{2}{r}\frac{\partial}{\partial r}\right)p(r,t) = \frac{1}{c^2}\frac{\partial^2}{\partial t^2}p(r,t)$$

This proves that the equation is valid.

5.3   Pressure Solutions of Wave Equation in Spherical Coordinate System

Solutions of the one-dimensional wave equation in the spherical coordinate system can be formulated as a real trigonometric function, as shown below:

$$rp(r,t) = A \cos{(\omega t - kr + \theta)} \qquad (R_{EP}) \qquad (5.3)$$

or:

$$p(r,t) = \frac{A}{r} \cos{(\omega t - kr + \theta)} \qquad (R_{EP}) \qquad (5.3)$$

Unlike plane waves, it is hard and unusual to create a symmetrical spherical backward or inward traveling wave. Therefore, only the forward or outward spherical waves are discussed in this course.

The following proves the outward traveling wave solution above. The wave equation in a spherically symmetric pressure field is described as:

$$\frac{\partial^2 p}{\partial r^2} + \frac{2}{r} \frac{\partial p}{\partial r} = \frac{1}{c^2} \frac{\partial^2 p}{\partial t^2}$$

The spherical wave equation above can be modified and arranged into the same form as the wave equation for a plane wave. For this purpose, we will define a new function φ:

$$\psi(r,t) \equiv rp(r,t)$$

Note that ψ could be either a real or a complex number, depending on the equation in which it is used. For example, if ψ is used in a complex equation, it is a complex number. Similarly, if it is used in a real equation, it is a real number.

Substituting the function $\psi(r,t)$ into the following equation will yield the spherical wave equation. The algebra is straightforward and is not shown here, but will be assigned as an exercise:

$$\frac{\partial^2}{\partial r^2} \psi(r,t) = \frac{1}{c^2} \frac{\partial^2}{\partial t^2} \psi(r,t)$$

Since the above equation has the same form as the one-dimensional plane wave equation, it has the same solutions as shown below:

$$\psi(r,t) = [A_{-c} \cos{(\omega t + kr)} - A_{-s} \sin{(\omega t + kr)}]$$
$$+ [A_{+c} \cos{(\omega t - kr)} - A_{+s} \sin{(\omega t - kr)}] \qquad (R_{IP})$$
$$= A_{+} \cos{(\omega t - kr + \theta_{+})} + A_{-} \cos{(\omega t + kr + \theta_{-})} \qquad (R_{EP})$$

where A_{-}, A_{+}, A_{-c}, A_{+c}, A_{-s}, A_{+s}, θ_{-}, and θ_{+} are real constants related by the following geometry relationships:

Geometric Relationship	
Geom 1	$A_\pm = \sqrt{A_{\pm c}^2 + A_{\pm s}^2}$
Geom 2	$\theta_\pm = \tan^{-1}\left(\dfrac{A_{\pm s}}{A_{\pm c}}\right)$
Geom 3	$A_{\pm c} = A_\pm \cos(\theta_\pm)$
Geom 4	$A_{\pm s} = A_\pm \sin(\theta_\pm)$

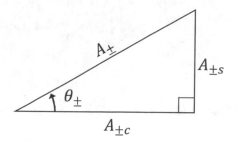

Since only forward traveling wave is considered in the spherical coordinates, the above equation can be simplified by defining the following new variables:

$$A_c \equiv A_{+c}$$

$$A_s \equiv A_{+s}$$

$$A \equiv A_+$$

$$\theta \equiv \theta_+$$

Therefore, the outward traveling wave in spherical coordinates becomes:

$$\psi(r,t) = A_c \cos(\omega t - kr) - A_s \sin(\omega t - kr) \qquad (R_{IP})$$
$$= A \cos(\omega t - kr + \theta) \qquad (R_{EP})$$

where A, A_c, A_s, and θ are real variables related by the following geometric relationships:

Geometric Relationship	
Geom 1	$A = \sqrt{A_c^2 + A_s^2}$
Geom 2	$\theta = \tan^{-1}\left(\dfrac{A_s}{A_c}\right)$
Geom 3	$A_c = A \cos(\theta)$
Geom 4	$A_s = A \sin(\theta)$

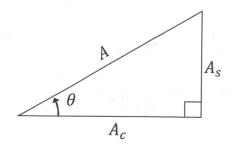

Transforming the new variable $\psi(r,t)$ back to its original form of pressure $p(r,t)$ yields:

$$rp(r,t) = \psi(r,t) = A \cos{(\omega t - kr + \theta)} \qquad\qquad (R_{EP})$$
$$= \frac{1}{2} \left[A e^{\,j(\omega t - kr + \theta)} + A e^{-j(\omega t - kr + \theta)} \right] \qquad (C_{EP})$$

5.4 Flow Velocity

Similar to the plane wave, the flow velocity can be obtained from sound pressure using Euler's force equation.

The one-dimensional Euler's force equation in the spherical coordinates can be obtained by replacing the vector gradient operator with the spherical coordinate gradient operator:

$$\nabla p = -\rho_o \frac{\partial}{\partial t} u \qquad\qquad \text{(Vector format)}$$

$$\rightarrow \frac{\partial}{\partial r} p(r,t) = -\rho_o \frac{\partial}{\partial t} u(r,t) \qquad\qquad \text{(Spherical coord)}$$

5.4.1 Flow Velocity in Real Format

Sound pressure $p(r,t)$ in the spherical coordinate system was derived in the previous section as:

$$rp(r,t) = A \cos{(\omega t - kr + \theta)}$$

Flow velocity $u(r,t)$ in spherical coordinates can be derived from Euler's force equation and expressed in real explicit phase $[R_{EP}]$ format as:

$$ru(r,t) = \frac{A}{\rho_o c \cos{\phi}} \cos{(\omega t - kr + \theta - \phi)} \qquad\qquad (R_{EP})$$

or:

$$u(r,t) = \frac{A}{r\rho_o c \cos{\phi}} \cos{(\omega t - kr + \theta - \phi)} \qquad\qquad (R_{EP})$$

The following will derive the above equation in real explicit phase $[R_{EP}]$ format:

Flow velocity expressed in real explicit phase $[R_{EP}]$ format can be derived independently using Euler's force equation:

$$\frac{\partial}{\partial r} p(r,t) \hat{e}_r = -\rho_o \frac{\partial}{\partial t} u(r,t)$$

For a given pressure $p(x,t)$, flow velocity $u(x,t)$ can be calculated as:

$$u(r,t) = -\frac{1}{\rho_o} \int \left[\frac{\partial}{\partial r} p(r,t) \right] dt$$

Flow velocity in real number format will be derived in this section:

$$ru(r,t) = \frac{A}{\rho_o c \cos \phi} \cos(\omega t - kr + \theta - \phi) \qquad (R_{EP})$$

Flow velocity formulated in real explicit phase $[R_{EP}]$ format is derived from pressure $p(r,t)$ as:

$$rp(r,t) = A \cos(\omega t - kr + \theta)$$

Substituting the above pressure $p(r,t)$ into the formula for flow velocity yields:

$$u(r,t) = \frac{-1}{\rho_o} \int \left[\frac{\partial}{\partial r} p(r,t) \right] dt = \frac{-A}{\rho_o} \int \left[\frac{\partial}{\partial r} \left[\frac{1}{r} \cos(\omega t - kr + \theta) \right] \right] dt$$

Carrying out the derivative with respect to r yields:

$$u(r,t) = \frac{-A}{\rho_o} \int \left[\frac{k}{r} \sin(\omega t - kr + \theta) - \frac{1}{r^2} \cos(\omega t - kr + \theta) \right] dt$$

Carrying out the integral for t yields:

$$u(r,t) = \frac{-A}{\rho_o} \left[\frac{k}{r} \left(\frac{-1}{\omega} \right) \cos(\omega t - kr + \theta) - \frac{1}{r^2} \left(\frac{1}{\omega} \right) \sin(\omega t - kr + \theta) \right]$$

$$= \frac{A}{r \rho_o} \frac{k}{\omega} \left[\cos(\omega t - kr + \theta) + \frac{1}{kr} \sin(\omega t - kr + \theta) \right]$$

Simplify the above equation and replace $\frac{\omega}{k}$ with c to arrive at:

$$u(r,t) = \frac{A}{r \rho_o c} \left[\cos(\omega t - kr + \theta) + \frac{1}{kr} \sin(\omega t - kr + \theta) \right]$$

Combining the cosine and sine functions into one cosine function yields:

$$u(r,t)=\frac{A}{r\rho_o c}\left[\cos(\omega t-kr+\theta)+\frac{1}{kr}\sin(\omega t-kr+\theta)\right]$$

$$=\frac{A}{r\rho_o ckr}\frac{1}{kr}[kr\cos(\omega t-kr+\theta)+\sin(\omega t-kr+\theta)]$$

$$=\frac{A}{r\rho_o c}\frac{\sqrt{1+(kr)^2}}{kr}\left[\frac{kr}{\sqrt{1+(kr)^2}}\cos(\omega t-kr+\theta)+\frac{1}{\sqrt{1+(kr)^2}}\sin(\omega t-kr+\theta)\right]$$

Define a new variable ϕ as:

$$\cos\phi=\frac{kr}{\sqrt{1+(kr)^2}};\quad \sin\phi=\frac{1}{\sqrt{1+(kr)^2}}$$

Then, $u(r,t)$ can be simplified as:

$$u(r,t)=\frac{A}{r\rho_o c}\frac{1}{\cos(\phi)}\left[\cos(\phi)\cos(\omega t-kr+\theta)+\sin(\phi)\sin(\omega t-kr+\theta)\right]$$

$$\rightarrow\quad ru(r,t)=\frac{A}{\rho_o c}\frac{1}{\cos\phi}\left[\cos(\omega t-kr+\theta-\phi)\right] \qquad (5.4)$$

where:

$$\phi=\tan^{-1}\left(\frac{1}{kr}\right)$$

Eq. (5.4) is the flow velocity of spherically symmetric acoustic wave, for a given pressure function, as described in Eq.(5.3).

5.4.2   Flow Velocity in Complex Format

Sound pressure $p(r,t)$ in the spherical coordinate system was derived in the previous ion as:

$$rp(r,t)=A\cos(\omega t-kr+\theta)$$

It can be formulated in a complex explicit phase $[C_{EP}]$ as shown below using Euler's formula:

$$rp(r,t) = A\frac{1}{2}\left[e^{j(\omega t - kr + \theta)} + e^{-j(\omega t - kr + \theta)}\right] \qquad (C_{EP})$$

The corresponding flow velocity in a complex exponential format as shown below can be derived using Euler's force equation:

$$ru(r,t) = \frac{A}{\rho_o c \cos\phi}\frac{1}{2}\left[e^{j(\omega t - kr + \theta - \phi)} + e^{-j(\omega t - kr + \theta - \phi)}\right] \qquad (C_{EP})$$

where:

$$\phi = \tan^{-1}\left(\frac{1}{kr}\right)$$

The derivation of the flow velocity formulated in complex explicit phase [C_{EP}] format is not derived here. The derivation is part of the homework in this chapter.

Example 5.2
What is the phase difference between the pressure and the flow velocity of a spherical wave at 3 [m] from the origin at 1000 [Hz]?

Use 340 $\left[\frac{m}{s}\right]$ for the speed of sound (c) in air.

Example 5.2 Solution The general form of pressure of a spherical wave is:

$$rp(r,t) = A \cos(\omega t - kr + \theta)$$

The formula of flow velocity is derived from pressure using Euler's force equation:

$$ru(r,t) = \frac{A}{\rho_o c \cos\phi} \cos(\omega t - kr + \theta - \phi)$$

where:

$$\phi = \tan^{-1}\left(\frac{1}{kr}\right)$$

Hence, by inspecting the exponents of pressure and flow velocity, the phase difference between pressure and flow velocity of a spherical wave is equal to ϕ, with flow velocity lagging behind the pressure.

For the given conditions:

$$f = 1000 \, [Hz]$$

$$r = 3 \ [m]$$

The wavenumber is:

$$k = \frac{\omega}{c} = \frac{2\pi \cdot f}{c} = \frac{2\pi \cdot 1000 \left[\frac{1}{s}\right]}{340 \left[\frac{m}{s}\right]} = 18.5 \ \left[\frac{1}{m}\right]$$

The phase difference is:

$$\phi = \tan^{-1}\left(\frac{1}{kr}\right) = \tan^{-1}\left(\frac{1}{18.5 \times 3}\right) = 0.018 \ [rad] = 1.03^{\circ}$$

5.5 RMS Pressure and Acoustic Intensity

The RMS pressure of a spherical outward wave is represented by the following equation:

$$p_{RMS}^2 \equiv \frac{1}{T} \int_0^T \left(\frac{A}{r} \cos\left(\omega t - kr + \theta\right)\right)^2 dt = \left(\frac{A}{r}\right)^2 \frac{1}{T} \int_0^T \cos^2(\omega t - kr + \theta) dt$$

$$= \frac{1}{2} \frac{A^2}{r^2}$$

The acoustic intensity of a given pressure:

$$p(t) = \frac{A}{r} \cos\left(\omega t - kr + \theta\right)$$

is:

$$I = \frac{1}{\rho_o c} p_{RMS}^2$$

where:

$$p_{RMS}^2 = \frac{1}{2} \frac{A^2}{r^2}$$

The following is the derivation of the above equations:

The acoustic intensity of spherical waves can be computed by taking the average power over a period of time similar to that of plane waves. Hence:

$$I \equiv \frac{1}{T} \int_0^T p(t)u(t)dt$$

$$= \frac{1}{T} \int_0^T \left[\frac{A}{r} \cos{(\omega t - kr + \theta)} \right] \left[\frac{A}{r\rho_o c \, \cos{(\phi)}} \cos{(\omega t - kr + \theta - \phi)} \right] dt$$

$$= \frac{A^2}{r^2 \rho_o c \, \cos{(\phi)}} \frac{1}{T} \int_0^T \cos{(\omega t - kr + \theta)}$$

$$\times [\cos{(\omega t - kr + \theta)} \cos{(\phi)} + \sin{(\omega t - kr + \theta)} \sin{(\phi)}]dt$$

Separate the above equation into two parts as:

$$I = I_1 + I_2$$

where:

$$I_1 = \frac{A^2 \cos{(\phi)}}{r^2 \rho_o c \, \cos{(\phi)}} \frac{1}{T} \int_0^T \cos^2{(\omega t - kr + \theta)}dt = \frac{1}{2} \frac{A^2}{r^2 \rho_o c}$$

$$I_2 = \frac{A^2 \sin{(\phi)}}{r^2 \rho_o c \, \cos{(\phi)}} \frac{1}{T} \int_0^T \cos{(\omega t - kr + \theta)} \sin{(\omega t - kr + \theta)}dt = 0$$

Therefore:

$$I = \frac{1}{2} \frac{A^2}{r^2 \rho_o c} = \frac{1}{\rho_o c} p_{RMS}^2$$

Hence, the intensity for spherical waves is in the same form as the intensity for plane waves. However, the magnitude P is different than that of plane wave and is given by $P = \frac{A}{r}$; the acoustic intensity of spherical waves is reduced by r^2 with relation to the distance from the source.

Example 5.3

Flow velocity magnitude U_0 of a spherical wave is given as 0.01 [m/s] at 0.5 m from the center of a source.

Assuming that the wavelength is $\lambda = 0.1$ [m], answer the following questions, and show units in the Meter-Kilogram-Second (MKS) system:

a) What is the particle velocity magnitude at any point in space?
b) What is the acoustic pressure magnitude at any point in space?
c) What is the acoustic intensity at 0.5 and 10 [m] from the center of the source?

Use 415 [rayls] for characteristic impedance ($\rho_o c$) and 340 [m/s] for the speed of sound (c).

Example 5.3 Solution

a)   The wavenumber k and angular frequency ω are:

$$k = \frac{2\pi}{\lambda} = \frac{2\pi}{0.1} \left[\frac{rad}{m}\right] = 20\pi \left[\frac{rad}{m}\right]$$

$$\omega = ck = 340 \left[\frac{m}{s}\right] 20\pi \left[\frac{rad}{m}\right] = 6800\pi \left[\frac{rad}{s}\right]$$

The general solution of velocity with the known wavenumber is:

$$u(r,t) = \frac{A}{r\rho_o c \cos \phi} \cos (\omega t - kr + \theta - \phi)$$

where:

$$\cos \phi = \frac{kr}{\sqrt{1 + (kr)^2}}$$

Comparing the velocity amplitude of the general solution to the given velocity magnitude $U_0(r = 0.5m){=}0.01 \left[\frac{m}{s}\right]$ yields:

$$\frac{A}{r\rho_o c \cos \phi} = 0.01$$

$$\rightarrow \frac{A}{0.5 \times 415 \times 0.99949} = 0.01$$

$$\rightarrow A = 2.0739$$

where:

$$\cos \phi = \frac{kr}{\sqrt{1 + (kr)^2}} = \frac{20\pi \times 0.5}{\sqrt{1 + (20\pi \times 0.5)^2}} = 0.99949$$

The flow velocity magnitude at r:

$$U_r(r) = \frac{A}{r\rho_o c \cos \phi} = \frac{A}{r\rho_o c} \frac{\sqrt{1 + (kr)^2}}{kr}$$

$$= \frac{2.0739}{r \cdot 415} \frac{\sqrt{1 + (20\pi r)^2}}{20\pi \cdot r}$$

$$= 7.9537 \cdot 10^{-5} \frac{\sqrt{1 + 3948 r^2}}{r^2} \ \left[\frac{m}{s}\right]$$

$$p(r,t) = \frac{2.0739}{r} \cos(\omega t - kr + \theta)[\qquad\qquad (5.\text{Pa})$$

The pressure magnitude at r: $P_r(r) = \frac{2.0739}{r} \ [Pa]$

b) The acoustic intensity at $r = 0.5 \ m$:

$$I = \frac{1}{\rho_o c} P_{RMS}^2 = \frac{1}{\rho_o c} \frac{1}{r^2} \frac{A^2}{2} = \frac{1}{415} \cdot \frac{1}{0.5^2} \cdot \frac{2.0739^2}{2} = 0.020728 \ [w/m^2]$$

The acoustic intensity at $r = 10 \ m$:

$$I = \frac{1}{\rho_o c} P_{RMS}^2 = \frac{1}{\rho_o c} \frac{1}{r^2} \frac{A^2}{2} = \frac{1}{415} \cdot \frac{1}{10^2} \cdot \frac{2.0739^2}{2} = 5.1820 \cdot 10^{-5} \ [w/m^2]$$

Example 5.4
If the field pressure created by a spherical source is $p(r = 10m, t) = 0.02 \cos(\omega t + 0.25\pi)[Pa]$ at $r = 10 \ [m]$ from the center of the source, calculate the surface velocity of the source if $a = 0.1 \ [m]$, $k = 0.125\pi$.

Use 415 $[rayls]$ for characteristic impedance $(\rho_o c)$ and 340 $[m/s]$ for the speed of sound (c). Show units in the Meter-Kilogram-Second (MKS) system.

Example 5.4 Solution
The general forms of acoustic pressure and velocity are:

$$p(r,t) = \frac{A}{r} \cos(\omega t - kr + \theta)$$

$$u(r,t) = \frac{A}{r\rho_o c \, \cos(\phi)} \cos(\omega t - kr + \theta - \phi)$$

where $\omega = ck = 340 \left[\frac{m}{s}\right] \times 0.125\pi = 42.5\pi = 133.52 \left[\frac{rad}{s}\right]$.
Solve for A and θ with the given pressure at $r = 10 \ [m]$ as:

$$p(r = 10m, , t) = \frac{A}{10} \cos(\omega t - 0.125\pi \cdot 10 + \theta) = 0.02 \cos(\omega t + 0.25\pi)$$

Then:

$$A = 10 \times 0.02 = 0.20; \qquad \theta = 1.25\pi + 0.25\pi = 1.5\pi$$

Substituting the calculated A and θ into velocity equation yields:

$$u(r,t) = \frac{A}{r\rho_o c \, \cos{(\phi)}} \cos{(\omega t - kr - \phi + \theta)}$$

$$= \frac{0.20}{r\rho_o c \, \cos{(\phi)}} \cos{(\omega t - kr - \phi + 1.5\pi)} \left[\frac{m}{s}\right]$$

Because the surface velocity of the sphere is the same as the velocity of the wave at the surface of the sphere, the velocity at $r = a = 0.1 \, [m]$ is:

$$u(a = 0.1, t) = \frac{0.20}{a\rho_o c \, \cos{(\phi)}} \cos{(\omega t - ka - \phi + 1.5\pi)} \left[\frac{m}{s}\right]$$

$$= \frac{0.20}{0.1 \cdot 415 \cdot 0.0392} \cos{(133.52\,t - 0.125\pi \cdot 0.1 - 0.488\pi + 1.5\pi)} \left[\frac{m}{s}\right]$$

$$= 0.123 \, \cos{(133.52\,t + \pi)} \left[\frac{m}{s}\right]$$

where:

$$ka = 0.125\pi \cdot 0.1$$

$$\phi = \tan^{-1}\left(\frac{1}{ka}\right) = \tan^{-1}\left(\frac{1}{0.125\pi \cdot 0.1}\right) = 1.53 \, rad = 0.488\pi \, rad = 87.75^{\circ}$$

$$\cos{(\phi)} = \cos{\left(87.75^{\circ}\right)} = 0.0392$$

$$\omega = ck = 340 \left[\frac{m}{s}\right] \times 0.125\pi = 42.5\pi = 133.52 \left[\frac{rad}{s}\right]$$

5.6   Specific Acoustic Impedance

The specific acoustic impedance, by definition, is calculated by dividing pressure by velocity. Based on this definition, we have two choices for formulating the specific acoustic impedance: (1) dividing a real number of pressure by a real number of velocity and (2) dividing a complex number (half of a complex conjugate pair) of a pressure by a complex number of a velocity. The two choices for formulating the specific acoustic impedance are summarized as follows:

Choice 1: Specific acoustic impedances formulated using real numbers of pressure and velocity.

The specific acoustic impedance can be obtained by substituting the pressure and flow velocity into the specific acoustic impedance equation.

The specific acoustic impedance in real explicit phase $[R_{EP}]$ format can be represented as:

$$z \equiv \frac{p}{u} = \frac{\frac{A}{r}\cos(\omega t - kr + \theta)}{\frac{A}{r\rho_o c \cos(\phi)}\cos(\omega t - kr + \theta - \phi)}$$

$$= \rho_o c \cos(\phi)\frac{\cos(\omega t - kr + \theta)}{\cos(\omega t - kr + \theta - \phi)} \qquad (R_{EP})$$

where:

$$\phi = \tan^{-1}\left(\frac{1}{kr}\right) \text{ and } \cos(\phi) = \frac{kr}{\sqrt{1+(kr)^2}}$$

Note that the specific acoustic impedances formulated using real numbers of pressure and velocity are real numbers. As shown in the equation above, the specific acoustic impedance (real number) is a function of space kr and time ωt.

Issues with Using Real Numbers of Pressure and Velocity (Choice 1)

For analyzing the sound pressures in connected pipes, equilibrium equations of pressure and flow (see Chap. 11 for details) use functions of space kr. Using any function of time ωt in an equilibrium equation is impossible for analyzing the sound pressures in connected pipes. For this reason, only specific impedances defined by complex numbers of pressure and velocity are used for the analysis in filter design (Chaps. 11 and 12).

Choice 2: Specific acoustic impedance formulated using complex numbers of pressure and velocity.

The specific acoustic impedance in a complex explicit phase $[C_{EP}]$ format can be represented as:

$$z \equiv \frac{p}{u} = \frac{\frac{A}{r}e^{j(\omega t - kr + \theta)}}{\frac{A}{r\rho_o c \cos(\phi)}e^{j(\omega t - kr + \theta - \phi)}}$$

$$= \rho_o c \cos(\phi)e^{j\phi} \qquad (C_{EP})$$

where:

$$\phi = \tan^{-1}\left(\frac{1}{kr}\right) \text{ and } \cos(\phi) = \frac{kr}{\sqrt{1+(kr)^2}}$$

Note that the specific acoustic impedances formulated using complex numbers of pressure and velocity are complex numbers. As shown in the equation above, the specific acoustic impedance (complex number) is the function of space kr only (time function ωt is canceling out). Therefore, the specific acoustic impedance defined by complex numbers of pressure and velocity is not a function of time ωt.

In addition, the specific acoustic impedance defined by real numbers of pressure and velocity (Choice 1) can be any real value varying from $-\infty$ to ∞. On the other hand, the specific acoustic impedances defined by complex numbers of pressure and velocity (Choice 2) have a finite number: when $kr \ll 1$, $\phi \cong \pi/2$, $z = \rho_o ckrj$ is an imaginary number; when $kr \gg 1$, $\phi \cong 0, z = \rho_o c$ is a real number and is the same as a plane wave.

5.7 Homework Exercises

Exercise 5.1

Acoustic pressure magnitude P_0 of a spherical wave is given as 4.149 [Pa] at 1 [m] from the center of a source. If the wavelength is 0.1 [m], answer the following questions, and show units in the Meter-Kilogram-Second (MKS) system:

a) What is the particle velocity magnitude at any point in space?
b) What is the acoustic pressure magnitude at any point in space?
c) What is the acoustic intensity at 1 and 25 [m] from the center of the source?
d) What is the phase difference between the pressure and flow velocity of the sphere wave at 1 and 25 [m]?

Use 415 [rayls] for the characteristic impedance ($\rho_o c$) of air.
(Answers):

a) The velocity magnitude $U_r = 0.0001591 \frac{\sqrt{1+3948r^2}}{r^2}$ [m/s]
b) The pressure magnitude at r: $P_r = \frac{4.149}{r}$ [Pa]
c) The acoustic intensity at $r = 1$ [m]: 0.02074 [w/m²]
   The acoustic intensity at $r = 25$ [m]: 0.00003318 [w/m²]
d) The phase difference at $r = 1$ [m]: $\phi_1 = 0.01591\ rad = 0.9118$ [Deg]
   The phase difference at $r = 25$ [m]: $\phi_{25} = 0.00064\ rad = 0.0365$ [Deg]

Exercise 5.2

The acoustic pressure magnitude from a spherical source, measured at 10 m from the center of the source, is 0.8 Pa. If the frequency of radiation is 250 Hz:

a) What is the surface velocity magnitude of the source if the source radius is 0.05 [m]?
b) What is the acoustic pressure magnitude on the surface of the source?

Use 415 [rayls] for the characteristic impedance ($\rho_o c$) of air and 340 $\left[\frac{m}{s}\right]$ for the speed of sound (c) in air.
(Answers): (a) 1.713 $\left[\frac{m}{s}\right]$; (b) 160[Pa]

Exercise 5.3

Flow velocity magnitude U_0 of a spherical wave is given as 0.02 [m/s] at 2 [m] from the center of a source. If the frequency is 1700 [Hz].

a)   What is the flow velocity magnitude at any point in space?
b)   What is the acoustic pressure magnitude at any point in space?
c)   What is the acoustic intensity at 5 [m] from the center of the source?

Use 415 [rayls] for the characteristic impedance $(\rho_0 c)$ of air and 340 $\left[\frac{m}{s}\right]$ for the speed of sound (c) in air.

(Answers): (a) $1.2731 * 10^{-3} \frac{\sqrt{1+100\pi^2 r^2}}{r^2}$ $\left[\frac{m}{s}\right]$; (b) $\frac{16.598}{r}$ $[Pa]$; (c) 0.0133 $\left[\frac{w}{m^2}\right]$

Exercise 5.4

Derive the complex flow velocity $u(r, t)$ from the following complex sound pressure $p(r, t)$ using Euler's force equation:

$$\text{Pressure}: \quad p(r,t) = \frac{A}{r} e^{j(\omega t - kr + \theta)} \qquad\qquad (C_{EP})$$

Hint: The flow velocity $u(r, t)$ can be obtained by finding the integral of the derivative of the pressure $p(r, t)$ in the Euler's force equation shown below:

$$\text{Euler's force equation}: \quad \frac{\partial}{\partial r} p(r,t) = -\rho_o \frac{\partial}{\partial t} u(r,t)$$

(Answers): Flow velocity : $u(r,t) = \frac{A}{r \rho_o c \cos\phi} e^{j(\omega t - kr + \theta - \phi)}$

Exercise 5.5

Derive the complex flow velocity $u(r, t)$ from the following complex sound pressure $p(r, t)$ using Euler's force equation:

$$\text{Pressure}: \quad p(r,t) = \frac{A}{r} e^{-j(\omega t - kr + \theta)} \qquad\qquad (C_{EP})$$

(Answers): Flow velocity : $u(r,t) = \frac{A}{r \rho_o c \cos\phi} e^{-j(\omega t - kr + \theta - \phi)}$

Chapter 6
Acoustic Waves from Spherical Sources

In the previous chapter, formulas for sound pressure, flow velocity, acoustic intensity, and specific acoustic impedance were formulated in spherical coordinates. Note that in the previous chapter, the formulas are just general solutions and are not based on any physical source. In this chapter, you will learn how to calculate sound pressure radiated from a spherical source of different radii. Note that this spherical source has physical quantities such as radius and surface vibration velocity.

The end goal of this chapter is to obtain the formula for calculating sound pressure radiating from a point source. Even though this formula is derived from a spherical source with physical quantities of radius and velocity, the radius is actually eliminated in this formula which makes it possible to model any shape for vibration surfaces. The formula for acoustic waves from a point source is summarized below and in the table on the next page:

$$p(r,t) = \frac{\rho_o ck}{4\pi r} Q_s \cos\left(\omega t - kr + \theta_o + \frac{\pi}{2}\right)$$

$$u(r,t) = \frac{k}{4\pi r \cos(\phi)} Q_s \cos\left(\omega t - kr + \theta_o + \frac{\pi}{2} - \phi\right)$$

where Q_s is the source strength and is defined as:

$$Q_s \equiv \int_s U_n ds = 4\pi a^2 U_a$$

This formula can be used for numerical calculations of sound pressure radiating from an arbitrary vibrating surface. The vibrating surface is numerically modeled as a large number of singular point sources. Each point source has a source strength representing the vibration power of its corresponding vibration surface. This technique is used in the project in this chapter.

© The Author(s), under exclusive license to Springer Nature Switzerland AG 2021
H. Lin et al., *Lecture Notes on Acoustics and Noise Control*,
https://doi.org/10.1007/978-3-030-88213-6_6

To obtain the formulas for point sources, we will follow the three formulas as described below:

Formula 1: Formula for a pulsating spherical source with an arbitrary radius a
Formula 2: Formula for a pulsating spherical source with a small radius ($a \ll 1$)
Formula 3: Formula for a point source with a zero radius ($a = 0$)

Each formula is developed based on the previous formula. For example, Formula 3 is based on Formula 2; Formula 2 is based on Formula 1; Formula 1 is based on the general spherical wave derived in the previous chapter.

The following is a summary of formulations of acoustic waves from a spherical source that will be derived in this chapter:

Source types	Pressure: $p(r,t)$	Velocity: $u(r,t)$
General spherical wave	$rp(r,t)$ $=A \cos(\omega t - kr + \theta)$ where r is the distance from the origin of the coordinate A and θ are determined by BC	$ru(r,t) = \frac{A}{\rho_o c \cos(\phi)} \cos(\omega t - kr + \theta - \phi)$ where $\phi = \tan^{-1}\left(\frac{1}{kr}\right) \to \cos(\phi) = \frac{kr}{\sqrt{1+(kr)^2}}$
Pulsating spherical source of radius a Based on BC: $u(a,t) = U_a \cos(\omega t + \theta_o)$	$rp(r,t)$ $=A \cos(\omega t - kr + \theta)$ where $A = U_a \rho_o c \cos(\phi_a) a$ $\theta = ka + \theta_o + \phi_a$ $\phi_a = \tan^{-1}\left(\frac{1}{ka}\right)$ $\to \cos(\phi_a) = \frac{ka}{\sqrt{1+(ka)^2}}$	$ru(r,t) = \frac{A}{\rho_o c \cos(\phi)} \cos(\omega t - kr + \theta - \phi)$ where $\phi = \tan^{-1}\left(\frac{1}{kr}\right)$
Small pulsating spherical source if $ka \ll 1$ Based on BC: $u(a,t) = U_a \cos(\omega t + \theta_o)$ $\to ka \approx 0$ $\to \phi_a \approx \frac{\pi}{2}$ $\to \cos(\phi_a) \approx ka$	$rp(r,t)$ $=A \cos(\omega t - kr + \theta)$ where $A \approx U_a \rho_o c\, ka^2$ $\theta \approx \theta_o + \frac{\pi}{2}$	$ru(r,t) = \frac{A}{\rho_o c \cos(\phi)} \cos(\omega t - kr + \theta - \phi)$ where $\phi = \tan^{-1}\left(\frac{1}{kr}\right)$ <table><tr><td>Near field: $kr \approx 0$ $\cos(\phi) \approx kr$ $\phi \approx \frac{\pi}{2}$ (out of phase)</td><td>Far field: $kr \approx \infty$ $\cos(\phi) \approx 1$ $\phi \approx 0$ (in phase)</td></tr></table>
Point source Based on BC: $u(a,t) = U_a \cos(\omega t + \theta_o)$ $Q_s \equiv \int_s U_n ds = 4\pi a^2 U_a$	$rp(r,t)$ $=A \cos(\omega t - kr + \theta)$ where $A = Q_s \frac{\rho_o c\, k}{4\pi}$ $\theta = \theta_o + \frac{\pi}{2}$	$ru(r,t) = \frac{A}{\rho_o c \cos(\phi)} \cos(\omega t - kr + \theta - \phi)$ where $\phi = \tan^{-1}\left(\frac{1}{kr}\right)$

6.1   Review of Pressure and Velocity Formulas for Spherical Waves

This section will use the spherical wave formulas from the previous section. A review of spherical wave formulas is given below.

The acoustic pressure and velocity of spherical waves in real number format and complex number format are:

$$p(r,t) \quad = \frac{A}{r} \cos\left(\omega t - kr + \theta\right) \tag{R_{EP}}$$
$$= \frac{A}{r} \frac{1}{2} \left[e^{\,j(\omega t - kr + \theta)} + e^{-j(\omega t - kr + \theta)} \right] \tag{C_{EP}}$$
$$u(r,t) \quad = \frac{A}{r \rho_o c \cos(\phi)} \cos\left(\omega t - kr + \theta - \phi\right) \tag{R_{EP}}$$
$$= \frac{A}{r \rho_o c \cos(\phi)} \frac{1}{2} \left[e^{\,j(\omega t - kr + \theta - \phi)} + e^{-j(\omega t - kr + \theta - \phi)} \right] \tag{C_{EP}}$$

where:

$$\tan(\phi) = \frac{1}{kr}; \quad \cos(\phi) = \frac{kr}{\sqrt{1 + (kr)^2}}; \quad \sin(\phi) = \frac{1}{\sqrt{1 + (kr)^2}}$$

For simplicity, this chapter will formulate in R_{EP} format only.

6.2   Acoustic Waves from a Pulsating Sphere

Sound pressure and velocity radiating from a pulsating spherical source will be formulated in this section:

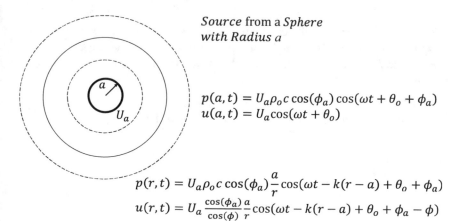

Source from a Sphere
with Radius a

$$p(a,t) = U_a \rho_o c \cos(\phi_a) \cos(\omega t + \theta_o + \phi_a)$$
$$u(a,t) = U_a \cos(\omega t + \theta_o)$$

$$p(r,t) = U_a \rho_o c \cos(\phi_a) \frac{a}{r} \cos(\omega t - k(r - a) + \theta_o + \phi_a)$$
$$u(r,t) = U_a \frac{\cos(\phi_a)}{\cos(\phi)} \frac{a}{r} \cos(\omega t - k(r - a) + \theta_o + \phi_a - \phi)$$

Let's assume that the spherical source of radius a is uniformly dilating and radiating spherical waves. The vibration $u(a, t)$ at the surface of this pulsating sphere ($r = a$) has an amplitude of U_a, an angular frequency of ω, and a phase of θ_o. The velocity at the surface of the sphere provides the boundary condition for solving the velocity at any point outside the sphere:

$$u(r = a, t) = U_a \cos(\omega t + \theta_o)$$

At the surface of the sphere ($r = a$), the velocity $u(x, t)$ of the radiated wave must equal to the velocity $u(a, t)$ of the surface of the sphere. Based on this boundary condition at the surface of the sphere, velocity and sound pressure at any point outside of the sphere can be derived and are shown below before validation:

$$rp(r, t) = A \cos(\omega t - kr + \theta)$$

$$ru(r, t) = \frac{A}{\rho_o c \cos(\phi)} \cos(\omega t - kr + \theta - \phi)$$

where:

$$A = U_a \rho_o c \cos(\phi_a) a$$

$$\theta = ka + \theta_o + \phi_a$$

$$\phi_a = \tan^{-1}\left(\frac{1}{ka}\right) \rightarrow \cos(\phi_a) = \frac{ka}{\sqrt{1 + (ka)^2}}$$

$$\phi = \tan^{-1}\left(\frac{1}{kr}\right) \rightarrow \cos(\phi) = \frac{kr}{\sqrt{1 + (kr)^2}}$$

The following section will derive the above pressure and velocity radiating from the sphere based on the boundary condition at the surface of the sphere.

From the previous section, the general solution for the velocity of the spherical wave is:

$$u(r, t) = \frac{A}{r \rho_o c \cos(\phi)} \cos(\omega t - kr + \theta - \phi)$$

where:

$$\tan(\phi) = \frac{1}{kr}$$

To calculate the unknown constants θ and A, substitute the boundary conditions of velocity on the surface ($r = a$) of the sphere into the above general solution as shown below:

at $r = a$: $u(a,t) = \dfrac{A}{a\rho_o c \cos(\phi_a)} \cos(\omega t - ka + \theta - \phi_a) = U_a \cos(\omega t + \theta_o)$

where the subscript a indicates the value of the associated constant at $r = a$ and then ϕ_a will be:

$$\phi_a = \tan^{-1}\left(\frac{1}{ka}\right)$$

The unknown constant A can be obtained by comparing the magnitude of the velocity evaluated at $r = a$ to the magnitude of the given velocity at $r = a$ as:

$$\frac{A}{a\rho_o c \cos(\phi_a)} = U_a$$

$$\rightarrow A = a\rho_o c \cos(\phi_a) U_a$$

This concludes the validation of formulas for velocity and sound pressure from a pulsating spherical source.

6.3 Acoustic Waves from a Small Pulsating Sphere

This section will study velocity and sound pressure of spherical waves radiated from a small spherical source. Formulas for velocity and sound pressure can be simplified when the wave is radiating from a small sphere ($a \ll 1$). Such a source is called a small spherical source. Since a is small, ka and ϕ_a become:

$$ka = \frac{\omega}{c} a \approx 0$$

$$\rightarrow \quad \phi_a = \tan^{-1}\left(\frac{1}{ka}\right) \approx \frac{\pi}{2}$$

$$\rightarrow \quad \cos(\phi_a) = \frac{ka}{\sqrt{1 + (ka)^2}} \approx ka$$

Substitute the above approximated values into the following equations:

$$p(r,t) = U_a \rho_o c \cos(\phi_a) \frac{a}{r} \cos(\omega t - k(r - a) + \theta_o + \phi_a)$$

$$u(r,t) = U_a \frac{\cos(\phi_a)}{\cos(\phi)} \frac{a}{r} \cos(\omega t - k(r - a) + \theta_o + \phi_a - \phi)$$

The pressure and velocity of the small spherical source are simplified into:

$$p(r,t) \cong U_a \rho_o cka \frac{a}{r} \cos\left(\omega t - kr + \theta_o + \frac{\pi}{2}\right)$$

$$u(r,t) \cong U_a \frac{ka}{\cos(\phi)} \frac{a}{r} \cos\left(\omega t - kr + \theta_o + \frac{\pi}{2} - \phi\right)$$

The error of this simplification for the small spherical source can be calculated by using the amplitude and phase. The errors of amplitude and phase are:

$$E_{amplitude} = \frac{ka - \cos(\phi_a)}{\cos(\phi_a)} = \sqrt{1 + (ka)^2} - 1$$

$$E_{phase} = \frac{\pi}{2} - (\phi_a + ka) = \frac{\pi}{2} - \left(\tan^{-1}\left(\frac{1}{ka}\right) + ka\right)$$

Define a total error of the approximation as:

$$E = \sqrt{E_{amplitude}^2 + E_{phase}^2}$$

Depending on the application, the threshold of the total error can be determined. For example, if $ka = 0.1$, $E = 0.005$; if $ka = 0.01$, $E = 0.00005$.

6.3.1   Near-Field Solutions of a Small Spherical Source (kr ≪ 1)

It is also of interest to study the behavior of sound pressure and velocity at short distances ($r \ll 1$) from the source or at very low frequencies ($\omega \ll 1$) of a wave.

When the solution (measurement point) is very close to the source, kr and ϕ are approximately given by:

$$kr = \frac{\omega}{c} r \approx 0$$

$$\phi = \tan^{-1}\left(\frac{1}{kr}\right) \approx \frac{\pi}{2}$$

$$\cos(\phi) = \frac{kr}{\sqrt{1 + (kr)^2}} \approx kr$$

Substituting the above approximated values back to the pressure and velocity equations induced by the pulsating small sphere yields:

$$p(r,t) \cong U_a \rho_o c\, ka \frac{a}{r} \cos\left(\omega t - kr + \theta_o + \frac{\pi}{2}\right)$$

$$u(r,t) \cong U_a \left(\frac{a}{r}\right)^2 \cos\left(\omega t - kr + \theta_o\right)$$

Hence, the pressure is nearly 90° out of phase with the velocity near the source.

6.3.2   Far-Field Solutions of a Small Spherical Source (kr ≫ 1)

It is also of interest to study the behavior of sound pressure and velocity at large distances ($r \gg 1$) from the source or at high frequencies ($\omega \gg 1$) of a wave. When the solution (measurement point) is far away from the source, kr and ϕ are approximated as:

$$kr = \frac{\omega}{c} r \approx \infty$$

$$\phi = \tan^{-1}\left(\frac{1}{kr}\right) \approx 0$$

$$\cos(\phi) = \frac{kr}{\sqrt{1 + (kr)^2}} \approx 1$$

Substituting the above approximated values back to the pressure and velocity relations induced by the pulsating small sphere yields:

$$p(r,t) \cong U_a \rho_o c\, ka \frac{a}{r} \cos\left(\omega t - kr + \theta_o + \frac{\pi}{2}\right)$$

$$u(r,t) \cong U_a ka \frac{a}{r} \cos\left(\omega t - kr + \theta_o + \frac{\pi}{2}\right)$$

A comparison of the above pressure and velocity shows that the pressure is in phase with the velocity at large distances from the source or at high frequencies of waves.

Example 6.1 (A Small Spherical Source)
An acoustic pressure $p(r,t)$ is created by a surface vibration of a spherical source with a radius $a = \frac{1}{100}$ [m]. Given the surface velocity of the sphere as:

$$u(r = a, t) = U_a \cos\left(2\pi ft - \frac{1}{2}\pi\right)\left[\frac{m}{s}\right] = \text{Re}\left(U_a e^{\, j\left(\omega t - \frac{1}{2}\pi\right)}\right)\left[\frac{m}{s}\right]$$

where the surface velocity is $U_a = 20$ [m/s] and frequency of radiation is $f = 680$ [Hz].

Assuming that this spherical source can be treated as a *small spherical source*, calculate (a) the flow velocity, (b) acoustic pressure, (c) intensity, and (d) power radiating from the small spherical source.

Use 415 [*rayls*] for characteristic impedance ($\rho_0 c$) and 340 [m/s] for the speed of sound. Show units in the Meter-Kilogram-Second (MKS) system.

Example 6.1 Solution

The angular frequency can be calculated as:

$$\omega = 2\pi f = 2\pi \times 680 \left[\frac{1}{s}\right] = 1360\pi \left[\frac{1}{s}\right]$$

The wavenumber can be calculated as:

$$k = \frac{\omega}{c} = \frac{2\pi f}{c} = \frac{2\pi \times 680 \left[\frac{1}{s}\right]}{340 \left[\frac{m}{s}\right]} = 4\pi \left[\frac{1}{m}\right]$$

Part (a)

Compare the given the surface velocity of the sphere:

$$u(r = a, t) = U_a \cos\left(2\pi ft - \frac{1}{2}\pi\right) \left[\frac{m}{s}\right]$$

to the boundary condition (BC) of a small pulsating spherical source in the summary table below:

Source types	Pressure: $p(r,t)$	Velocity: $u(r,t)$	
Small pulsating spherical source $a \ll 1$	$rp(r,$ $t) \cong A \cos(\omega t - kr + \theta)$ where	$ru(r,t) \cong \frac{A}{\rho_o c \cos(\phi)} \cos(\omega t - kr + \theta - \phi)$ where $\phi = \tan^{-1}\left(\frac{1}{kr}\right) \to \cos(\phi) = \frac{kr}{\sqrt{1+(kr)^2}}$	
Based on BC: $u(a,t)=$ $U_a \cos(\omega t + \theta_o)$ $\to ka \approx 0$ $\to \phi_a \approx \frac{\pi}{2}$ $\to \cos(\phi_a) \approx ka$	$A = U_a \rho_o c \, ka^2$ $\theta = \theta_o + \frac{\pi}{2}$	Near field: $\cos(\phi) \approx kr$ Remarks: • p-u out of phase by $\frac{\pi}{2}$ spherical wave	Far field: $\cos(\phi) \approx 1$ Remarks: • p-u in phase, $\therefore \phi = 0$ spherical wave

Comparing the given BC to the formulas in the summary table above gives:

$$a = \frac{1}{100} \, [m]$$

$$U_a = 20[m/s]$$

$$\theta_o = -\frac{\pi}{2}$$

The formula of velocity from a small source as shown is shown in the summary table as:

$$ru(r,t) \cong \frac{A}{\rho_o c \cos{(\phi)}} \cos{(\omega t - kr + \theta - \phi)}$$

where

$$A = U_a \rho_o c \, ka^2 = 20 \cdot 415 \cdot 4\pi \cdot 0.01^2 = 10.4$$

$$\theta = \theta_o + \frac{\pi}{2} = -\frac{\pi}{2} + \frac{\pi}{2} = 0$$

Therefore:

$$ru(r,t) \cong \frac{10.4}{415 \cos{(\phi)}} \cos{(\omega t - kr - \phi)}$$

$$\rightarrow u(r,t) \cong \frac{0.0251}{r \cos{(\phi)}} \cos{(1360\pi t - 4\pi r - \phi)} \left[\frac{m}{s}\right]$$

where:

$$\phi = \tan^{-1}\left(\frac{1}{kr}\right)$$

Part (b)
The formula for acoustic pressure from a small source is shown in the summary table as:

$$p(r,t) \cong \frac{A}{r} \cos{(\omega t - kr + \theta)}[Pa]$$

$$\cong \frac{10.4}{r} \cos{(1360\pi t - 4\pi r)} \, [Pa]$$

Part (c)
The formula of the intensity of spherical waves is derived in Chap. 5 as:

$$I(r) = \frac{1}{2\rho_o c} \frac{A^2}{r^2} = \frac{1}{2 \times 415} \frac{10.4^2}{r^2} = \frac{0.130}{r^2} \left[\frac{w}{m^2}\right]$$

Part (d)

The formula of the sound power of spherical waves is derived in Chap. 5 as:

$$w = \frac{2\pi A^2}{\rho_o c} = \frac{2\pi \times 10.4^2}{415} = 1.64 \ [w]$$

6.4 Acoustic Waves from a Point Source

6.4.1 Point Sources Formulated with Source Strength

The small spherical source in the previous section can be further reduced to a point
source. When a small spherical source is reduced to a point source, the radius and
velocity of the surface of the sphere are eliminated and replaced by acoustic source
strength.

The difference between a small spherical source and a point source in the
formulation of a radiation wave is that in small spheres, the radiation wave is
formulated with radius and vibration at the surface of a small sphere. In point
sources, the radiation wave is formulated with an acoustic source strength.

Unlike small spheres, point sources do not have a physical body. A point source is
a hypothetical source that is an acoustic source strength that can radiate spherical
waves of any frequency. Since the pressure and velocity from a small spherical
sphere can be presented in terms of source strength Q_s as:

$$p(r,t) = \frac{\rho_o c k}{4\pi r} Q_s \cos\left(\omega t - kr + \theta_o + \frac{\pi}{2}\right)$$

$$u(r,t) = \frac{k}{4\pi r \cos(\phi)} Q_s \cos\left(\omega t - kr + \theta_o + \frac{\pi}{2} - \phi\right)$$

where θ_o is a given phase and Q_s is source strength and is defined as:

$$Q_s \equiv \int_s U_n ds = 4\pi a^2 U_a$$

Q_s represents the volume of fluid flowing into the acoustic medium from the
source. The definition of the source strength factor removes the explicit dependence
of pressure on source size as well as the source surface velocity and replaces it with
the volume of fluid flow into the acoustic medium. This abstraction of the source also

removes the requirement that the source surface must be spherical. It only assumes that the radiation wave from the source is omnidirectional regardless of its shape. Hence, point sources are mathematical tools that can collectively represent very complex waves radiating from general geometrics.

6.4.2 Flow Rate as Source Strength

Source strength Q_s is defined as:

$$Q_s \equiv \oint_s U_n ds = 4\pi a^2 U_a$$

where s is an arbitrary closed surface enclosing the source, U_n is the amplitude of the velocity at the normal direction to the closed surface, and U_a is the amplitude of velocity at the surface of this pulsating sphere ($r = a$) introduced in Sect. 6.2 and shown below for reference:

$$u(r, t) \cong U_a \left(\frac{a}{r}\right)^2 \cos\left(\omega t - kr + \theta_o\right)$$

Note that Q_s is also implied as an amplitude of the quantity.

According to the above definition, the source strength, Q_s, is the total volume of all particles passing through the closed surface per unit time. When the air density ρ_0 is constant, the total volume passing through a closed surface multiplying ρ_0 becomes the total mass passing through a closed surface. Therefore, based on the definition, the source strength, Q_s, is also the flow rate of the source:

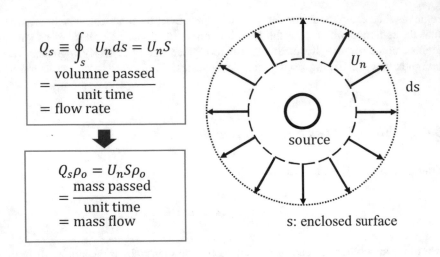

It can be shown that the source strength is independent of the closed surface. This means that the source strength is the same for every closed surface:

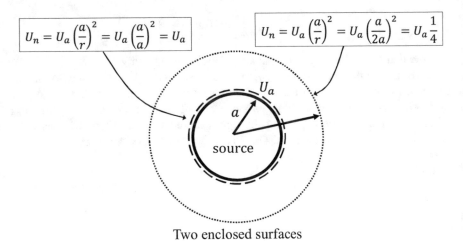

Two enclosed surfaces

The following example will demonstrate this statement by calculating the source strength by integrating over the surface at a distance of (i) one radius ($1a$) and (ii) two radii ($2a$):

(i) s *at* $r = 1a$ (at the surface of the pulsating sphere, $r = a$):

$$U_n = U_a \left(\frac{a}{r}\right)^2 = U_a \left(\frac{a}{a}\right)^2 = U_a$$

$$\rightarrow Q_s \equiv \oint_{s \text{ at } 1a} U_n ds = \oint_{s \text{ at } 1a} U_a \left(\frac{a}{a}\right)^2 ds = U_a \left(\frac{a}{a}\right)^2 4\pi(1a)^2 = 4\pi a^2 U_a$$

(ii) s at $r = 2a$ (at a distance of two times the radius of the sphere, $r = 2a$):

$$U_n = U_a \left(\frac{a}{r}\right)^2 = U_a \left(\frac{a}{2a}\right)^2 = U_a \frac{1}{4}$$

$$\rightarrow Q_s \equiv \oint_{s \text{ at } 2a} U_n ds = \oint_{s \text{ at } 2a} U_a \left(\frac{a}{2a}\right)^2 ds = U_a \left(\frac{a}{2a}\right)^2 4\pi(2a)^2 = 4\pi a^2 U_a$$

It thus follows that the source strength is the same at the surface of the sphere and one radius distance away from the surface of the sphere.

6.4.3   Point Source in an Infinite Baffle

A point source is placed in an infinite baffle as shown in the figure below:

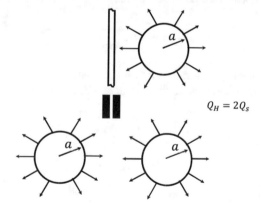

$Q_H = 2Q_s$

Point source in the infinite baffle

The source strength of the hemispherical point source is expressed as

$$Q_H = 2Q_s$$

Since the surface area is half of the complete sphere, the field pressure for a hemispherical point source Q_H is:

$$p(r,t) = \frac{\rho_o c k}{2\pi r} Q_H \cos\left(\omega t - kr + \theta_o + \frac{\pi}{2}\right)$$

$$u(r,t) = \frac{k}{2\pi r \cos(\phi)} Q_H \cos\left(\omega t - kr + \theta_o + \frac{\pi}{2} - \phi\right)$$

Therefore, it requires only half of the source strength to produce the same pressure field as a full spherical pressure field. In other words, if the same source strength is put on the half space, half of the source strength will be reflected to the half space and produce twice the pressure field in the full spherical space created by the same source strength.

6.5   Acoustic Intensity and Sound Power

Sound intensity I is the acoustic intensity, formulated in the previous section as:

$$I(r) \equiv \frac{1}{T}\int_0^T (pu)dt = \frac{1}{2\rho_o c}\frac{A^2}{r^2} = \frac{\rho_o c}{2}\left(\frac{k}{4\pi}\right)^2 \frac{Q_s^2}{r^2}$$

where A was calculated for a point source as:

$$A = Q_s \frac{\rho_o c\, k}{4\pi}$$

For a spherical source with radius $r = a$, acoustic intensity I is constant on the surface of the sphere. Therefore, sound power can be formulated without integration:

$$w \equiv \int_s I(r)\, ds$$

$$= 4\pi a^2 I(r = a)$$

$$= 4\pi a^2 \frac{1}{2}\frac{A^2}{r^2\rho_o c} = \frac{2\pi A^2}{\rho_o c} = \frac{\rho_o c k^2}{8\pi} Q_s^2 [w]$$

Even though the above formula for sound power is formulated at $r = a$, this formula is valid for any other choice of surface for integration. It makes sense because the sound power is the total radiation energy of the source and will not vary with the choices of the surface for integration. This can be checked by comparing the above sound power, at $r = a$, to the sound power at $r = 2a$ as shown below:

$$w \equiv \int_s I \, ds = 4\pi(2a)^2 I(r = 2a) = 4\pi(2a)^2 \frac{1}{2} \frac{A^2}{(2a)^2 \rho_o c} = \frac{2\pi A^2}{\rho_o c}$$

Example 6.2 (A Point Source)

An acoustic pressure $p(r, t)$ is created by a surface vibration of a spherical source with a radius $a = \frac{1}{100}$ [m]. Given the surface velocity of the sphere as:

$$u(r = a, t) = U_a \cos\left(2\pi f t - \frac{1}{2}\pi\right)\left[\frac{m}{s}\right] = \text{Re}\left(U_a e^{j\left(\omega t - \frac{1}{2}\pi\right)}\right)\left[\frac{m}{s}\right]$$

where the surface velocity is $U_a = 20$ [m/s] and frequency of radiation is $f = 680$ [Hz].

Assuming that this spherical source can be treated as *a point source*, calculate (a) the source strength, (b) flow velocity, (c) acoustic pressure, (d) intensity, and (e) power radiating from the point source.

Use 415 [rayls] for characteristic impedance ($\rho_o c$) of air and 340 [m/s] for the speed of sound in air. Show units in the Meter-Kilogram-Second (MKS) system.

Example 6.2 Solution

Similar to Example 6.1, the angular frequency can be calculated as:

$$\omega = 2\pi f = 2\pi \times 680 \left[\frac{1}{s}\right] = 1360\pi \left[\frac{1}{s}\right]$$

And the wave number can be calculated as:

$$k = \frac{\omega}{c} = \frac{2\pi f}{c} = \frac{2\pi \times 680 \left[\frac{1}{s}\right]}{340 \left[\frac{m}{s}\right]} = 4\pi \left[\frac{1}{m}\right]$$

Compare the given surface velocity of the sphere:

$$u(r = a, t) = U_a \cos\left(2\pi f t - \frac{1}{2}\pi\right)\left[\frac{m}{s}\right]$$

to the boundary condition of a small pulsating spherical source in the summary table below:

Source types	Pressure: $p(r, t)$	Velocity: $u(r, t)$
Point source Based on BC: $u(a,t) = U_a \cos(\omega t + \theta_o)$ $Q_s = 4\pi a^2 U_a$	$rp(r, t) = A \cos(\omega t - kr + \theta)$ where $A = Q_s \frac{\rho_o c k}{4\pi}$; $\theta = \theta_o + \frac{\pi}{2}$	$ru(r, t) = \frac{A}{\rho_o c \cos(\phi)} \cos(\omega t - kr + \theta - \phi)$ where $\phi = \tan^{-1}\left(\frac{1}{kr}\right)$; $\cos(\phi) = \frac{kr}{\sqrt{1 + (kr)^2}}$

Comparing the given boundary condition to the formulas in the summary table above gives:

$$a = \frac{1}{100} \ [m]$$

$$U_a = 20 [m/s]$$

$$\theta_o = -\frac{\pi}{2}$$

Part (a)
When this source is modeled as a point source, the source strength is given by:

$$Q_s = 4\pi a^2 U_a = 4\pi \cdot 0.01^2 \cdot 20 = 0.0251 \ \left[\frac{m^3}{s}\right]$$

Part (b)
The formula of the velocity from a point source is shown in the summary table as:

$$u(r,t) \cong \frac{A}{r \, \rho_o c \cos(\phi)} \cos(\omega t - kr + \theta - \phi)$$

where:

$$A = Q_s \frac{\rho_o c \, k}{4\pi} = 0.0251 \frac{415 \times 4\pi}{4\pi} = 10.4$$

$$\theta = \theta_o + \frac{\pi}{2} = -\frac{\pi}{2} + \frac{\pi}{2} = 0$$

Therefore:

$$u(r,t) \cong \frac{A}{r \, \rho_o c \cos(\phi)} \cos(\omega t - kr + \theta - \phi)$$

$$= \frac{10.4}{415 \, r \cos(\phi)} \cos(1360\pi t - 4\pi r - \phi)$$

$$= \frac{0.0251}{r \, \cos(\phi)} \cos(1360\pi t - 4\pi r - \phi)$$

where:

$$\phi = \tan^{-1}\left(\frac{1}{kr}\right)$$

Part (c)

The formula for the acoustic pressure from a point source is shown in the summary table as:

$$p(r,t) = \frac{A}{r} \cos(\omega t - kr + \theta)$$

$$= \frac{10.4}{r} \cos(1360\pi t - 4\pi r)$$

Part (d)

Acoustic intensity:

$$I(r) = \frac{\rho_o c}{2} \left(\frac{k}{4\pi}\right)^2 \frac{Q_s^2}{r^2} \left[\frac{w}{m^2}\right]$$

$$= \frac{415}{2} \left(\frac{4\pi}{4\pi}\right)^2 \frac{0.0251^2}{r^2} \left[\frac{w}{m^2}\right] = \frac{0.131}{r^2} \left[\frac{w}{m^2}\right]$$

Part (e)

Sound power:

$$w = \frac{\rho_o c k^2}{8\pi} Q_s^2 [w] = \frac{415 \times (4\pi)^2}{8\pi} 0.0251^2 [w] = 1.65 \ [w]$$

Remarks

The results show that the acoustic quantities calculated from a small spherical formulation (Example 6.1) and a point source formulation (Example 6.2) are identical, as expected.

For a small sphere, the constant A is calculated by comparing the magnitudes of the velocity:

$$\frac{A}{r\rho_o c \cos(\phi)} = \frac{k}{4\pi r \cos(\phi)} Q_s$$

to arrive at:

$$A = \frac{\rho_o c k}{4\pi} Q_s = \frac{415 \cdot 4\pi}{4\pi} \cdot 0.0251 = 10.4$$

This shows that the constant A calculated from a point source is the same or close to the value calculated from a small spherical source. Therefore, we can conclude that this spherical source can be modeled as a point source.

Example 6.3

The rectangular plate has the width L_x and the height L_y with the following given dimensions:

$$L_x = 0.8 \; [m]$$

$$L_y = 0.6 \; [m]$$

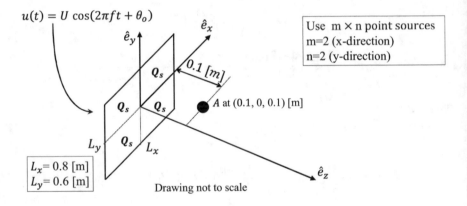

Assume that this plate vibrates at frequency f and the amplitude U of surface velocity is:

$$f = 340 \; [Hz]$$

$$U = 20 \; \left[\frac{m}{s}\right]$$

$$\theta_o = 0 \; [rad]$$

Therefore, the surface velocity is:

$$u(t) = U \; \cos(2\pi f t) = 20 \; \cos(2\pi * 340 * t) \; \left[\frac{m}{s}\right]$$

We will equally divide the rectangular plate into m sections in the x-direction and n sections in the y-direction. The values of m and n are:

$$m = 2; \qquad n = 2$$

Calculate (a) sound pressure and (b) flow velocity (three components) generated by surface vibration of the plate at Point A (0.1, 0, 0.1) [m].

Use 415 [*rayls*] for characteristic impedance ($\rho_0 c$) of air and 340 [*m/s*] for the speed of sound in air. Show units in the Meter-Kilogram-Second (MKS) system.

You can use the function POINTSOURCE (in Sect. 6.6) for calculating sound pressure and flow velocity of a vibration plate. Or you can implement this function using your preferred tool such as Python or Excel spreadsheet.

Example 6.3 Solution
We will equally divide the rectangular plate into m sections in the x-direction and n sections in the y-direction. The values of m and n are:

$$m = 2; \qquad n = 2$$

Each divided piece has the same dimension as:

$$\Delta x = \frac{L_x}{m} = \frac{0.8}{2} \ [m] = 0.4 \ [m]$$

$$\Delta y = \frac{L_y}{n} = \frac{0.6}{2} \ [m] = 0.3 \ [m]$$

$$Q_s = \Delta x \times \Delta y \times U = 0.4 \times 0.3 \times 20 = 2.4 \ \left[\frac{m^3}{s}\right]$$

Due to the hemispherical point source, the source strength is doubled as:

$$Q_H = Q_s \times 2$$

Use a time resolution of $\Delta t = 0.0001$ [s] for the time history plot.

The locations of the point sources are user-defined arguments in the main program as:

```
src(1,:)=[-0.2, 0.15, 0, 2.4]; % sx,sy,sz,Qs
src(2,:)=[ 0.2, 0.15, 0, 2.4]; % sx,sy,sz,Qs
src(3,:)=[-0.2,-0.15, 0, 2.4]; % sx,sy,sz,Qs
src(4,:)=[ 0.2,-0.15, 0, 2.4]; % sx,sy,sz,Qs
```

The location of Point A is also a user-defined argument in the main program as:

```
x=0.1; y=0 ; z=0.1; % [m] Point A
```

The function POINTSOURCE.m and the main program VibrationPlate.m of this example can be found on Moodle.

The outputs of the first 11 time steps are:

time [s]	pressure [Pa]	velocity x [m/s]	velocity y [m/s]	velocity z [m/s]
0.0000	13902.74	−6.87	0.00	15.36
0.0001	13738.71	−5.46	0.00	16.93
0.0002	12950.06	−3.80	0.00	17.74
0.0003	11572.66	−1.97	0.00	17.73
0.0004	9669.12	−0.05	0.00	16.92
0.0005	7325.99	1.87	0.00	15.34
0.0006	4649.79	3.71	0.00	13.07
0.0007	1762.19	5.38	0.00	10.19
0.0008	−1205.52	6.81	0.00	6.86
0.0009	−4118.43	7.92	0.00	3.21
0.0010	−6844.10	8.68	0.00	-0.58

The output of the first 51 time steps can be plotted as:

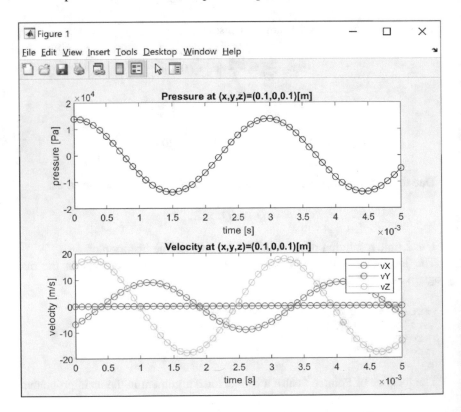

6.6 Computer Program

Function POINTSOURCE

The function POINTSOURCE calculates the pressure and velocity induced by a point source with a source strength Q_s. The input and output arguments of the function are as follows:

Input Arguments

```
sx: x coordinate of the source
sy: y coordinate of the source
sz: z coordinate of the source
Qs: Source Strength
x: x coordinate of the calculation point
y: y coordinate of the calculation point
z: z coordinate of the calculation point
k: Wavenumber
LoC: Characteristic Impedance
w: Frequency
t: time
tho: theda of the source
```

Output Arguments

```
p: pressure
vX: x component of velocity
vY: y component of velocity
vZ: z component of velocity
```

This function is based on the formulas of pressure and velocity of a point source in Chap. 6. The formulas are shown in the table below. This function in MATLAB code is provided in Appendix 1 and can be implemented with your preferred tool such as Python or Excel spreadsheet.

Source types	Pressure: $p(r,t)$	Velocity: $u(r,t)$
Point source Based on BC: $u(a,t) = U_a \cos(\omega t + \theta_o)$ $Q_s \equiv \int_s U_n ds = 4\pi a^2 U_a$	$rp(r,t)$ $= A \cos(\omega t - kr + \theta)$ where $A = Q_s \frac{\rho_o c\, k}{4\pi}$ $\theta = \theta_o + \frac{\pi}{2}$	$ru(r,t) = \frac{A}{\rho_o c \cos(\phi)} \cos(\omega t - kr + \theta - \phi)$ where $\phi = \tan^{-1}\left(\frac{1}{kr}\right)$

MATLAB Code (Function)

```
function [p,vX,vY,vZ]=POINTSOURCE(sx,sy,sz,Qs,x,y,z,k,loC,w,t,tho)
% calculate position vector from the source to the calculation point
rX=x-sx;
rY=y-sy;
rZ=z-sz;
% distance from the point source to (x,y,z)
r=sqrt(rX^2+rY^2+rZ^2);
% calculate the pressure
% formula for calculating pressure
% p=(A/r)*cos(w*t-k*r+theta_a)
% where
% A=Qs*loC*k/(4*pi)
% tha=tho+pi/2
A=Qs*loC*k/(4*pi);
tha=tho+pi/2;
p=(1/r)*A*cos(w*t-k*r+tha);
%
% calculate the velocity vector
% velocity of point source in the summary table
% v=(A/r)/(loC*k*cos_phi)*cos(w*t-k*r+tha-phi)
% where
% A=Qs*loC*k/(4*pi)
% tha=tho+pi/2;
% cos_phi=kr/sqrt(1+kr^2)
% phi=atan2(1,kr)
phi=atan2(1,k*r);
cos_phi=(k*r)/sqrt(1+(k*r)^2);
v=(A/r)/(loC*cos_phi)*cos(w*t-k*r+tha-phi);
% calculate components of velocity in the x, y, and z directions
vX=v*(rX/r);
vY=v*(rY/r);
vZ=v*(rZ/r);
end
```

MATLAB Code (Main Code)

```
function vibrationPlate
% Example to use function "POINTSOURCE"
clear all
%----------------------------------------------%
% Define time array
%----------------------------------------------%
TimeIncrement=0.0001; % the time increment [s]
nTime=51;          % the number of time data
%----------------------------------------------%
% Define Source Strength
%----------------------------------------------%
% define the properties of air
c= 340;   % [m/s]
```

```
loC=415;   % [rayl]
f= 340;    % [Hz]
% calculate the angular frequency (w) and wave number (k)
w=2*pi*f;  % the angular frequency [rad/s]
k=w/c;     % the wave number [rad/m]
tho=0;     % theda_o: phase of the source
% define sx,sy,sz,Qs
iCase=2;
if iCase==1
% Case 1: Lx=0.8; Ly=0.6; m=2; n=2
src(1,:)=[-0.2, 0.15, 0, 2.4]; % sx,sy,sz,Qs
src(2,:)=[ 0.2, 0.15, 0, 2.4]; % sx,sy,sz,Qs
src(3,:)=[-0.2,-0.15, 0, 2.4]; % sx,sy,sz,Qs
src(4,:)=[ 0.2,-0.15, 0, 2.4]; % sx,sy,sz,Qs
elseif iCase==2
% Case 2: Lx=0.9; Ly=0.6; m=3; n=3
src(1,:)= [-0.3, 0.2, 0, 1.8]; % sx,sy,sz,Qs
src(2,:)= [ 0 , 0.2, 0, 1.8]; % sx,sy,sz,Qs
src(3,:)= [ 0.3, 0.2, 0, 1.8]; % sx,sy,sz,Qs
src(4,:)= [-0.3, 0,  0, 1.8]; % sx,sy,sz,Qs
src(5,:)= [ 0 , 0,  0, 1.8]; % sx,sy,sz,Qs
src(6,:)= [ 0.3, 0,  0, 1.8]; % sx,sy,sz,Qs
src(7,:)= [-0.3,-0.2, 0, 1.8]; % sx,sy,sz,Qs
src(8,:)= [ 0 ,-0.2, 0, 1.8]; % sx,sy,sz,Qs
src(9,:)= [ 0.3,-0.2, 0, 1.8]; % sx,sy,sz,Qs
end
%-----------------------------------------------%
% Define the Location of the measurement points
%-----------------------------------------------%
x=0;   y=0 ; z=0.1; % [m] Point A
% x=0;   y=0 ; z=10; % [m] Point B
% x=0.1; y=0 ; z=0.1; % [m] Point C
% x=0.1; y=0 ; z=10; % [m] Point D
%-----------------------------------------------%
% Calculation the Pressure and Velocity
%-----------------------------------------------%
for kTime=1:1:nTime
   time=TimeIncrement*(kTime-1);
   % reset the pressure and velocity to zero for each time step
   pSum=0;
   vzSum=0;
   vxSum=0;
   vySum=0;
   for kSrc=1:1:size(src,1)
       sx=src(kSrc,1);
       sy=src(kSrc,2);
       sz=src(kSrc,3);
       Qs=src(kSrc,4);
       % calculate the pressure and velocity
       [p,vx,vy,vz]=POINTSOURCE(sx,sy,sz,Qs*2,x,y,z,k,loC,w,time,
tho);
       % superpose the pressure and velocity
       pSum=pSum+p;
```

```
        vzSum=vzSum+vz;
        vxSum=vxSum+vx;
        vySum=vySum+vy;
    end
    % record the pressure and velocity in arrays
    tArr(kTime,1)=time;
    pArr(kTime,1)=pSum;
    vxArr(kTime,1)=vxSum;
    vyArr(kTime,1)=vySum;
    vzArr(kTime,1)=vzSum;
end
%-----------------------------------------------%
% Plotting
%-----------------------------------------------%
% plot the pressure
figure(1)
subplot(2,1,1)
plot(tArr,pArr(:,1),'k-o')
title(strcat('Pressure at (x,y,z)=(', num2str(floor(x*1000)/
1000),',',',...
    num2str(floor(y*1000)/1000),',',',num2str(floor(z*1000)/1000),')
[m]'))
xlabel('time [s]')
ylabel('pressure [Pa]')
% plot velocity
subplot(2,1,2)
plot(tArr,[vxArr(:,1),vyArr(:,1),vzArr(:,1)],'-o')
title(strcat('Velocity at (x,y,z)=(',num2str(floor(x*1000)/
1000),',',',...
    num2str(floor(y*1000)/1000),',',',num2str(floor(z*1000)/1000),')
[m]'))
xlabel('time [s]')
ylabel('velocity [m/s]')
legend('vX','vY','vZ','Location','NorthEast')
end
```

6.7   Project

A rectangular plate has a width L_x and a height L_y with the following given dimensions:

$$L_x = 0.9\,[m]; L_y = 0.6\,[m]$$

Assume that this plate vibrates at a frequency f with a phase θ_o and an amplitude U as:

$$f = 340 \ [Hz]; \ U = 30 \ \left[\frac{m}{s}\right]; \theta_o = 0 \ [rad]$$

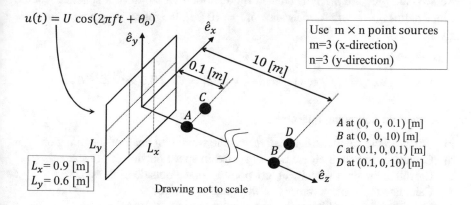

Divide the rectangular plate into m sections in the x-direction and n sections in the y-direction as:

$$m = 3; \qquad\qquad n = 3$$

Use a time resolution of $\Delta t = 0.0001 \ [s]$ for the time history plot.

Use the function POINTSOURCE.m and the main program VibrationPlate.m to complete this project.

Write your group report and submit it as a single pdf file that includes the following items:

6.8   Objective

1) Procedures
2) Your completed MATLAB script.
3) Output figures of the sound pressure at points A, B, C, and D
4) Output figures of the flow velocity at points A, B, C, and D
5) Comparison of the pressures at points A and B
6) Comparison of velocities at points A and C
7) Conclusion of the simulation results

Hint: This project is an extension of Example 6.3. You can use Example 6.3 as a reference for this project.

6.9   Homework Exercises

Exercise 6.1

An acoustic pressure $p(r, t)$ is created by a surface vibration of a spherical source with a radius $a = \frac{1}{2}$ [m]. At a distance of $r = 10$ [m], the following sound pressure is given:

$$p(r = 10, t) = 0.02 \cos\left(\frac{1}{8}\pi ct + \frac{1}{4}\pi\right)[Pa] = \mathrm{Re}\left(0.02 e^{j\left(\frac{1}{8}\pi ct + \frac{1}{4}\pi\right)}\right) [Pa]$$

Answer the following questions:

a) What are the angular frequency ω and the wave number k of the radiation?
b) Calculate the sound pressure at any point in space outside of the sphere.
c) Calculate the flow velocity at any point in space outside of the sphere.
d) Calculate the surface velocity of the spherical source.
e) Calculate the sound intensity at any point outside of the sphere.
f) Calculate the sound power radiated from the source.
g) What are the period and wavelength of the radiation?

Use 415 [*rayls*] for the characteristic impedance ($\rho_0 c$) of air and 340 [*m/s*] for the speed of sound in air. Show units in the Meter-Kilogram-Second (MKS) system.
(Answers):

$$\omega = \frac{1}{8}\pi c \left[\frac{rad}{sec}\right]; \quad k = \frac{1}{8}\pi \left[\frac{rad}{m}\right]$$

$$p(r, t) = \frac{1}{r} 0.2 \times \cos\left(\frac{1}{8}\pi ct - \frac{1}{8}\pi r + 1.5\pi\right)[Pa]$$

$$u(r, t) = \frac{1}{r}\frac{0.2}{\rho_0 c \cos(\phi)} \cos\left(\frac{1}{8}\pi ct - \frac{1}{8}\pi r + 1.5\pi - \phi\right)\left[\frac{m}{s}\right];$$

where $\cos(\phi) = \dfrac{\pi r}{8\sqrt{1+\left(\frac{\pi r}{8}\right)^2}}$; $\phi = \tan^{-1}\left(\frac{8}{\pi r}\right)$

$$u(t) = 0.005 \cos\left(\frac{1}{8}\pi ct + 0.9992\pi\right)\left[\frac{m}{s}\right]$$

$$I(r) = \frac{4.82 \times 10^{-5}}{r^2}\left[\frac{w}{m^2}\right]$$

$$w = 6.06 \times 10^{-4}[w]$$

$$T = \frac{16}{c}[s]; \lambda = 16[m]$$

Exercise 6.2 (Reverse of Exercise 6.1)

A sound pressure $p(r, t)$ is created by a surface vibration of a spherical source with a radius $a = \frac{1}{2}$ [m]. Assume that the surface velocity of the sphere is given by:

$$u(r = a, t) = 0.005 \ \cos\left(\frac{1}{8}\pi c t + 0.9992 \ \pi\right)\left[\frac{m}{sec}\right]$$

$$= \operatorname{Re}\left(0.005 \ e^{\ j\left(\frac{1}{8}\pi c t + 0.9992 \ \pi\right)}\right)\left[\frac{m}{s}\right]$$

Answer the following questions:

a) What are the angular frequency ω and the wave number k of the radiation?
b) Calculate the sound pressure at any point in space outside of the sphere.
c) Calculate the flow velocity at any point in space outside of the sphere.
d) Calculate the sound pressure at $r = 10$ [m].
e) Calculate the sound intensity at any point outside of the sphere.
f) Calculate the sound power radiated from the source.

Use 415 [rayls] for the characteristic impedance ($\rho_0 c$) of air and 340 [m/s] for the speed of sound in air. Show units in the Meter-Kilogram-Second (MKS) system.

(Answers):

$$\omega = \frac{1}{8}\pi c \left[\frac{rad}{sec}\right]; \quad k = \frac{1}{8}\pi \left[\frac{rad}{m}\right]$$

$$p(r, t) = \frac{1}{r} 0.2 \times \ \cos\left(\frac{1}{8}\pi c t - \frac{1}{8}\pi r + 1.5\pi\right)[Pa]$$

$$u(r, t) = \frac{1}{r}\frac{0.2}{\rho_0 c \cos(\phi)} \ \cos\left(\frac{1}{8}\pi c t - \frac{1}{8}\pi r + 1.5\pi - \phi\right)\left[\frac{m}{s}\right]$$

where $\cos(\phi) = \dfrac{\pi \ r}{8\sqrt{1+\left(\frac{\pi \ r}{8}\right)^2}}; \quad \phi = \tan^{-1}\left(\frac{8}{\pi \ r}\right)$

$$p(r = 10m, t) = 0.02 \cos\left(\frac{1}{8}\pi c t + \frac{1}{4}\pi\right)[Pa]$$

$$I(r) = \frac{4.82 \times 10^{-5}}{r^2}\left[\frac{w}{m^2}\right]$$

$$w = 6.06 \times 10^{-4}[w]$$

Exercise 6.3

A sound pressure is created by the pulsating surface of a sphere with a radius $a = \frac{1}{2}$ [m]. Assume that the surface velocity of the sphere is:

$$u(r = a, t) = 4 \, \cos\left(2\pi ct - \frac{\pi}{2}\right)\left[\frac{m}{sec}\right]$$

where c is the speed of the sound. At a distance $r = 5$ [m] from the center of the sphere, calculate the following:

a) Sound pressure amplitude
b) Sound intensity
c) Sound power (or sound pressure)

Note: This pulsating sphere *cannot be treated* as a small sphere source or a point source.

Use 415 [*rayls*] for the characteristic impedance ($\rho_0 c$) of air and 340 [*m/s*] for the speed of sound in air. Show units in the Meter-Kilogram-Second (MKS) system.

(Answers): (a) 158.18 [*Pa*]; (b) 30.15 $\left[\frac{w}{m^2}\right]$; (c) 9470.5 [*w*]

Exercise 6.4 (Point Source)
The sound pressure $p(r, t)$ is created by a surface vibration of a spherical source with the radius $a = \frac{1}{20}$ [m]. Assume that the surface velocity of the sphere is given by:

$$u(r = a, t) = 0.5 \, \cos\left(\frac{1}{8}\pi ct + \pi\right)\left[\frac{m}{sec}\right]$$

$$= \text{Re}\left(0.5 \, e^{\, j\left(\frac{1}{8}\pi ct + \pi\right)}\right)\left[\frac{m}{s}\right]$$

Treat this small spherical source as a point source to answer the following questions:

a) What are the angular frequency ω and the wave number k of the radiation?
b) Calculate the source strength at $r = a$ and $r = 2a$ from the center of the sphere.
c) Use the source strength to calculate flow velocity $u(r, t)$ at any point in space.
d) Use the source strength to calculate sound pressure $p(r, t)$ at any point in space.
e) Calculate the sound intensity at any point.
f) Calculate the sound power radiated from the source.

Use 415 [*rayls*] for the characteristic impedance ($\rho_0 c$) of air and 340 [*m/s*] for the speed of sound in air. Show units in the Meter-Kilogram-Second (MKS) system.

(Answers):

$$\omega = \frac{1}{8}\pi c\left[\frac{rad}{sec}\right]; \text{wave number} : k = \frac{1}{8}\pi\left[\frac{rad}{m}\right]$$

a) $Q_s = 0.0157 \left[\frac{m^3}{s}\right]$; $Q_s = 0.0157 \left[\frac{m^3}{s}\right]$

$$u(r,t) = \frac{1}{r} \frac{0.2}{\rho_o c \cos(\phi)} \cos\left(\frac{1}{8}\pi ct - \frac{1}{8}\pi r + 1.5\pi - \phi\right)\left[\frac{m}{s}\right];$$

where, $\cos(\phi) = \frac{\pi r}{8\sqrt{1+\left(\frac{\pi r}{8}\right)^2}}$; $\phi = \tan^{-1}\left(\frac{8}{\pi r}\right)$

$$p(r,t) = \frac{1}{r}0.2 \times \cos\left(\frac{1}{8}\pi ct - \frac{1}{8}\pi r + 1.5\pi\right)[Pa]$$

$$I(r) = \frac{5.0 \times 10^{-5}}{r^2}\left[\frac{w}{m^2}\right]$$

$$w = 6.3 \times 10^{-4}[w]$$

Exercise 6.5 (Point Source)
A sound pressure $p(r,t)$ is created by a surface vibration of a small spherical source with radius $a = \frac{1}{25}$ [m]. Assume the surface velocity of the sphere is:

$$u(r = a, t) = 3 \cos\left(\frac{1}{4}\pi ct - \frac{\pi}{2}\right)\left[\frac{m}{sec}\right]$$

Treat this small spherical source as a point source to answer the following questions:

d) Calculate the source strength of the spherical source.
e) Calculate sound pressure $p(r,t)$ at any point in space.
f) Calculate sound intensity at any point in space.
g) Calculate sound power radiated from the source.

Use 415 [rayls] for the characteristic impedance ($\rho_0 c$) of air and 340 [m/s] for the speed of sound in air. Show units in the Meter-Kilogram-Second (MKS) system.
  (Answers):

a) $Q_s = 0.0603 \left[\frac{m^3}{s}\right]$
b) $p(r,t) = \frac{1.5645}{r} \cos\left(\frac{\pi c}{4}t - \frac{\pi}{4}r\right)[Pa]$
c) $I(r) = \frac{0.0029}{r^2}\left[\frac{w}{m^2}\right]$
g) $w = 0.0371[w]$

Chapter 7
Resonant Cavities

In the previous chapters, when we calculated the sound pressure radiated from a vibrating surface, we only considered forward waves but not returning waves. This is because we have not considered any reflecting surfaces yet.

From this chapter forward, we will study sound waves that are bouncing inside cavities, waveguides, and pipes. If we consider the pre-reflection waves as forward waves, the post-reflection waves will be backward waves. The resulting wave of the addition of a pre-reflection forward wave and its post-reflection backward wave is a standing wave because the amplitudes of the pre-reflection wave and its post-reflection wave are (almost) the same.

In resonant cavities, the standing waves caused by the pre-reflection forward wave and the post-reflection backward wave will cause the air in the cavity to be resonant at certain frequencies. The resonance of air in cavities, if not considered properly, can be a serious design flaw. On the other hand, the resonance of cavities, if understood correctly, can be used as vibration absorbers in filter designs (Chapter 13).

In this chapter, natural (resonant) frequencies and their corresponding mode shapes of resonant cavities will be formulated. The discretized natural frequencies and mode shapes will be derived from standing wave solutions with constraints on boundary conditions. It can be difficult to fully comprehend and visualize standing waves in rectangular cavities because they propagate in 3D space. It is easier to first understand and formulate standing waves in 1D than in 3D. For this reason, we will first study the 1D standing waves between two walls and then extend the formulas to 2D and 3D.

A summary of formulas of 1D standing waves between two walls is listed below and will be derived in Section 7.1:

The natural frequency of an eigenmode (l, m, n) in an enclosure is:

© The Author(s), under exclusive license to Springer Nature Switzerland AG 2021
H. Lin et al., *Lecture Notes on Acoustics and Noise Control*,
https://doi.org/10.1007/978-3-030-88213-6_7

$$f_{lmn} = \frac{c}{2\pi} k_{lmn}$$

where k_{lmn} (combined wavenumber) is related to k_{xl}, k_{ym}, k_{zn} as:

$$k_{lmn}^2 = k_{xl}^2 + k_{ym}^2 + k_{zn}^2$$

The component wavenumbers k_{xl}, k_{ym}, and k_{zn} are the discretized wavenumbers of the eigenmode (l, m, n) and are related to circular frequencies as:

$$k_{xl} = \frac{\pi}{L_x} l \rightarrow f_{xl} = \frac{c}{2\pi} k_{xl}$$

$$k_{ym} = \frac{\pi}{L_y} m; \quad \rightarrow f_{ym} = \frac{c}{2\pi} k_{ym}$$

$$k_{zn} = \frac{\pi}{L_z} n \rightarrow f_{zn} = \frac{c}{2\pi} k_{zn}$$

Four combinations of boundary conditions of a pipe will be discussed near the end of this chapter. The natural frequencies of the four cases of boundary conditions are summarized as:

	Left end	Right end	Wavenumber	Natural frequency	Wavelength
Case 1	CLOSED	CLOSED	$k_l = \frac{\pi}{L} l$	$f_l = \frac{c}{2L} l$	$\lambda_l = 2L \frac{1}{l}$
Case 2	OPEN	OPEN	$k_l = \frac{\pi}{L} l$	$f_l = \frac{c}{2L} l$	$\lambda_l = 2L \frac{1}{l}$
Case 3	OPEN	CLOSED	$k_l = \frac{\pi}{2L}(2l-1)$	$f_l = \frac{c}{4L}(2l-1)$	$\lambda_l = 4L \frac{1}{(2l-1)}$
Case 4	CLOSED	OPEN	$k_l = \frac{\pi}{2L}(2l-1)$	$f_l = \frac{c}{4L}(2l-1)$	$\lambda_l = 4L \frac{1}{(2l-1)}$

7.1   1D Standing Waves Between Two Walls

Between two parallel walls, standing waves exist as mode shapes of air movement between two walls. The standing wave solutions of the acoustic wave equation can be obtained by solving the boundary conditions imposed by the rigid walls.

In a previous chapter, the standing wave solutions were constructed by the addition of two traveling waves with the same amplitude ($A_+ = A_- = A$) that moved in opposite directions as:

$$\begin{aligned} p_s(x,t) &= p_+(x,t) + p_-(x,t) \\ &= A \cos(\omega t - kx + \theta_+) + A \cos(\omega t + kx + \theta_-) \\ &= 2A \cos(\omega t + \theta_t) \cos(kx + \theta_x) \end{aligned}$$

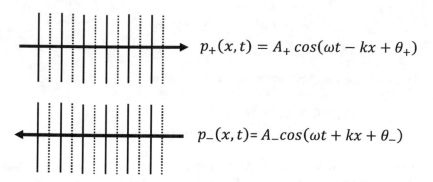

$$p_+(x,t) = A_+ \cos(\omega t - kx + \theta_+)$$

$$p_-(x,t) = A_- \cos(\omega t + kx + \theta_-)$$

In this chapter, standing waves $p_s(x, t)$ are obtained directly by solving the acoustic wave equation.

The one-dimensional wave equation in Cartesian coordinate is:

$$\nabla^2 p = \frac{1}{c^2} \frac{\partial^2 p}{\partial t^2}$$

$$\rightarrow \frac{\partial^2 p}{\partial x^2} = \frac{1}{c^2} \frac{\partial^2 p}{\partial t^2}$$

The differential equation is separable if a solution can be cast in the following form that satisfies the acoustic wave equation:

$$p(x, t) = T(t)X(x)$$

Substituting the above solution into the acoustic wave equation above yields the two separate ordinary differential equations (ODEs):

$$\frac{1}{c^2} \frac{T''(t)}{T(t)} = \frac{X''(x)}{X(x)} = -k^2$$

where k is a new variable that relates the spatial and temporal parts of the differential equations. The new variable k gives us the two separate ordinary differential equations (ODEs) as:

$$\frac{1}{c^2} \frac{T''(t)}{T(t)} = -k^2$$

$$\frac{X''(x)}{X(x)} = -k^2$$

With the new variable k, the solutions of the temporal and the spatial differential equations are:

$$T(t) = A_t \cos (\omega t + \theta_t)$$
$$X(x) = A_x \cos (kx + \theta_x)$$

Therefore, the standing wave solution of the 1D wave equation is:

$$p_s(x,t) = T(t)X(x)$$
$$= A_t A_x \cos (\omega t + \theta_t) \cos (k_x x + \theta_x)$$

where the unknown constants A_t, A_x, θ_t, and θ_x are calculated using the boundary conditions from the two rigid walls. Since $T(t)$ and $X(x)$ are both real numbers, the standing wave functions are real.

Using Euler's force equation, the velocity of the one-dimensional standing wave was derived in a previous chapter and is:

$$u(x,t) = -\frac{1}{\rho_o} \int \left[\frac{\partial}{\partial x} p(x,t) \right] dt$$

$$\rightarrow u_s(x,t) = \frac{1}{\rho_o c} A_t A_x \sin (\omega t + \theta_t) \sin (kx + \theta_x)$$

7.2   Natural Frequencies and Mode Shapes in a Pipe

If the boundary condition at the end of a pipe is either open or closed, there will be four combinations of boundary conditions of a single pipe. The four combinations of the boundary conditions of a single pipe are summarized in the table below:

	Left end	Right end	Wavenumber	Natural frequency	Wavelength
Case 1	CLOSED	CLOSED	$k_l = \frac{\pi}{L}l$	$f_l = \frac{c}{2L}l$	$\lambda_l = 2L\frac{1}{l}$
Case 2	OPEN	OPEN	$k_l = \frac{\pi}{L}l$	$f_l = \frac{c}{2L}l$	$\lambda_l = 2L\frac{1}{l}$
Case 3	OPEN	CLOSED	$k_l = \frac{\pi}{2L}(2l-1)$	$f_l = \frac{c}{4L}(2l-1)$	$\lambda_l = 4L\frac{1}{(2l-1)}$
Case 4	CLOSED	OPEN	$k_l = \frac{\pi}{2L}(2l-1)$	$f_l = \frac{c}{4L}(2l-1)$	$\lambda_l = 4L\frac{1}{(2l-1)}$

Case 1 and Case 2 in the table above are demonstrated in the examples listed below. You can practice Case 3 and Case 4 in the homework exercises listed below:

Case 1: Example 7.1
Case 2: Example 7.2
Case 3: Homework Exercise 7.1
Case 4: Homework Exercise 7.2

Example 7.1 (CLOSED-CLOSED PIPE)
A pipe with CLOSED-CLOSED boundary conditions has a finite length L as shown
below. Set the left end of the pipe as x=0:

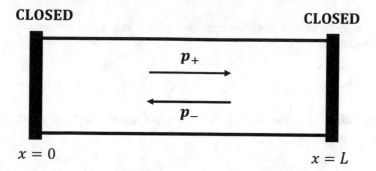

$$x = 0 \qquad\qquad\qquad\qquad x = L$$

(a) Derive the following formulas for wavenumber and natural frequency of eigenmode l:

$$k_l = \tfrac{\pi}{L} l,$$

$$f_l = \frac{c}{2L} l$$

(b) Plot the mode shapes of the first three natural frequencies.

Example 7.1 Solution
Part (a)
   Natural Frequencies of a CLOSED-CLOSED PIPE
   The closed end represents a fixed wall that requires the flow velocity to be zero at
all times.
   The boundary condition at the left wall (CLOSED):

$$u_s(x, t)|_{x=0} = 0$$

$$\rightarrow u_s(x = 0, t) = \frac{1}{\rho_0 c} A_t A_x \sin{(\omega t + \theta_t)} \sin{(kx + \theta_x)} \Big|_{x=0} = 0$$

The above boundary condition requires that the velocity is zero at $x = 0$ at any
time. Therefore:

$$\sin{(kx + \theta_x)}|_{x=0} = 0 \quad \rightarrow \quad \sin{(0 + \theta_x)} = 0 \quad \rightarrow \quad \theta_x = 0$$

Note that the sine function above is simplified from $\sin(kx + \theta_x)$ to $\sin(\theta_x)$ by
letting $x = 0$ at the left wall.
   The boundary condition at the right wall (CLOSED):

$$u_s(x, t)|_{x=L} = 0$$

$$\rightarrow u_s(x = L, t) = \frac{1}{\rho_0 c} A_t A_x \sin(\omega t + \theta_t) \sin(kL) = 0$$

The above boundary condition requires that the velocity is zero at $x = L$ at any time. Therefore:

$$\sin(kx)|_{x=L} = 0 \quad \rightarrow \quad k_l L = l\pi \quad \rightarrow \quad k_l = \frac{l\pi}{L}, \quad l \text{ is an integer}$$

Substituting k_l into the velocity yields the final one-dimensional standing waves between two rigid walls:

$$u_s(x, t) = \frac{1}{\rho_0 c} A_t A_x \sin(\omega t + \theta_t) \sin(k_l x)$$

$$p_s(x, t) = A_t A_x \cos(\omega t + \theta_t) \cos(k_l x)$$

where the wavenumber of eigenmode l is:

$$k_l = \frac{\pi}{L} l$$

Therefore, the natural frequency of eigenmode l is:

$$f_l = \frac{c}{2\pi} k_l = \frac{c}{2L} l$$

And the corresponding wavelength is:

$$\lambda_l = \frac{2\pi}{k_l} = 2L \frac{1}{l}$$

Part (b)

Mode Shapes of a CLOSED-CLOSED PIPE

The discretized velocity, pressure, and wavelength of the eigenmode l are relisted here for plotting the mode shapes:

$$u_s(x, t) = \frac{1}{\rho_0 c} A_t A_x \sin(\omega t + \theta_t) \sin(k_l x)$$

$$p_s(x, t) = A_t A_x \cos(\omega t + \theta_t) \cos(k_l x)$$

$$k_l = \frac{\pi}{L} l$$

$$\lambda_l = \frac{2\pi}{k_l} = \frac{2L}{l}$$

The wavelength of each eigenmode is helpful for plotting the mode shapes, and the wavelengths of the first three eigenmodes are shown below:

$$\lambda_1 = 2L; \quad \lambda_2 = L; \quad \lambda_3 = \frac{2}{3}L;$$

Based on the wavelengths above and zero velocities on the edge, we can plot the mode shape of the mode $(l, m) = (2, 3)$ as follows:

The first two mode shapes of velocity and pressure of a CLOSED-CLOSED pipe are shown below:

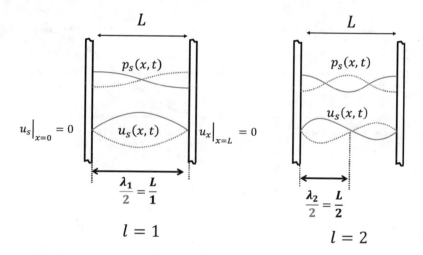

Note that the zero velocity corresponds to the maximum pressure (zero slope) based on Euler's force equation.

Also, the mode shapes of flow velocity are typically not plotted in 2D and 3D because flow velocities are vectors. It will require more than one figure to show a mode shape of flow velocities: one figure for each direction. Therefore, mode shapes of sound pressures are commonly used.

The first three mode shapes of the pressure of a CLOSED-CLOSED pipe are shown below:

Mode 1: $(l = 1)$

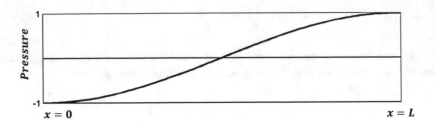

$$\lambda_l = \frac{2\pi}{k_l} = \frac{2L}{l} \rightarrow \lambda_1 = 2L$$

Mode 2: ($l = 2$)

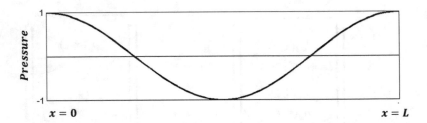

$$\lambda_l = \frac{2\pi}{k_l} = \frac{2L}{l} \rightarrow \lambda_2 = L$$

Mode 3: ($l = 3$)

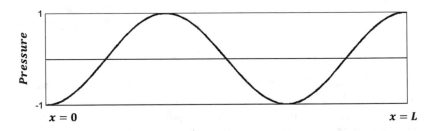

$$\lambda_l = \frac{2\pi}{k_l} = \frac{2L}{l} \rightarrow \lambda_3 = \frac{2}{3}L$$

Example 7.2 (OPEN-OPEN PIPE)
A pipe with OPEN-OPEN boundary conditions has a finite length L as shown below. Set the left end of the pipe as x=0. Determine the first three natural frequencies and plot their corresponding mode shapes:

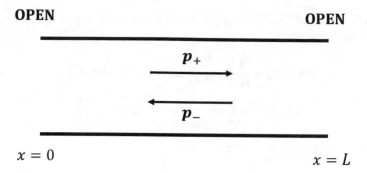

Example 7.2 Solutions

The boundary condition of an *open end* requires that the sound pressure be zero at all times. Note that the boundary condition of a *closed end* requires that the flow velocity be zero at all times.

The boundary condition at the left wall (OPEN):

$$p_s(x, t)|_{x=0} = 0$$

$$\rightarrow p_s(x = 0, t) = A_t A_x \cos(\omega t + \theta_t) \cos(kx + \theta_x)|_{x=0} = 0$$

The above boundary condition requires that the pressure is zero at $x = 0$ at any time. Therefore:

$$\cos(kx + \theta_x)|_{x=0} = 0 \quad \rightarrow \quad \cos(0 + \theta_x) = 0 \quad \rightarrow \quad \theta_x = \frac{\pi}{2}$$

The boundary condition at the right wall (OPEN):

$$p_s(x, t)|_{x=L} = 0$$

$$\rightarrow p_s(x = L_x, t) = A_t A_x \cos(\omega t + \theta_t) \cos\left(kx + \frac{\pi}{2}\right) = 0$$

The above boundary condition requires that the velocity is zero at $x = L$ at any time. Therefore:

$$\cos\left(kx + \frac{\pi}{2}\right)\bigg|_{x=L} = 0$$

$$\rightarrow \quad k_l L + \frac{\pi}{2} = \frac{\pi}{2}(2l + 1)$$

$$\rightarrow \quad k_l = \frac{\pi}{L} l, \quad l \text{ is an integer}$$

Substituting k_l into the velocity yields the final one-dimensional standing waves between two rigid walls:

$$u_s(x,t) = \frac{1}{\rho_0 c} A_t A_x \sin(\omega t + \theta_t) \sin\left(k_l x + \frac{\pi}{2}\right)$$

$$p_s(x,t) = A_t A_x \cos(\omega t + \theta_t) \cos\left(k_l x + \frac{\pi}{2}\right)$$

where the wavenumber of the eigenmode l is:

$$k_l = \frac{\pi}{L}l, \quad l = 1, 2, \ldots$$

Therefore, the natural frequency of the eigenmode l is:

$$f_l = \frac{c}{2\pi} k_l = \frac{c}{2L}l$$

And the corresponding wavelength is:

$$\lambda_l = \frac{2\pi}{k_l} = 2L\frac{1}{l}$$

Mode Shapes of an OPEN-OPEN PIPE:

The first three mode shapes of the pressure of an OPEN-OPEN pipe are shown below:

Mode 1: ($l = 1$)

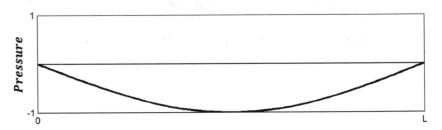

$$\lambda_l = \frac{2\pi}{k_l} = \frac{2L}{l} \rightarrow \frac{\lambda_1}{2} = \frac{L}{1}$$

Mode 2: ($l = 2$)

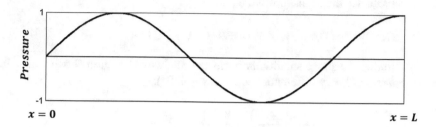

$$\lambda_l = \frac{2\pi}{k_l} = \frac{2L}{l} \rightarrow \frac{\lambda_2}{2} = \frac{L}{2}$$

Mode 3: ($l = 3$)

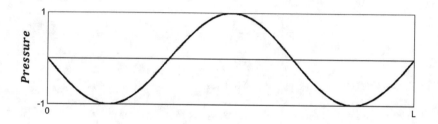

$$\lambda_l = \frac{2\pi}{k_l} = \frac{2L}{l} \rightarrow \frac{\lambda_3}{2} = \frac{L}{3}$$

Note that mode shapes in a pipe are standing waves which are equivalent to two traveling waves with the same amplitude and traveling in opposite directions.

7.3   2D Boundary Conditions Between Four Walls

7.3.1   2D Standing Wave Solutions of the Wave Equation

The two-dimensional wave equation in Cartesian coordinate is:

$$\nabla^2 p = \frac{1}{c^2} \frac{\partial^2 p}{\partial t^2}$$

$$\rightarrow \quad \frac{\partial^2 p}{\partial x^2} + \frac{\partial^2 p}{\partial y^2} = \frac{1}{c^2} \frac{\partial^2 p}{\partial t^2}$$

The differential equation is separable if a solution can be cast in the following form that satisfies the acoustic wave equation:

$$p_s(x, y, t) = T(t)X(x)\, Y(y)$$

Substituting the above solution into the acoustic wave equation above yields the three separate ordinary differential equations (ODEs):

$$\frac{1}{c^2}\frac{T''(t)}{T(t)} = \frac{X''(x)}{X(x)} + \frac{Y''(y)}{Y(y)} = -k_x^2 - k_y^2 = -k^2$$

where k is a new variable that relates the spatial and temporal parts of the differential equations. The new variable k is separated into two independent variables k_x and k_y for two independent differential equations in two independent directions:

$$\frac{1}{c^2}\frac{T''(t)}{T(t)} = -k^2$$

$$\frac{X''(x)}{X(x)} = -k_x^2$$

$$\frac{Y''(y)}{Y(y)} = -k_y^2$$

where the new variable k can be considered a combined wavenumber that related to the wavenumbers k_x and k_y as:

$$k^2 = k_x^2 + k_y^2$$

With the new variables k, k_x, and k_y, the solutions of the temporal and the spatial differential equations are:

$$T(x) = A_t \cos\left(\omega t + \theta_t\right)$$

$$X(x) = A_x \cos\left(k_x x + \theta_x\right)$$

$$Y(x) = A_y \cos\left(k_y y + \theta_y\right)$$

Therefore, the standing wave solution of the 2D wave equation is:

$$p_s(x, y, t) = T(t)X(x)Y(y)$$
$$= A_t A_x A_y \cos\left(\omega t + \theta_t\right)\cos\left(k_x x + \theta_x\right)\cos\left(k_y y + \theta_y\right)$$

According to Euler's force equation, the velocity of the 2D standing wave is:

$$\vec{u}_s(x,y,t) = -\frac{1}{\rho_o} \int [\nabla p_s(x,y,t)] dt$$

$$= -\frac{1}{\rho_o} \int \left[\left(\hat{e}_x \frac{\partial}{\partial x} + \hat{e}_y \frac{\partial}{\partial y} \right) p(x,y,t) \right] dt$$

$$= u_x(x,y,t) \hat{e}_x + u_y(x,y,t) \hat{e}_y$$

where:

$$u_x(x,y,t) = \frac{1}{\rho_0 c} \frac{k_x}{k} A_t A_x A_y \sin(\omega t + \theta_t) \sin(k_x x + \theta_x) \cos(k_y y + \theta_y)$$

$$u_y(x,y,t) = \frac{1}{\rho_0 c} \frac{k_y}{k} A_t A_x A_y \sin(\omega t + \theta_t) \cos(k_x x + \theta_x) \sin(k_y y + \theta_y)$$

Note that the pressure is a scaler, and the velocity is a vector as indicated in the equation above.

The detailed derivation of the 2D standing wave velocity above is shown in Appendix 7.7.1.

7.3.2   2D Nature Frequencies Between Four Walls

Assume these two walls are oriented parallel to each other with the surface normal to the x-axis (y-axis). Also, assume the left (bottom) wall is located at $x = 0$ $(y = 0)$ and the right (top) wall is located at $= L_x$ $(y = L_y)$ as shown below:

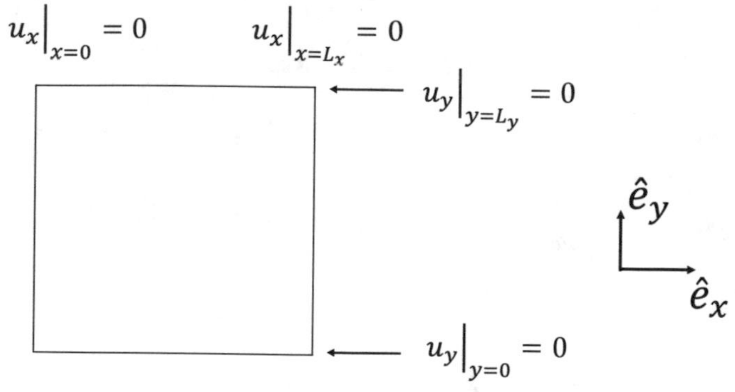

$$u_x \Big|_{x=0} = 0 \qquad u_x \Big|_{x=L_x} = 0$$

$$u_y \Big|_{y=L_y} = 0$$

$$\hat{e}_y$$

$$\hat{e}_x$$

$$u_y \Big|_{y=0} = 0$$

The general 2D standing wave pressure and velocity are shown below:

$$p_s(x, y, t) = A_t A_x A_y \cos (\omega t + \theta_t) \cos (k_x x + \theta_x) \cos (k_y y + \theta_y)$$

$$\vec{u}_s(x, y, t) = u_x(x, y, t)\widehat{e}_x + u_y(x, y, t)\widehat{e}_y$$

where:

$$u_x(x, y, t) = \frac{1}{\rho_0 c} \frac{k_x}{k} A_t A_x A_y \sin (\omega t + \theta_t) \sin (k_x x + \theta_x) \cos (k_y y + \theta_y)$$

$$u_y(x, y, t) = \frac{1}{\rho_0 c} \frac{k_y}{k} A_t A_x A_y \sin (\omega t + \theta_t) \cos (k_x x + \theta_x) \sin (k_y y + \theta_y)$$

The boundary condition at the left wall (CLOSED at $x = 0$):

$$u_x(x, y, t)|_{x=0} \rightarrow \theta_x = 0$$

The boundary condition at the right wall (CLOSED at $x = L_x$):
$$u_x(x, y, t)|_{x=L_x} = \sin (k_{xl} L_x) = 0, \quad k_{xl} = \frac{\pi}{L_x} l, \qquad l = 1, 2, \ldots$$
The boundary condition at the bottom wall (CLOSED at $y = 0$):

$$u_y(x, y, t)\big|_{y=0} \rightarrow \theta_y = 0$$

The boundary condition at the top wall (CLOSED at $y = L_y$):

$$u_y(x, y, t)\big|_{y=L_y} = \sin (k_{ym} L_y) = 0, \qquad k_{ym} = \frac{\pi}{L_y} m, \quad m = 1, 2, \ldots$$

Therefore, the discretized flow velocity becomes:

$$\vec{u}_s(x, y, t) = u_x(x, y, t)\widehat{e}_x + u_y(x, y, t)\widehat{e}_y$$

$$u_x(x, y, t) = \frac{1}{\rho_0 c} \frac{k_x}{k} A_t A_x A_y \sin (\omega t + \theta_t) \sin (k_{xl} x) \cos (k_{ym} y)$$

$$u_y(x, y, t) = \frac{1}{\rho_0 c} \frac{k_y}{k} A_t A_x A_y \sin (\omega t + \theta_t) \cos (k_{xl} x) \sin (k_{ym} y)$$

where:

$$k_{xl} = \frac{\pi}{L_x} l; \qquad k_{ym} = \frac{\pi}{L_y} m; \qquad l, m = 1, 2, \ldots$$

And the discretized sound pressure becomes:

$$p_s(x, y, t) = A_t A_x A_y \cos(\omega t + \theta_t) \cos(k_{xl}x) \cos(k_{ym}y)$$

The combined wavenumber k_{lm} of an eigenmode (l, m) are related to the discretized wavenumbers k_{xl} and k_{ym} as:

$$k_{lm}^2 = k_{xl}^2 + k_{ym}^2$$

The above relationship between the combined wavenumber k_{lm} and the discretized wavenumbers k_{xl} and k_{ym} will be explained in this chapter and derived in the next chapter.

The combined frequency f_{lm} of an eigenmode (l, m) is related to the combined wavenumber k_{lm} as:

$$f_{lm} = \frac{c}{2\pi} k_{lm}$$

7.3.3   2D Mode Shapes Between Four Walls

The boundary conditions imposed by four rigid walls require that the flow velocity is zero on the four walls. The mode shape of the flow velocity for the mode $(l, m) = (2, 3)$ is shown below as an example.

The formulas of the discretized wavelength shown below can be used for plotting mode shapes:

$$\lambda_{xl} = \frac{2\pi}{k_{xl}} = 2L_x \frac{1}{l}; \qquad \lambda_{ym} = \frac{2\pi}{k_{ym}} = 2L_y \frac{1}{m}$$

where:

$$k_{xl} = \frac{\pi}{L_x} l; \qquad k_{ym} = \frac{\pi}{L_y} m; \qquad l, m = 1, 2, \ldots$$

In the x-direction, the half wavelength of $l = 2$ is:

$$\frac{\lambda_{x2}}{2} = \frac{L_x}{2}$$

In the y-direction, the half wavelength of $m = 3$ is:

$$\frac{\lambda_{y3}}{2} = \frac{L_y}{3}$$

Based on the wavelengths above and zero velocities on the edge, we can plot the mode shape of the mode $(l, m) = (2, 3)$ as:

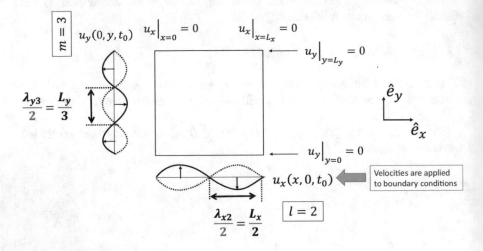

Usually, we only plot the mode shapes of pressure instead of the mode shapes of velocity because (1) pressure is scaler and velocity is vector and (2) pressure is more straightforward than velocity to be measured.

Based on Euler's equation, zero velocity results in a maximum amplitude of pressure on fixed boundaries where the slope of the pressure is zero ($\frac{\partial p}{\partial x} = -\rho_o \frac{\partial v_f}{\partial t} = 0$):

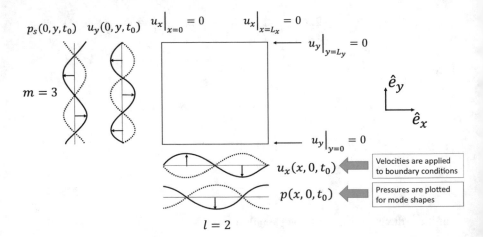

Locations at zero pressures are shown in nodal lines, as shown in the figure below. The mode shapes are indicated with plus and minus signs indicating peaks and valleys, respectively, and are separated by nodal lines where sound pressure is zero at all times, as shown in the figure below:

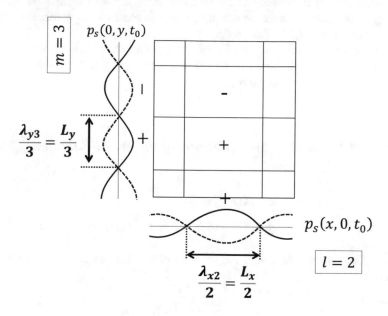

In the figure above, the pressure along the x-axis (y-axis) on the bottom (left) wall where y=0 (x=0) can be written as:

$$p_s(x, 0, t) = A_t A_x A_y \cos(\omega t + \theta_t) \cos(k_{xl} x)$$

$$p_s(0, y, t) = A_t A_x A_y \cos(\omega t + \theta_t) \cos(k_{ym} y)$$

where:

$$k_{xl} = \frac{\pi}{L_x} l; \qquad k_{ym} = \frac{\pi}{L_y} m; \qquad l, m = 1, 2, \ldots$$

7.4   3D Boundary Conditions of Rectangular Cavities

It can be difficult to plot and visualize 3D standing waves in a rectangular cavity on paper. However, it is possible to assume that there are no standing waves in an arbitrary dimension and draw the standing waves in other two dimensions as shown in Examples 7.3 and 7.4 in this section.

On the other hand, formulations for the 3D standing waves in a rectangular cavity can be extended from the formulas of 1D and 2D standing waves as the following (this is tedious but straightforward):

7.4.1   3D Standing Wave Solutions of the Wave Equation

$$p_s(x, y, z, t) = T(t)X(x)Y(y)Z(z)$$
$$= A_t A_x A_y A_z \cos(\omega t + \theta_t) \cos(k_x x) \cos(k_y y) \cos(k_z z)$$

From Euler's force equation, the velocity of the 3D standing wave solution is given by:

$$\vec{u}_s(x, y, z, t) = u_x(x, y, z, t)\widehat{e}_x + u_y(x, y, z, t)\widehat{e}_y + u_z(x, y, z, t)\widehat{e}_z$$

where:

$$u_x(x, y, z, t) = \frac{1}{\rho_0 c} \frac{k_x}{k} A_t A_x A_y A_z \sin(\omega t + \theta_t) \sin(k_x x) \cos(k_y y) \cos(k_z z)$$

$$u_y(x, y, z, t) = \frac{1}{\rho_0 c} \frac{k_y}{k} A_t A_x A_y A_z \sin(\omega t + \theta_t) \cos(k_x x) \sin(k_y y) \cos(k_z z)$$

$$u_z(x, y, z, t) = \frac{1}{\rho_0 c} \frac{k_z}{k} A_t A_x A_y A_z \sin(\omega t + \theta_t) \cos(k_x x) \cos(k_y y) \sin(k_z z)$$

The procedures for deriving the 3D velocity from the 3D pressure are similar to the procedures for deriving the 2D velocity from the 2D pressure. It is repetitive and is not shown here.

7.4.2   3D Natural Frequencies and Mode Shapes

The mode shapes of a rectangular cavity with six rigid walls can be calculated from the above pressure and velocity formulas with boundary conditions imposed by the

six rigid walls. Assume that the cavity is oriented at the origin of the coordinate, as
shown in the figure below:

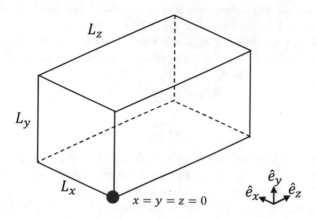

The boundary conditions on all rigid walls require that the flow velocity is zero.
By constraining the velocity at the fixed boundaries will discretize the wavenumber
similar to the 2D standing waves as:

$$k_{xl} = \frac{\pi}{L_x}l; \quad k_{ym} = \frac{\pi}{L_y}m; \quad k_{zn} = \frac{\pi}{L_z}n; \qquad l, m, n = 1, 2, \ldots$$

Therefore, the discretized flow velocity vector and pressure are shown below:

$$\vec{u}_s(x, y, z, t) = u_x(x, y, z, t)\hat{e}_x + u_y(x, y, z, t)\hat{e}_y + u_z(x, y, z, t)\hat{e}_z$$

where:

$$u_x(x, y, t) = \frac{1}{\rho_0 c}\frac{k_x}{k}A_t A_x A_y A_z \sin(\omega t + \theta_t) \sin(k_{xl}x) \cos(k_{ym}y) \cos(k_{zn}z)$$

$$u_y(x, y, t) = \frac{1}{\rho_0 c}\frac{k_y}{k}A_t A_x A_y A_z \sin(\omega t + \theta_t) \cos(k_{xl}x) \sin(k_{ym}y) \cos(k_{zn}z)$$

$$u_z(x, y, t) = \frac{1}{\rho_0 c}\frac{k_z}{k}A_t A_x A_y A_z \sin(\omega t + \theta_t) \cos(k_{xl}x) \cos(k_{ym}y) \sin(k_{zn}z)$$

and:

$$p_s(x, y, z, t) = A_t A_x A_y A_z \cos(\omega t + \theta_t) \cos(k_{xl} x) \cos(k_{ym} y) \cos(k_{zn} z)$$

The standing wave of the pressure is shown below. The derivation for these mode shapes is similar to the derivation for the 2D mode shapes.

The combined wavenumber k_{lmn} and the combined frequency f_{lmn} of an eigen-mode (l, m, n) are related to the discretized wavenumbers k_{xl} and k_{ym} and k_{zn} as:

$$k_{lmn}^2 = k_{xl}^2 + k_{ym}^2 + k_{zn}^2$$

$$f_{lmn} = \frac{c}{2\pi} k_{lmn}$$

The boundary conditions imposed by six rigid walls require that the flow velocity normal to the wall be zero on the six walls. A pattern of mode shape can be defined by the nodal lines where the pressure is zero. As described for the 2D pressure in the previous section, the nodal lines can be determined by the standing waves of the sound pressure. An example of a mode shape and a detailed explanation of the nodal lines can be found at the end of the following example:

Example 7.3 (Cavity Resonant Frequencies)
A rectangular cavity has the dimensions $(L_x, L_y, L_z) = (5, 8, 10)$ [m].

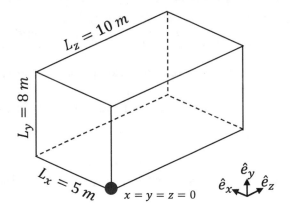

(a) Calculate all resonant frequencies of this cavity that are lower than 50 [Hz]. Indicate the associated mode number. For example, denote f_{213} for eigenmode $(l, m, n) = (2, 1, 3)$ corresponding to the dimensions (L_x, L_y, L_z).

(b) Plot the mode shape of the eigenmode $(l, m, n) = (0, 4, 7)$ using the pressure nodal lines, and indicate the peaks and valleys with "+" and "-" signs, respectively.

Example 7.3 Solution
Part (a)

The frequency (combined) of standing waves in the cavities can be formulated using the following relationship of eigenmodes:

$$f_{lmn} = \frac{\omega_{lmn}}{2\pi} = \frac{c}{2\pi} k_{lmn} = \frac{c}{2} \left[\left(\frac{l}{L_x} \right)^2 + \left(\frac{m}{L_y} \right)^2 + \left(\frac{n}{L_z} \right)^2 \right]^{\frac{1}{2}}$$

where the combined wavenumber k_{lmn} and combined frequency f_{lmn} of an eigenmode (l, m, n) are related to the discretized wavenumbers k_{xl} and k_{ym} and k_{zn} as:

$$k_{lmn}^2 = k_{xl}^2 + k_{ym}^2 + k_{zn}^2$$

$$k_{xl} = \frac{\pi}{L_x} l; k_{ym} = \frac{\pi}{L_y} m; k_{zn} = \frac{\pi}{L_z} n$$

The formula above shows that the larger the dimension (L_x, L_y or L_z) and the smaller the mode (l, m or n), the lower the modal frequency f_{lmn}.

As a result, the first few lower modal frequencies can be calculated using combinations of the first few modes. Lower modal frequencies correspond to larger dimensions. For this reason, if the dimensions of the cavity are defined from short to long, the lower frequency mode will exist in the z-direction (the longest dimension).

The following procedures demonstrate a searching method for finding all the resonant frequencies of this cavity lower than 50 [Hz].

Step 1:

Set $l=0$ and $m=0$. Next, increase n by 1 from 1. Finally, calculate the combined frequency using the following formula until the calculated frequency is higher than the maximum prescribed frequency (50 Hz):

$$f_{lmn} = \frac{\omega_{lmn}}{2\pi} = \frac{c}{2} \left[\left(\frac{l}{L_x} \right)^2 + \left(\frac{m}{L_y} \right)^2 + \left(\frac{n}{L_z} \right)^2 \right]^{\frac{1}{2}}$$

Step 2:

Set $l=0$ and $m=1$. Next, increase n by 1 from 0. Finally, calculate the combined frequency using the above formula until the calculated frequency is higher than the prescribed frequency (50 [Hz]).

Step 3:

Repeat Step 2 by increasing m by 1 until the calculated frequency is higher than the prescribed frequency maximum.

Step 4:

Increase l by 1 and set $m=0$ and $n=0$. Then, repeat Step 2 and Step 3 above until the calculated frequency is higher than the prescribed frequency maximum:

l	m	n	Frequency [Hz]	Note
0	0	1	17	< 50 (OK)
0	0	2	34	< 50 (OK)
0	0	3	51	> 50 (NOK)
Stop and increase m by 1				
0	1	0	21.25	< 50 (OK)
0	1	1	27.21	< 50 (OK)
0	1	2	40.09	< 50 (OK)
0	1	3	~~55.25~~	> 50 (NOK)
Stop and increase m by 1				
0	2	0	42.50	< 50 (OK)
0	2	1	45.77	< 50 (OK)
0	2	2	~~54.43~~	> 50 (NOK)
Stop and increase m by 1				
0	3	0	~~63.75~~	> 50 (NOK)
Stop and increase l by 1				
1	0	0	34	< 50 (OK)
1	0	1	38.01	< 50 (OK)
1	0	2	48.08	< 50 (OK)
1	0	3	~~61.30~~	> 50 (NOK)
Stop and increase m by 1				
1	1	0	40.09	< 50 (OK)
1	1	1	43.55	< 50 (OK)
1	1	2	~~52.57~~	> 50 (NOK)
Stop and increase m by 1				
1	2	0	~~54.43~~	> 50 (NOK)
Stop and increase l by 1				
2	0	0	~~68.00~~	> 50 (NOK)
Stop				

Frequencies lower than 50 [Hz] can be calculated by hand or in MATLAB or Excel as follows (partial list):

$$f_{lmn} = f_{001} = \frac{340}{2}\left(\left(\frac{0}{5}\right)^2 + \left(\frac{0}{8}\right)^2 + \left(\frac{1}{10}\right)^2\right)^{\frac{1}{2}}[Hz] = 170 \cdot \frac{1}{10}[Hz] = 17[Hz]$$

$$f_{lmn} = f_{002} = \frac{340}{2}\left(\left(\frac{0}{5}\right)^2 + \left(\frac{0}{8}\right)^2 + \left(\frac{2}{10}\right)^2\right)^{\frac{1}{2}}[Hz] = 170 \cdot \frac{2}{10}[Hz] = 34\ [Hz]$$

$$f_{lmn} = f_{003} = \frac{340}{2}\left(\left(\frac{0}{5}\right)^2 + \left(\frac{0}{8}\right)^2 + \left(\frac{3}{10}\right)^2\right)^{\frac{1}{2}}[Hz] = 170 \cdot \frac{3}{10}[Hz] = 51\ [Hz]$$

$$f_{lmn} = f_{010} = \frac{340}{2}\left(\left(\frac{0}{5}\right)^2 + \left(\frac{1}{8}\right)^2 + \left(\frac{0}{10}\right)^2\right)^{\frac{1}{2}}[Hz] = 170 \cdot \frac{1}{8}[Hz] = 21.25\ [Hz]$$

$$f_{lmn} = f_{011} = \frac{340}{2}\left(\left(\frac{0}{5}\right)^2 + \left(\frac{1}{8}\right)^2 + \left(\frac{1}{10}\right)^2\right)^{\frac{1}{2}}[Hz] = 27.21[Hz]$$

$$f_{lmn} = f_{012} = \frac{340}{2}\left(\left(\frac{0}{5}\right)^2 + \left(\frac{1}{8}\right)^2 + \left(\frac{2}{10}\right)^2\right)^{\frac{1}{2}}[Hz] = 40.09\ [Hz]$$

$$f_{lmn} = f_{111} = \frac{340}{2}\left(\left(\frac{1}{5}\right)^2 + \left(\frac{1}{8}\right)^2 + \left(\frac{1}{10}\right)^2\right)^{\frac{1}{2}}[Hz] = 43.55\ [Hz]$$

Part (b)

The pressure nodal lines can be determined by the following three properties of sound pressure:

Property #1: The relationship between wavelength and length of the cavity

The formulas of the discretized wavelengths shown below can be used for plotting the mode shapes:

$$\frac{\lambda_{ym}}{2} = \frac{\pi}{k_{ym}} = \frac{L_y}{m}; \qquad \frac{\lambda_{zn}}{2} = \frac{\pi}{k_{zn}} = \frac{L_z}{n}$$

where:

$$k_{ym} = \frac{\pi}{L_y}m; \qquad k_{zn} = \frac{\pi}{L_z}n; \qquad m, n = 1, 2, \ldots$$

In the x-direction, the half wavelength of $m = 4$ is:

$$\frac{\lambda_{ym}}{2} = \frac{L_y}{m} = \frac{L_y}{4}$$

In the y-direction, the half wavelength of $n = 7$ is:

$$\frac{\lambda_{zn}}{2} = \frac{L_z}{n} = \frac{L_z}{7}$$

Property #2: Boundary conditions of fixed surface

Pressure is maximum at fixed boundaries because the flow velocity on a fixed surface is zero. We can use this property to align the harmonic function (cosine or sine function) to the boundaries.

Property #3: Multiplication of spatial functions

The peaks and valleys can be determined by the multiplication of signs of $p_s(y, 0, t_0)$ in y-direction and $p_s(0, z, t_0)$ in z-direction because standing wave solutions are the result of multiplication of temporal function $T(t)$ and spatial functions $X(x)$, $Y(y)$, and $Z(z)$ as:

$$p(x, y, z, t) = T(t)X(x)\, Y(y)\, Z(z)$$
$$= P_{lmn} \cos(\omega t + \theta_t) \cos(k_{xl}x) \cos(k_{ym}y) \cos(k_{zn}z)$$

Therefore, the pressure nodal lines and valleys of eigenmode $(l, m, n) = (0, 4, 7)$ can be drawn as follows:

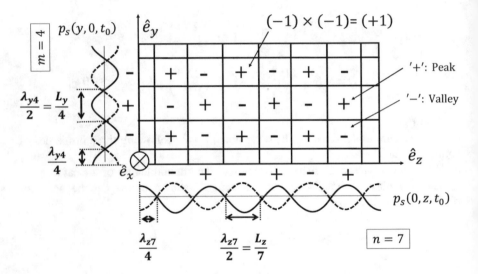

The index l is zero in the eigenmode $(l, m, n) = (0, 4, 7)$ which means that this eigenmode has no mode shape in the x-direction. In other words, the sound pressure of this eigenmode is the same everywhere in the x-direction. Therefore, the standing wave of the sound pressure can be reduced to:

$$p(x, y, z, t) = A_t A_x A_y A_z \cos(\omega t + \theta_t) \cos(k_{xl}x) \cos(k_{ym}y) \cos(k_{zm}z)$$
$$\rightarrow p(y, z, t) = A_t A_x A_y A_z \cos(\omega t + \theta_t) \cos(k_{ym}y) \cos(k_{zm}z)$$

The pressure at y and z directions can be presented as:

$$p(y, z = 0, t) = A_t A_x A_y A_z \cos(\omega t + \theta_t) \cos(k_{ym}y)$$
$$p(y = 0, z, t) = A_t A_x A_y A_z \cos(\omega t + \theta_t) \cos(k_{zm}z)$$

where:

$$k_{ym} = \frac{\pi}{L_y}m = \frac{\pi}{L_y}4$$

$$k_{zn} = \frac{\pi}{L_z}n = \frac{\pi}{L_z}7$$

Example 7.4 (Standing Wave Amplitude)

The dimensions of a rectangular cavity are given as $(L_x, L_y, L_z) = (10, 5, 2)$ [m]. Assume that this rectangular cavity is positioned at the origin.

If the RMS pressure of mode $(1,1,0)$ is 1 [Pa] at $x = 2$ [m] and $y = 1$ [m], what is the pressure magnitude of this mode?

Example 7.4 Solutions

The standing wave pressure solution of the wave equation is:

$$p(x, y, z, t) = T(t)X(x)\,Y(y)\,Z(z)$$
$$= P_{lmn}\cos{(\omega t + \theta_t)}\cos{(k_{xl}x)}\cos{(k_{ym}y)}\cos{(k_{zn}z)}$$

[R_{EP}] where k_{xl}, k_{ym}, and k_{zn} are the discretized values that satisfy the boundary conditions of the rigid walls:

$$k_{xl} = \frac{\pi}{L_x}l \quad \rightarrow k_{x1} = \frac{\pi}{10}*1\left[\frac{rad}{m}\right]$$

$$k_{ym} = \frac{\pi}{L_y}m \rightarrow k_{y1} = \frac{\pi}{5}*1\left[\frac{rad}{m}\right]$$

$$k_{zn} = \frac{\pi}{L_z}n \quad \rightarrow k_{z0} = \frac{\pi}{2}*0\left[\frac{rad}{m}\right]$$

Therefore, the pressure at the given location ($x = 2$ [m], $y = 1$ [m], and $z = 0$ [m]) is:

$$p(1, 1, 0, t) = P_{lmn}\cos{(\omega t + \theta_t)}\cos{\left(\frac{\pi}{10}*2\right)}\cos{\left(\frac{\pi}{5}*1\right)}\cos{(0*0)}$$
$$= P_{lmn}\cos{\left(\frac{\pi}{10}*2\right)}\cos{\left(\frac{\pi}{5}*1\right)}\cos{(\omega t + \theta_t)}$$
$$= A\cos{(\omega t + \theta_t)}$$

Note that in the above equation, A is the amplitude of $\cos(\omega t + \theta_t)$:

$$A = P_{lmn}\cos{\left(\frac{\pi}{10}*2\right)}\cos{\left(\frac{\pi}{5}*1\right)}$$

Amplitude A is related to the RMS pressure of the standing wave (from the section of one-dimensional plane waves):

$$p_{RMS} = \frac{1}{\sqrt{2}} |A|$$

Solve for P_{lmn} using the given RMS pressure ($p_{RMS} = 1$ [Pa]) and the location ($x = 2$ [m], $y = 1$ [m], and $z = 0$ [m]) as shown below:

$$1 \; [Pa] = \frac{1}{\sqrt{2}} |A| = \frac{1}{\sqrt{2}} P_{lmn} \left| \cos\left(\frac{\pi}{5}\right) \cos\left(\frac{\pi}{5}\right) \right| = \frac{1}{\sqrt{2}} P_{lmn} \cos^2\left(\frac{\pi}{5}\right)$$

$$P_{lmn} = \frac{\sqrt{2}}{\cos^2\left(\frac{\pi}{5}\right)} \; [Pa]$$

7.5   Homework Exercises

Exercise 7.1 (OPEN-CLOSED PIPE)
A pipe with OPEN-CLOSED boundary conditions has a finite length L as shown below. Set the left end of the pipe as $x = 0$:

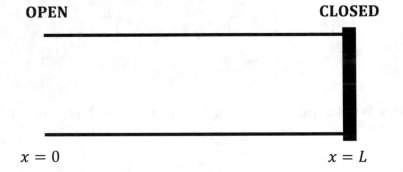

(a) Show that the formulas for the wavenumber and natural frequency of the eigenmode l are:

$$k_l = \frac{\pi}{2L}(2l - 1)$$

$$f_l = \frac{c}{4L}(2l - 1)$$

(b) If $L = 0.5$ [m], determine the first three natural frequencies, and plot their corresponding mode shapes. Use 340 [m/s] for the speed of sound.

Use 340 [m/s] for the speed of sound (c) in air. Show units in the Meter-Kilogram-Second (MKS) system.

(Answers) (b) 170 [Hz], 510 [Hz], 850 [Hz]

Exercise 7.2 (CLOSE-OPEN PIPE)
A pipe with OPEN-CLOSED boundary conditions has a finite length L as shown below. Set the left end of the pipe as $x = 0$:

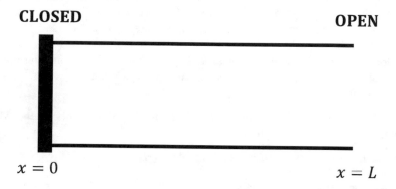

CLOSED   **OPEN**

$x = 0$ $x = L$

(a) Show that the formulas for the wavenumber and natural frequency of the eigenmode l are:

$$k_l = \tfrac{\pi}{2L}(2l - 1), \quad f_l = \tfrac{c}{4L}(2l - 1)$$

(b) If $L = 0.5$ [m], determine the first three natural frequencies, and plot their corresponding mode shapes. Use 340 [m/s] for the speed of sound.

Use 340 [m/s] for the speed of sound (c) in air. Show units in the Meter-Kilogram-Second (MKS) system.

(Answers) (b) 170 [Hz], 510 [Hz], 850 [Hz]

Exercise 7.3 (3D Rectangular Cavity)
The dimensions of a rectangular cavity are $(L_x, L_y, L_z) = (2, 5, 10)[m]$.

(a) Calculate all resonant frequencies of this cavity that are lower than 60 Hz.
(b) Plot the mode shape of the eigenmode $(l, m, n) = (0, 3, 5)$ using the pressure nodal lines, and indicate the peaks and valleys with "+" and "-" signs, respectively.
(c) Plot the mode shape of the eigenmode $(l, m, n) = (0, 5, 8)$ using the pressure nodal lines, and indicate the peaks and valleys with "+" and "-" signs, respectively.

Use 340 [m/s] for the speed of sound (c) in air. Show units in the Meter-Kilogram-Second (MKS) system.

(Answers): (a) 17 [Hz], 34 [Hz], 38.01 [Hz], 48.08 [Hz], 51 [Hz]

Exercise 7.4 (3D Rectangular Cavity)
The dimensions of a rectangular cavity are $(L_x, L_y, L_z) = (3, 4, 10)[m]$.

(a) Calculate all resonant frequencies of this cavity that are lower than 50 [Hz].
(b) Plot the mode shape of the eigenmode $(l, m, n) = (0, 2, 3)$ using the pressure nodal lines, and indicate the peaks and valleys with "+" and "-" signs, respectively.
(c) Plot the mode shape of the eigenmode $(l, m, n) = (0, 5, 8)$ using the pressure nodal lines, and indicate the peaks and valleys with "+" and "-" signs, respectively.

Use 340 [m/s] for the speed of sound (c) in air. Show units in the Meter-Kilogram-Second (MKS) system.

(Answers): (a) 17 [Hz], 34 [Hz], 42.5 [Hz], 45.8 [Hz]

Exercise 7.5 (3D Rectangular Cavity)
The dimensions of a rectangular cavity are $(L_x, L_y, L_z) = (4, 6, 10)[m]$. Use 340 [m/s] for the speed of sound.

(a) Calculate the three lowest resonant frequencies of this cavity. Indicate the eigenmode. State the mode numbers.
(b) Plot the mode shape of the eigenmode $(l, m, n) = (0, 3, 5)$ using the pressure nodal lines, and indicate the peaks and valleys with "+" and "-" signs, respectively.

(Answer): (a) $f_{001} = 17$ [Hz]; $f_{010} = 28.33$ [Hz]; $f_{011} = 33.04$ [Hz]

Exercise 7.6 (Standing Waves, Complex) (Optional)
Derive the 2D velocity vector $\vec{u}_s(x, y, t)$ of the plane standing wave from the 2D acoustic pressure using Euler's force equation. Use complex number format in the derivation:

$$p(x, y, t) = \frac{1}{8} A_t A_x A_y \left(e^{j(\omega t + \theta_t)} + e^{-j(\omega t + \theta_t)} \right)$$

$$\cdot \left(e^{j(k_x x + \theta_x)} + e^{-j(k_x x + \theta_x)} \right) \cdot \left(e^{j(k_y y + \theta_y)} + e^{-j(k_y y + \theta_y)} \right)$$

$$\vec{u}_s(x, y, t) = -\frac{n_x \widehat{e}_x}{8 \rho_0 c} A_t A_x A_y \left(e^{j(\omega t + \theta_t)} - e^{-j(\omega t + \theta_t)} \right)$$

$$\cdot \left(e^{j(k_x x + \theta_x)} - e^{-j(k_x x + \theta_x)} \right) \cdot \left(e^{j\left(k_y y + \theta_y\right)} + e^{-j\left(k_y y + \theta_y\right)} \right)$$

$$- \frac{n_y \widehat{e}_y}{8 \rho_0 c} A_t A_x A_y \left(e^{j(\omega t + \theta_t)} - e^{-j(\omega t + \theta_t)} \right)$$

$$\cdot \left(e^{j(k_x x + \theta_x)} + e^{-j(k_x x + \theta_x)} \right) \cdot \left(e^{j\left(k_y y + \theta_y\right)} - e^{-j\left(k_y y + \theta_y\right)} \right)$$

Use Euler's force equation:

$$\nabla p(x, y, t) = -\rho_o \frac{\partial}{\partial t} \vec{u}_S(x, y, t)$$

$$\vec{u}_S(x, y, t) = -\frac{\widehat{e}_x}{\rho_o} \int \left[\frac{\partial}{\partial x} p(x, y, t) \right] dt - \frac{\widehat{e}_y}{\rho_o} \int \left[\frac{\partial}{\partial y} p(x, y, t) \right] dt$$

Chapter 8
Acoustic Waveguides

In the previous chapter, resonance in rectangular cavities was formulated as standing waves as a result of the addition of a forward traveling wave (pre-reflection) and a backward traveling wave (post-reflection). In this chapter, we will study the propagation of plane waves in acoustic waveguides, as shown on the right-hand side of the figure below. In acoustic waveguides, traveling waves propagate along waveguides and have no returning waves; standing waves exist in lateral directions.

We can treat acoustic waveguides as resonant cavities with the two ends removed (see the figure below). Because there is no reflecting wave on the two ends, the wave in the waveguide will propagate without returning waves (as shown on the right-hand side of the figure below):

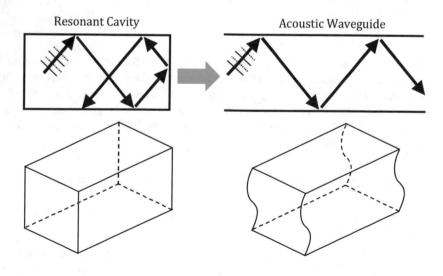

© The Author(s), under exclusive license to Springer Nature Switzerland AG 2021   197
H. Lin et al., *Lecture Notes on Acoustics and Noise Control*,
https://doi.org/10.1007/978-3-030-88213-6_8

The propagation of plane waves in acoustic waveguides is in the form of resonance (standing waves in the transverse direction) and propagation (traveling waves in the axial direction). Based on what we learned from the previous chapters, we can treat a standing wave (shown in the upper left of the figure below) as the addition of two traveling waves with the same amplitudes and traveling in opposite directions. By adding these two traveling waves (in the transverse direction) back to the traveling wave (in the axial direction), we can reconstruct the propagating wave in the acoustic waveguide as shown in the upper right of the figure below:

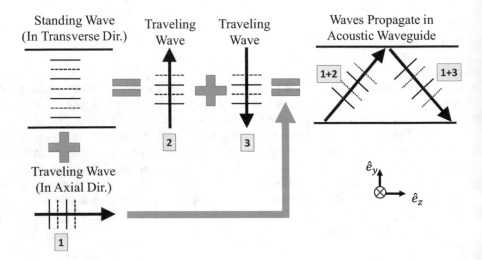

After the two walls in the z-direction are removed, the four wavenumber vectors (in 2D resonant cavities as shown in the previous chapter) become two wavenumber vectors because there is no reflection in the z-direction as shown in the figure above. Without the two walls in the z-direction, the wavenumbers in the y-direction are still discretized (as in 2D resonant cavities), but the wavenumbers in the z-direction become continuous numbers as:

$$k_{ym} = \frac{\pi}{L_y} m \qquad k_{ym} \text{ is DISCRETIZED where m is an INTEGER}$$

$$k_{zn} = \frac{\pi}{L_z} n = k_z \qquad k_z \text{ is CONTINOUS since } L_z \text{ can be ANY LENGTH}$$

In a waveguide, because L_z can be considered as a very large number, and based on the formula above, k_{zn} is continuous. The formulas shown above are the most important formulas of this chapter and will be derived in detail in this chapter.

Based on the formulas shown above, topics such as cutoff frequency, echoes, and sound distortion in waveguides will be discussed in this chapter.

8.1   2D Traveling Wave Solutions

8.1.1   Definition of Wavenumber Vectors

We have derived the relationship between the combined wavenumber k and the component wavenumbers k_x and k_y in Chap. 7:

$$k^2 = k_x^2 + k_y^2$$

Substituting the relationship between the wavenumber k and the wavelength λ:

$$k = \frac{2\pi}{\lambda}$$

into the first equation gives the relationship between the combined wavelength λ and the component wavelengths λ_x and λ_y as:

$$\frac{1}{\lambda^2} = \frac{1}{\lambda_x^2} + \frac{1}{\lambda_y^2}$$

The relationship above can also be proved geometrically with trigonometric relationships:

$$\cos\theta = \frac{\lambda}{\lambda_x}; \ \sin\theta = \frac{\lambda}{\lambda_y}$$

and the trigonometric identity:

$$\cos^2\theta + \sin^2\theta = 1 \rightarrow 1 = \frac{\lambda^2}{\lambda_x^2} + \frac{\lambda^2}{\lambda_y^2} \rightarrow \frac{1}{\lambda^2} = \frac{1}{\lambda_x^2} + \frac{1}{\lambda_y^2}$$

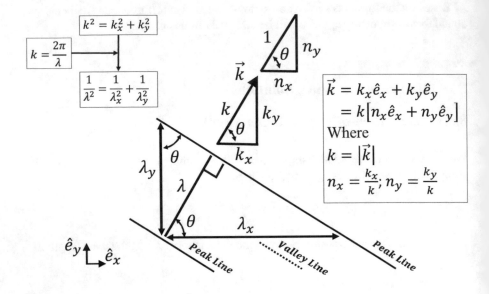

The combined wavenumber is a vector that presents the propagation direction and the magnitude of a 2D propagating plane wave as:

$$\vec{k} = k_x\widehat{e}_x + k_y\widehat{e}_y$$
$$= k(n_x\widehat{e}_x + n_y\widehat{e}_y)$$

where:

$$n_x = \frac{k_x}{k}; n_y = \frac{k_y}{k}$$

where $\widehat{e}_x$ and $\widehat{e}_y$ are the base vectors of the Cartesian coordinate systems and k_x and k_y are vector components of $\vec{k}$.

Note that n_x and n_y indicate the direction of the wavenumber vector $\vec{k}$ and have the following properties:

$$n_x = \cos\theta; n_y = \sin\theta$$
$$\rightarrow n_x^2 + n_y^2 = 1$$

8.1.2 *Wavenumber Vectors in 2D Traveling Wave Solutions*

The sound pressure and flow velocity of 1D plane waves were derived in Chap. 4 and are listed below for reference:

$$p_\pm(x,t) = A_\pm \cos\left(\omega t \mp kx + \theta_\pm\right)$$

$$u_\pm(x,t) = \pm\frac{1}{\rho_o c}p_\pm(x,t)$$

The formulas above will be transferred to 2D plane waves using the wavenumber vector we learned in the previous section:

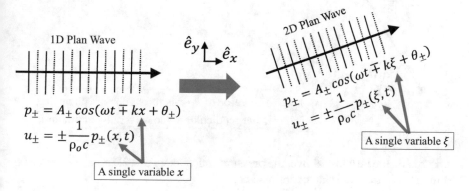

Note that the formulas of pressure and velocity on the left-hand side of the figure are for the plane wave traveling along the x-axis. The right-hand side of the figure shows the same plane wave rotated by an angle. The rotated plane wave becomes 2D and can be formulated with a new variable ξ. The formulas of pressure and velocity of the 2D plane wave are the same as the 1D plane wave except using the different variable names – x and ξ as:

$$p_\pm(\xi,t) = A_\pm \cos\left(\omega t \mp k\xi + \theta_\pm\right)$$

$$u_\pm(\xi,t) = \pm\frac{1}{\rho_o c}p_\pm(\xi,t) \quad \text{along traveling direction}$$

The new variable ξ is the result of the dot product of the wavenumber vector $\vec{k}\left(k_x, k_y\right)$ and the position vector $\vec{r}\left(x,y\right)$ as:

$$\xi = \frac{\vec{k} \cdot \vec{r}}{\left|\vec{k}\right|}$$

$$= \frac{(k_x, k_y) \cdot (x, y)}{\left|\vec{k}\right|} = \frac{k_x x + k_y y}{\sqrt{k_x^2 + k_y^2}} = \frac{k_x x + k_y y}{k}$$

$$= n_x x + n_y y$$

And the velocity of the plane wave above is along the traveling direction $\vec{k}$:

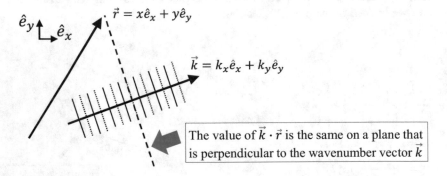

$$\hat{e}_y \quad \hat{e}_x$$

$$\vec{r} = x\hat{e}_x + y\hat{e}_y$$

$$\vec{k} = k_x \hat{e}_x + k_y \hat{e}_y$$

The value of $\vec{k} \cdot \vec{r}$ is the same on a plane that is perpendicular to the wavenumber vector $\vec{k}$

Based on the formulas above, the pressure and velocity of a 2D plane wave can be formulated in the Cartesian coordinate as:

$$p_{\pm}(x, y, t) = A_{\pm} \cos{(\omega t \mp k\xi + \theta_{\pm})}$$
$$= A_{\pm} \cos{\left[\omega t \mp k(n_x x + n_y y) + \theta_{\pm}\right]}$$
$$= A_{\pm} \cos{\left[\omega t \mp (k_x x + k_y y) + \theta_{\pm}\right]}$$
$$= A_{\pm} \cos{\left[\omega t \mp \vec{k} \cdot \vec{r} + \theta_{\pm}\right]}$$

where:

$$k = \left|\vec{k}\right|$$

$$n_x = \frac{k_x}{k}; n_y = \frac{k_y}{k}$$

The 2D flow velocity $\vec{u}_{\pm}$ of the 2D plane wave can be derived using Euler's force equation as the following:

$$\vec{u}_\pm(x, y, t) = \frac{-1}{\rho_o} \int [\nabla p_\pm(x, y, t)] dt$$

$$= \pm \frac{n_x \widehat{e}_x + n_y \widehat{e}_y}{\rho_o c} p_\pm(x, y, t)$$

Note that the sound pressure $p_\pm$ is a scalar but the flow velocity $\vec{u}_\pm$ is a vector because of the gradient operator ∇.

The detailed derivation of the backward flow velocity $\vec{u}_-$ is shown in Example 8.1. The derivation of the forward flow velocity $\vec{u}_+$ is one of the homework problems.

Example 8.1: (Derivation of 2D Plane Traveling Waves: Backward)

Derive the 2D backward flow velocity vector $\vec{u}_-(x, y, t)$ of a 2D plane backward traveling wave from the 2D acoustic pressure $p_-(x, y, t)$ using Euler's force equation:

$$p_-(x, y, t) = A_- \cos \left[\omega t + \left(k_x x + k_y y \right) + \theta \right]$$

$$\vec{u}_-(x, y, t) = -\left(\frac{n_x \widehat{e}_x + n_y \widehat{e}_y}{\rho_o c} \right) p_-(x, y, t)$$

where:

$$n_x = \frac{k_x}{\sqrt{k_x^2 + k_y^2}}, \text{and } n_y = \frac{k_y}{\sqrt{k_x^2 + k_y^2}}$$

Euler's force equation: $\nabla p(x, y, t) = -\rho_o \frac{\partial}{\partial t} \vec{u}(x, y, t)$

Example 8.1: Solution

Backward flow velocity $\vec{u}_-(x, y, t)$ is a vector and can be calculated from acoustic pressure based on Euler's force equation:

$$\nabla p(x, y, t) = -\rho_o \frac{\partial}{\partial t} \vec{u}(x, y, t)$$

$$\rightarrow \vec{u}(x, y, t) = -\frac{1}{\rho_o} \int [\nabla p(x, y, t)] dt$$

Euler's force equation for the 2D wave equation in Cartesian coordinate is formatted as:

$$\vec{u}(x, y, t) = -\frac{1}{\rho_o} \int \left[\left(\widehat{e}_x \frac{\partial}{\partial x} + \widehat{e}_y \frac{\partial}{\partial y} \right) p(x, y, t) \right] dt$$

Remarks

- Euler's force equation is valid for both forward and backward waves.
- Pressure $p(x, y, t)$ is a scalar function and velocity $\vec{u}(x, y, t)$ is a vector function.

Therefore, the backward velocity can be calculated from pressure as:

$$\vec{u}_-(x, y, t) = -\frac{1}{\rho_o} \int \left[\left(\hat{e}_x \frac{\partial}{\partial x} + \hat{e}_y \frac{\partial}{\partial y} \right) p_-(x, y, t) \right] dt$$

$$= -\frac{\hat{e}_x}{\rho_o} \int \left[\frac{\partial}{\partial x} p_-(x, y, t) \right] dt - \frac{\hat{e}_y}{\rho_o} \int \left[\frac{\partial}{\partial y} p_-(x, y, t) \right] dt$$

Separate the velocity vector into two vector components, $\hat{e}_x u_{x-}(x, y, t)$ and $\hat{e}_y u_{y-}(x, y, t)$:

$$\vec{u}_-(x, y, t) = \hat{e}_x u_{x-}(x, y, t) + \hat{e}_y u_{y-}(x, y, t)$$

where:

$$u_{x-}(x, y, t) = -\frac{1}{\rho_o} \int \left[\frac{\partial}{\partial x} p_-(x, y, t) \right] dt = -\left(\frac{k_x}{\rho_o \omega} \right) p_-(x, y, t)$$

$$u_{y-}(x, y, t) = -\frac{1}{\rho_o} \int \left[\frac{\partial}{\partial y} p_-(x, y, t) \right] dt = -\left(\frac{k_y}{\rho_o \omega} \right) p_-(x, y, t)$$

$$p_-(x, y, t) = A_- \cos \left(\omega t + \left(k_x x + k_y y \right) + \theta \right)$$

$$\frac{\partial}{\partial x} p_-(x, y, t) = A_- \frac{\partial}{\partial x} \cos \left(\omega t + \left(k_x x + k_y y \right) + \theta \right)$$

$$= -A_- k_x \sin \left(\omega t + \left(k_x x + k_y y \right) + \theta \right)$$

$$u_{x-}(x, y, t) = -\int \left[\frac{\partial}{\partial x} p_-(x, y, t)\right] dt$$

$$= A_- k_x \int \left[\sin\left(\omega t + \left(k_x x + k_y y\right) + \theta\right)\right] dt$$

$$= -A_- \left(\frac{k_x}{\omega}\right) \cos\left(\omega t + \left(k_x x + k_y y\right) + \theta\right)$$

$$= -\left(\frac{k_x}{\omega}\right) p_-(x, y, t)$$

Similarly:

$$u_{y-}(x, y, t) = -\int \left[\frac{\partial}{\partial y} p_-(x, y, t)\right] dt = -\left(\frac{k_y}{\omega}\right) p_-(x, y, t)$$

Combining two vector components back to one velocity vector yields

$$\vec{u}_-(x, y, t) = \widehat{e}_x u_{x-}(x, y, t) + \widehat{e}_y u_{y-}(x, y, t)$$

$$= -\widehat{e}_x \left(\frac{k_x}{\rho_o \omega}\right) p_-(x, y, t) - \widehat{e}_y \left(\frac{k_y}{\rho_o \omega}\right) p_-(x, y, t)$$

$$= -\frac{1}{\rho_o \omega} \left[k_x \widehat{e}_x + k_y \widehat{e}_y\right] p_-(x, y, t)$$

$$= -\frac{1}{\rho_o \omega} \sqrt{k_x^2 + k_y^2} \left[\frac{k_x}{\sqrt{k_x^2 + k_y^2}} \widehat{e}_x + \frac{k_y}{\sqrt{k_x^2 + k_y^2}} \widehat{e}_y\right] p_-(x, y, t)$$

$$= -\frac{k}{\rho_o \omega} \left[n_x \widehat{e}_x + n_y \widehat{e}_y\right] p_-(x, y, t)$$

where:

$$k \equiv \left|\vec{k}\right| = \sqrt{k_x^2 + k_y^2}$$

$$n_x = \frac{k_x}{\sqrt{k_x^2 + k_y^2}}, \text{ and } n_y = \frac{k_y}{\sqrt{k_x^2 + k_y^2}}$$

Replacing ω and k with c yields the final form:

$$\vec{u}_-(x, y, t) = -\frac{k}{\rho_0 \omega} \left[n_x \widehat{e}_x + n_y \widehat{e}_y \right] p_-(x, y, t) = -\frac{1}{\rho_0 c} \left[n_x \widehat{e}_x + n_y \widehat{e}_y \right] p_-(x, y, t)$$

$$= -\left(\frac{n_x \widehat{e}_x + n_y \widehat{e}_y}{\rho_0 c} \right) A_- \cos\left(\omega t + \left(k_x x + k_y y \right) + \theta \right)$$

$$= -\left(\frac{n_x \widehat{e}_x + n_y \widehat{e}_y}{\rho_0 c} \right) p_-(x, y, t)$$

8.2 Wavenumber Vectors in Resonant Cavities

In Chap. 7, the natural frequency f_{lmn} of the eigenmode (l, m, n) in resonant cavities was calculated without referring to the wavenumber vector based on the formulas below:

$$f_{lmn} = \frac{\omega_{lmn}}{2\pi} = \frac{c}{2\pi} k_{lmn}$$

where:

$$k_{lmn}^2 = k_{xl}^2 + k_{ym}^2 + k_{zn}^2$$

and:

$$k_{xl} = \frac{\pi}{L_x} l; k_{ym} = \frac{\pi}{L_y} m; k_{zn} = \frac{\pi}{L_z} n$$

In Chap. 8, we will relate the natural frequencies to the wavenumber vectors in resonant cavities.

In this section, the formulas of wavenumber vectors will be introduced. In the next section, the physics of wavenumber vectors as traveling waves in resonant cavities will be explained.

The discretized wavenumber vectors $\vec{k}_{lmn}$ of an eigenmode (l, m, n) can be calculated using the formulas developed in the previous chapter as:

$$\vec{k}_{lmn} = k_{xl} \widehat{e}_x + k_{ym} \widehat{e}_y + k_{zn} \widehat{e}_z$$

where:

$$k_{xl} = \frac{\pi}{L_x} l; k_{ym} = \frac{\pi}{L_y} m; k_{zn} = \frac{\pi}{L_z} n$$

and:

$$k_{lmn}^2 = k_{xl}^2 + k_{ym}^2 + k_{zn}^2$$

$$f_{lmn} = \frac{\omega_{lmn}}{2\pi} = \frac{c}{2\pi} k_{lmn} = \frac{c}{2\pi} \left| \vec{k}_{lmn} \right|$$

For example, the wavenumber vector and natural frequency of the eigenmode $(l, m, n) = (0, 4, 7)$ in terms of the cavity dimensions (L_x, L_y, L_z) are:

$$\vec{k}_{047} = \left(k_{x0}, k_{y4}, k_{z7} \right) = \left(0, \frac{\pi}{L_y} 4, \frac{\pi}{L_z} 7 \right)$$

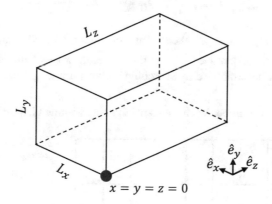

The discretized wavenumber vectors $\vec{k}_{lmn}$ can be visualized with a wavenumber grid.

In an example of a special case with the eigenmode $=0$, uniform pressure in the x-direction, the discretized wavenumber vectors $\vec{k}_{047}$ of the eigenmode $(l, m, n) = (0, 4, 7)$ can be visualized by sketching a wavenumber grid as:

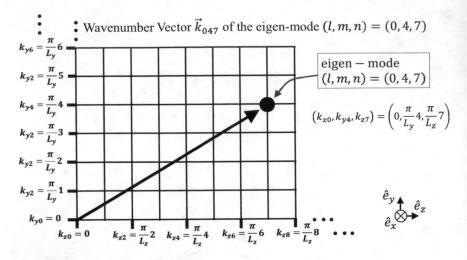

In another example of a special case with the eigenmode $=0$, uniform pressure in the y-direction , the discretized wavenumber vectors $\vec{k}_{205}$ of the eigenmode $(l, m, n) = (2, 0, 5)$ can be visualized by sketching a wavenumber grid as:

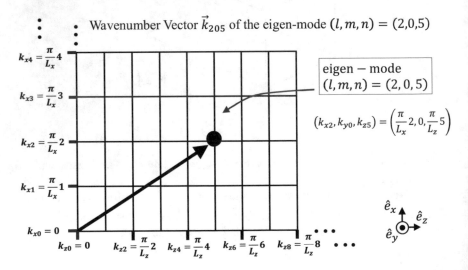

An example of a general case with an eigenmode with nonzero indexes, $l \neq 0$, $m \neq 0$, and $n \neq 0$, is $(l, m, n) = (2, 3, 6)$. The wavenumber vector can be drawn in 3D coordinates as:

Discretized Wavenumber Vector $\vec{k}_{236}$ of the eigen-mode $(l, m, n) = (2, 3, 6)$

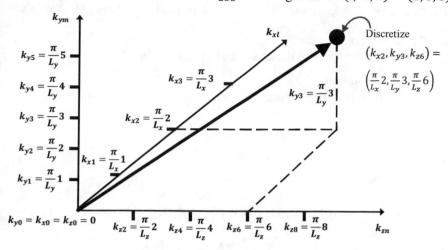

The discretized wavenumber vector has a tail located at the origin and a head (the black dot) in the discretized grid as shown in the figure above. The formula of the discretized wavenumber vector is:

$$\vec{k}_{lmn} = k_{xl}\hat{e}_x + k_{ym}\hat{e}_y + k_{zn}\hat{e}_z$$

where:

$$k_{xl} = \frac{\pi}{L_x}l; \; k_{ym} = \frac{\pi}{L_y}m; \; k_{zn} = \frac{\pi}{L_z}n$$

The natural frequency of resonant cavities is related to the wavenumber vector as:

$$f_{lmn} = \frac{\omega_{lmn}}{2\pi} = \frac{c}{2\pi}k_{lmn}$$
$$= \frac{c}{2\pi}\left|\vec{k}_{lmn}\right| = \frac{c}{2\pi}\sqrt{k_{xl}^2 + k_{ym}^2 + k_{zn}^2}$$

where k_{lmn} is the length of the discretized wavenumber vector $\vec{k}_{lmn}$ as:

$$k_{lmn} = \left|\vec{k}_{lmn}\right| = \sqrt{k_{xl}^2 + k_{ym}^2 + k_{zn}^2}$$

or:

$$k_{lmn}^2 = k_{xl}^2 + k_{ym}^2 + k_{zn}^2$$

Example 8.2: (Calculations of Wavenumber Vectors and Natural Frequencies)
The dimensions of a resonant cavity are $L_x = \frac{1}{3}$ [m] and $L_y = L_z = \frac{1}{2}$ [m]. Calculate:

(a) The wavenumber vector $\vec{k}_{lmn}$
(b) The natural frequency of the eigenmode $(l, m, n) = (0, 4, 7)$

Example 8.2: Solution
(a) The wavenumber vector of the eigenmode $(l, m, n) = (0, 4, 7)$ is:

$$\vec{k}_{lmn} = k_{xl}\widehat{e}_x + k_{ym}\widehat{e}_y + k_{zn}\widehat{e}_z$$

where:

$$k_{xl} = \frac{\pi}{L_x}l = \frac{\pi}{\frac{1}{3}\,[m]}0 = 0$$

$$k_{ym} = \frac{\pi}{L_y}m = \frac{\pi}{\frac{1}{2}\,[m]}4 = 8\pi\left[\frac{1}{m}\right]$$

$$k_{zn} = \frac{\pi}{L_z}n = \frac{\pi}{\frac{1}{2}\,[m]}7 = 14\pi\left[\frac{1}{m}\right]$$

(b) The natural frequency of the eigenmode $(l, m, n) = (0, 4, 7)$ is:

$$f_{lmn} = \frac{\omega_{lmn}}{2\pi} = \frac{c}{2\pi}k_{lmn} = \frac{c}{2\pi}\left|\vec{k}_{lmn}\right|$$

$$= \frac{c}{2\pi}\sqrt{k_{xl}^2 + k_{ym}^2 + k_{zn}^2} = \frac{c}{2\pi}\sqrt{(0)^2 + (8\pi)^2 + (14\pi)^2}\left[\frac{1}{m}\right]$$

$$= \frac{c}{2}\sqrt{(0)^2 + (8)^2 + (14)^2}\left[\frac{1}{m}\right] = \sqrt{65}\,c\left[\frac{1}{m}\right]$$

8.3   Traveling Waves in Resonant Cavities

In a 1D plane wave, the 1D standing wave can be constructed by the addition of two traveling waves with the same amplitudes and traveling in opposite directions.

In a resonant cavity, a 2D mode shape of resonant can be separated into two sets of 2D standing waves (plane wave). Each set of the 2D standing wave can be constructed by the addition of two 2D traveling waves with the same amplitudes and traveling in opposite directions. Therefore, a 2D mode shape of resonant can be constructed by four 2D traveling waves.

Assume that the dimensions of a rectangular resonant cavity are given as (L_x, L_y, L_z):

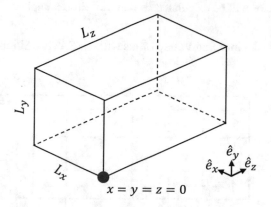

$$x = y = z = 0$$

In Example 7.2, we have plotted nodal lines and indicated peaks and valleys with plus and minus signs, as shown below:

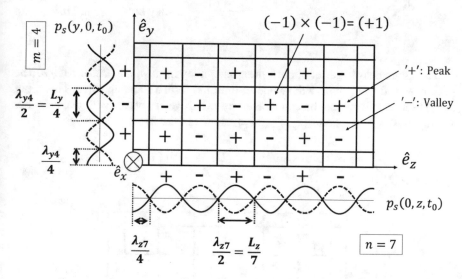

A set of peak lines and valley lines represent a set of the 2D standing wave constructed by two 2D traveling waves $(k_y \widehat{e}_y + k_z \widehat{e}_z)$ and $(-k_y \widehat{e}_y - k_z \widehat{e}_z)$ as shown in the figure below. Because it is a 2D standing wave, the peak lines and the valley lines will switch after half the cycle. The peak lines can be drawn by connecting some adjacent plus signs. The valley lines can be drawn by connecting some adjacent minus signs.

The figure below is a set of peak lines and valley lines with a *negative slope*. The peak lines and valley lines will switch in time and form a standing wave. This set of the 2D standing wave can be constructed by two traveling waves with wavenumber vectors $(k_y\hat{e}_y + k_z\hat{e}_z)$ and $(-k_y\hat{e}_y - k_z\hat{e}_z)$ as shown in the figure below. The actual wavenumber vector will be calculated in part c of this example:

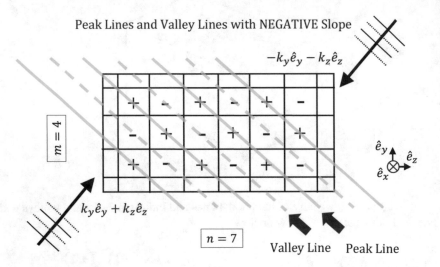

The figure below is a set of peak lines and valley lines with a *positive slope*. The peak lines and valley lines will switch in time and form a standing wave. This set of the 2D standing wave can be constructed by two traveling waves with wavenumber vectors $(-k_y\hat{e}_y + k_z\hat{e}_z)$ and $(k_y\hat{e}_y - k_z\hat{e}_z)$, shown in the figure below:

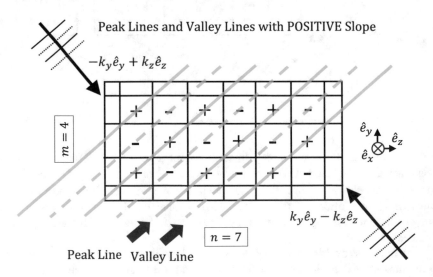

The four wavenumber vectors $\left(k_y\hat{e}_y + k_z\hat{e}_z\right)$, $\left(-k_y\hat{e}_y - k_z\hat{e}_z\right)$, $\left(-k_y\hat{e}_y + k_z\hat{e}_z\right)$, and $\left(k_y\hat{e}_y - k_z\hat{e}_z\right)$ can be shown in one figure as:

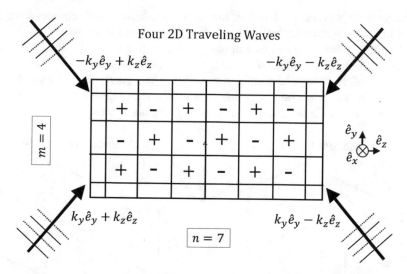

Inside this enclosed cavity, the peak and valley will switch in time and form a resonant mode shape.

These four wavenumber vectors represent the wave propagation of the four 2D traveling waves as a result of waves reflecting on the surface of the cavity as shown in the figure below:

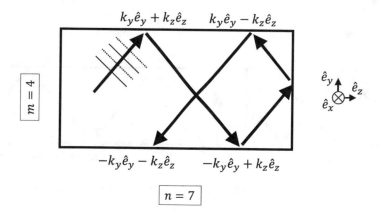

8.4   Wavenumber Vectors in Acoustic Waveguides

In the acoustic waveguide shown in the right figure below, the acoustic wave can travel in the z-direction but will reflect on all four sides of the waveguide.

Waveguides are elongated rectangular cavities with no closed walls in the axial direction. We can construct a waveguide by removing the two walls in the z-direction of an elongated rectangular cavity. The figure below shows that the two walls in the z-direction of a resonant cavity are removed to allow acoustic waves to propagate in the z-direction in an acoustic waveguide.

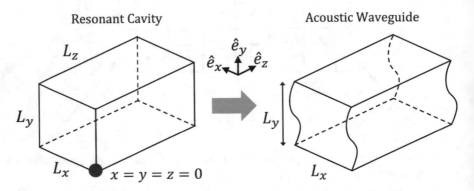

After the two walls in the z-direction are removed, the four wavenumber vectors become two wavenumber vectors because there is no reflection in the z-direction as shown in the figure below:

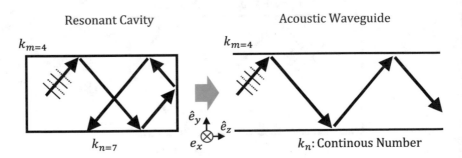

With the two walls in the z-direction, the wavenumbers in both the y-direction and z-direction are discretized as:

$$k_{ym} = \frac{\pi}{L_y}m \quad k_{ym} \text{ is DISCRETIZED where m is an INTEGER}$$

$$k_{zn} = \frac{\pi}{L_z}n \qquad k_{zn} \text{ is DISCRETIZED where n is an INTEGER}$$

Without the two walls in the z-direction, the wavenumbers in the y-direction are still discretized, but the wavenumbers in the z-direction become continuous numbers as:

$$k_{ym} = \frac{\pi}{L_y}m \qquad k_{ym} \text{ is DISCRETIZED where m is an INTEGER}$$

$$k_{zn} = \frac{\pi}{L_z}n = k_z \qquad k_z \text{ is CONTINOUS since } L_z \text{ can be ANY LENGTH}$$

In a waveguide, because L_z can be considered as a very large number, and based on the formula above, k_{zn} is continuous:

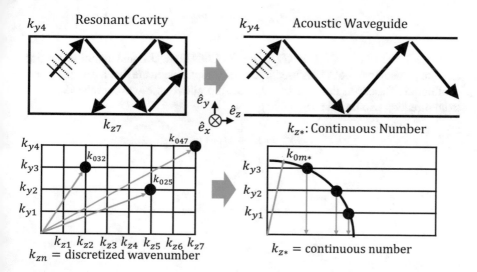

Because the wavenumber k_z changes from being discretized numbers to being continuous numbers, a pure tone (single-frequency sound) will propagate along the axial direction with any transverse eigenmode of the waveguide.

Example 8.3: (Wavenumber Vectors in a Waveguide)

Assume a waveguide has a rectangular cross-section of $L_x = \frac{1}{3}$ [m] and $L_y = \frac{1}{2}$ [m].

(a) Calculate all the possible transverse eigenmodes that can carry an 850 [Hz] pure tone in the waveguide. State the wavenumber vectors as $\vec{k}_{lmn} = k_{xl}\widehat{e}_x + k_{ym}\widehat{e}_y + k_{z*}\widehat{e}_z$

(b) Calculate the cutoff frequency of the waveguide.

Example 8.3: Solution
Part (a)

Similar to the searching method for calculating the natural frequencies which are lower than a given frequency in the previous chapter, a searching method will be used to calculate all of the possible wavenumber vectors that can carry a pure tone.

Calculations of Wavenumber Vectors by Using a Searching Method
The relationship between the combined wavenumber and the component wavenumber is:

$$k^2 = k_{xl}^2 + k_{ym}^2 + k_{zn}^2$$

where k is the combined wavenumber k_{lmn} and:

$$k_{xl} = \frac{\pi}{L_x} l; k_{ym} = \frac{\pi}{L_y} m; k_{zn} = \frac{\pi}{L_z} n$$

Since L_z can be ANY LENGTH, k_{zn} is CONTINOUS and is not discretized anymore. A star symbol "*" is used in the subscript to indicate that a wavenumber is continuous. Based on this and the relationship above, the wavenumber k_{z*} in the axial direction can be calculated using:

$$k_{z*}^2 = k^2 - k_{xl}^2 - k_{ym}^2$$

where:

$$k_{xl} = \frac{\pi}{L_x} l = \frac{\pi}{\frac{1}{3} [m]} l$$

$$k_{ym} = \frac{\pi}{\frac{1}{2} [m]} m$$

and k is the wavenumber of the 850 [Hz] pure tone as:

$$k = \frac{\omega}{c} = \frac{2\pi f}{c} = \frac{2\pi \times 850 [\text{Hz}]}{340 \left[\frac{m}{s}\right]} = 5\pi \left[\frac{1}{m}\right]$$

The following rule can be used to determine if an eigenmode exists to carry a pure tone in a waveguide:

If k_{z*}^2 is greater than zero, the eigenmode exists. If k_{z*}^2 is smaller than or equal to zero, the eigenmode does not exist.

Use the searching method to find all of the eigenmodes (l, m) with their wavenumber k_{z*}^2 being greater than zero as the following:

l	m	$k_{xl} \left[\frac{1}{m}\right]$	$k_{ym} \left[\frac{1}{m}\right]$	$k_{z*}^2 \left[\frac{1}{m^2}\right]$	Note
0	0	0	0	$25\pi^2$	Direct wave
0	1	0	$2\pi * 1$	$21\pi^2$	$k_{z*}^2 > 0$ (eigenmode exists)

0	2	0	$2\pi * 2$	$9\pi^2$	$k^2_{z*} > 0$ (eigenmode exists)
0	3	0	$2\pi * 3$	$-11\pi^2$	$k^2_{z*} \leq 0$ 0 (eigenmode does *not* exist)

Stop and increase l by 1					
1	0	$3\pi * 1$	0	$16\pi^2$	$k^2_{z*} > 0$ (eigenmode exist)
1	1	$3\pi * 1$	$2\pi * 1$	$12\pi^2$	$k^2_{z*} > 0$ (eigenmode exist)
1	2	$3\pi * 1$	$2\pi * 2$	0	$k^2_{z*} \leq 0$ (eigenmode does *not* exist)
Stop and increase m by 1					
2	0	$3\pi * 2$	0	$-11\pi^2$	$k^2_{z*} \leq 0$ (eigenmode does *not* exist)
Stop					

A summary of all the possible transverse eigenmodes found in the table is:

$$\vec{k}_{00*} = 5\pi\widehat{e}_z (\text{direct wave})$$

$$\vec{k}_{01*} = 2\pi\widehat{e}_y + \sqrt{21}\pi\widehat{e}_z$$

$$\vec{k}_{02*} = 4\pi\widehat{e}_y + 3\pi\widehat{e}_z$$

$$\vec{k}_{10*} = 3\pi\widehat{e}_x + 4\pi\widehat{e}_z$$

$$\vec{k}_{11*} = 3\pi\widehat{e}_x + 2\pi\widehat{e}_y + \sqrt{12}\pi\widehat{e}_z$$

The detailed calculations are shown below:
$(l, m, n) = (0, 0, *)$: ... this is the direct wave:

$$k^2_{z*} = k^2 - 0 - 0 = 25\pi^2 \left[\frac{1}{m}\right]$$

where k is the wavenumber of an 850 [Hz] pure tone as:

$$k = \frac{\omega}{c} = \frac{2\pi f}{c} = \frac{2\pi \times 850[\text{Hz}]}{340\left[\frac{m}{s}\right]} = 5\pi \left[\frac{1}{m}\right]$$

$$(l, m, n) = (0, 1, *) : k^2_{z*} = k^2 - k^2_{x0} - k^2_{y1} = (5\pi)^2 - 0 - (2\pi)^2 = 21\pi^2$$

$$(l, m, n) = (0, 2, *) : k^2_{z*} = k^2 - k^2_{x0} - k^2_{y2} = (5\pi)^2 - 0 - (4\pi)^2 = 9\pi^2$$

$$(l, m, n) = (0, 3, *) : k^2_{z*} = k^2 - k^2_{x0} - k^2_{y3} = (5\pi)^2 - 0 - (6\pi)^2 = -11\pi^2$$

$$(l, m, n) = (1, 0, *) : k^2_{z*} = k^2 - k^2_{x1} - k^2_{y0} = (5\pi)^2 - (3\pi)^2 - 0 = 16\pi^2$$

$$(l, m, n) = (1, 1, *) : k^2_{z*} = k^2 - k^2_{x1} - k^2_{y1} = (5\pi)^2 - (3\pi)^2 - (2\pi)^2 = 12\pi^2$$

$$(l, m, n) = (1, 2, *) : k_{z*}^2 = k^2 - k_{x1}^2 - k_{y2}^2 = (5\pi)^2 - (3\pi)^2 - (4\pi)^2 = 0$$

$$(l, m, n) = (2, 0, *) : k_{z*}^2 = k^2 - k_{x2}^2 - k_{y0}^2 = (5\pi)^2 - (6\pi)^2 - 0 = -11\pi^2$$

Summary of Wavenumber Vectors in Waveguides

For the eigenmodes having a mode number $l = 0$, there are two eigenmodes $(l, m, n) = (0, 1, *)$ *and* $(0, 2, *)$ in y-direction. These two eigenmodes have the following properties:

$$k_{xl} = 0.$$

k_{ym} is discretized due to the boundary condition: $k_{ym} = \frac{\pi}{L_y} m$.

k_{z*} is continued since L_z can be any length: $k_{z*} = \frac{\pi}{L_z} n$.

$k = \left| \vec{k}_{lmn} \right|$ is the combined wavenumber : $k = \frac{\omega}{c}$.

Because k_{z*} is continue, eigenmodes can be constructed by the combination of a discretized k_{ym} and an arbitrary k_{z*} that satisfies:

$$k = k_{x0}^2 + k_{ym}^2 + k_{z*}^2$$

Therefore, the possible eigenmodes are intersections (black dots) between a circle with a radius k and horizontal lines with a height k_{ym} as shown in the figure below:

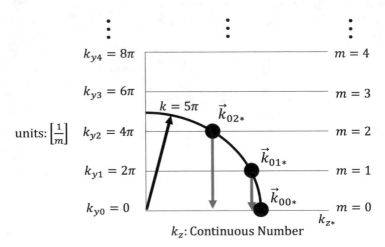

Visualization of Wavenumber Vectors by a Wavenumber Grid
Because:

$$k^2 = k_{xl}^2 + k_{ym}^2 + k_{zn}^2$$

all the possible eigenmodes (k_{xl}, k_{ym}) for positive k_{zn}^2 can be drawn as black dots inside a circle with a radius of $k = 5\pi$ in the figure below:

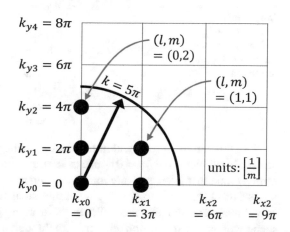

This method provides a quick way to find all the possible eigenmodes but does not provide the wavenumber vectors. The wavenumber vectors still need to be calculated after identifying the eigenmodes.

Part (b)

The cutoff frequency corresponds to the lowest wavenumber in both the x- and y-direction. In this example, $k_{x1} = 3\pi$ and $k_{y1} = 2\pi$. Therefore, the lowest wavenumber is:

$$k_{y1} = 2\pi$$

The corresponding cutoff frequency is:

$$\rightarrow f_{y10} = \frac{\omega_{y10}}{2\pi} = \frac{c}{2\pi} k_{y1} = \frac{340 \left[\frac{m}{s}\right]}{2\pi} 2\pi \left[\frac{1}{m}\right] = 340 \ [Hz]$$

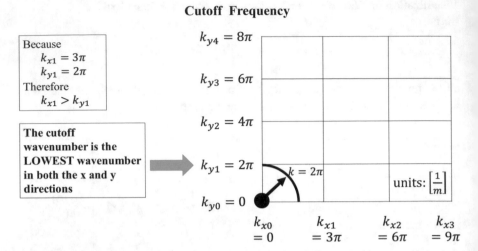

Note that any pure tone with a frequency smaller than the cutoff frequency will have only a direct wave ($m = 0$) since the corresponding wavenumber is too small to form a circle that intercepts any transverse eigenmode k_{ym} with $m > 0$.

As a result, any pure tone with a frequency smaller than the cutoff frequency have only direct wave and no echoes. When there are no echoes for any frequencies, the sound will have no distortion after passing through a waveguide and will be suitable for transmitting signals or sounds.

8.5   Traveling Waves in Acoustic Waveguides

The traveling time of a wave from one end of the waveguide to the other end depends on the eigenmode which carries the pure tone. Larger mode numbers have larger propagation angles and will result in longer traveling times. On the other hand, smaller mode numbers have smaller propagation angles and will result in shorter traveling times. When the pure tone is not carried by any mode (m>0), it will travel parallel to the axial direction and result in the shortest possible traveling time.

The traveling time for the wave (direct or reflected) to arrive at the other end of the waveguide can be formulated as:

$$\text{Traveling Time} = \frac{\text{Distance}}{\text{Velocity}}$$

where:

$$\text{Velocity} = c\frac{k_{z*}}{k}$$

Example 8.4: (Traveling Time in Waveguides)

An 850 $[Hz]$ pure tone is traveling through a waveguide. Assume the waveguide is 340 $[m]$ long and has a rectangular cross-section of $L_x = \frac{1}{3}$ $[m]$ and $L_y = \frac{1}{2}$ $[m]$.

(a) Calculate the traveling time of the pure tone moving in the axial direction as a direct wave ($l = m = 0$ mode) to arrive at the other end of the waveguide along the z-axis.

(b) Calculate the traveling times of the pure tone carried by all the eigenmodes to arrive at the other end of the waveguide. Use eigenmodes calculated in Example 8.3.

Example 8.4: Solution
Part (a)

The traveling time for the direct wave to arrive at the other end of the waveguide is:

$$\rightarrow \text{Traveling Time} = \frac{\text{Distance}}{\text{Velocity}} = \frac{340 \ [m]}{340 \ \left[\frac{m}{s}\right]} = 1 \ [s]$$

Part (b)

The eigenmodes calculated in Example 8.3 are:

$$\vec{k}_{00*} = 5\pi\widehat{e}_z \quad \text{(direct wave)}$$

$$\vec{k}_{01*} = 2\pi\widehat{e}_y + \sqrt{21}\pi\widehat{e}_z$$

$$\vec{k}_{02*} = 4\pi\widehat{e}_y + 3\pi\widehat{e}_z$$

$$\vec{k}_{10*} = 3\pi\widehat{e}_x + 4\pi\widehat{e}_z$$

$$\vec{k}_{11*} = 3\pi\widehat{e}_x + 2\pi\widehat{e}_y + \sqrt{12}\pi\widehat{e}_z$$

Traveling time of $(l, m, n) = (0, 0, *)$ for the direct wave is:

$$time = \frac{distance}{c\frac{k_{z*}}{k}} = \frac{340 \ [m]}{340 \ \left[\frac{m}{s}\right]\frac{5\pi}{5\pi}} = 1 \ [s]$$

Traveling time of $(l, m, n) = (0, 1, *)$ for the first eigenmode in the y-direction is:

$$time = \frac{distance}{c \frac{k_{z*}}{k}} = \frac{340 \, [m]}{340 \, \left[\frac{m}{s}\right] \frac{\sqrt{21}\pi}{5\pi}} = \frac{5}{\sqrt{21}} \, [s]$$

Traveling time of $(l, m, n) = (0, 2, *)$ for the second eigenmode in the y-direction is:

$$time = \frac{distance}{c \frac{k_{z*}}{k}} = \frac{340 \, [m]}{340 \, \left[\frac{m}{s}\right] \frac{3\pi}{5\pi}} = \frac{5}{3} \, [s]$$

Traveling time of $(l, m, n) = (1, 0, *)$ for the first eigenmode in the x-direction is:

$$time = \frac{distance}{c \frac{k_{z*}}{k}} = \frac{340 \, [m]}{340 \, \left[\frac{m}{s}\right] \frac{4\pi}{5\pi}} = \frac{5}{4} \, [s]$$

Traveling time of $(l, m, n) = (1, 1, *)$ is:

$$time = \frac{distance}{c \frac{k_{z*}}{k}} = \frac{340 \, [m]}{340 \, \left[\frac{m}{s}\right] \frac{\sqrt{12}\pi}{5\pi}} = \frac{5}{\sqrt{12}} \, [s]$$

This pure tone will travel in this waveguide with five modes. The difference in travel times between different eigenmodes and the direct wave will create echoes of the sound. A sound is usually containing many different frequencies.

When a sound is made of multiple frequencies, since the traveling times are different for different frequencies, even by the same eigenmode, the sound transmitted in the waveguide will be distorted.

Example 8.5: (Cutoff Frequency of Waveguides)
This example shows how to calculate the frequency range of an *undisturbed* pure tones in a narrow waveguide. Assume that a narrow waveguide has a square cross-section with the dimensions of $L_x = L_y = 0.0085 \, [m]$.

Example 8.5: Solution
Since the dimensions in the x- and y-direction are the same, the cutoff frequencies (the lowest first transverse first mode) are the same for both directions.

The transverse frequency of the first mode can be calculated as:

$$f_{y1} = \frac{\omega_{y1}}{2\pi} = \frac{c}{2\pi} k_{y1} = \frac{340 \left[\frac{m}{s}\right]}{2\pi} \frac{\pi}{0.0085 \, [m]} 1 = 20000 \, [Hz]$$

where the wavenumber k_{y1} is calculated using:

$$k_{ym} = \frac{\pi}{L_y} m; \, m = 1$$

The transverse frequency of the first mode is called the cutoff frequency of the waveguide because it is the maximum frequency that could travel in the waveguide without an echo. Any frequency above the cutoff frequency will disturb the direct sound due to the echo carried by the transverse eigenmode, as shown in Sect. 8.3.

Therefore, the non-echo effect frequency range of this waveguide is 0–20 [kHz]. And the frequency range of sounds that will not disturb the direct sound due to the echo of the sound carried by the transverse eigenmode of the waveguide is 0–20 [kHz].

Remarks: Stethoscopes Are Waveguides with a Very High Cutoff Frequency
Typical stethoscopes have diameters around 0 .0085 [m]. Based on the example above, the cutoff frequency of stethoscopes is around 20 [kHz].

Since the human audible range is 20 [Hz] to 20 [kHz] and there are no echoes with frequencies below 20 [kHz] in this waveguide, any sound waves such as heartbeats can be clearly heard on the other side of a stethoscope. Frequencies above the cutoff frequency of 20 [kHz] will still result in echoes, but these echoes cannot be heard by a human. Stethoscopes are designed with a diameter less than 8.5 [mm] in order to eliminate/disable echo sound with frequencies below the cutoff frequency of 20 [kHz].

8.6 Homework Exercises

Exercise 8.1: (2D Traveling Wave Velocity: Forward)
Derive the 2D forward flow velocity vector $\vec{u}_+(x, y, t)$ of a 2D plane forward traveling wave from the 2D acoustic pressure $p_+(x, y, t)$ using Euler's force equation:

$$p_+(x, y, t) = A_+ \cos\left(\omega t - \left(k_x x + k_y y\right) + \theta\right)$$

$$\vec{u}_+(x, y, t) = \frac{n_x \hat{e}_x + n_y \hat{e}_y}{\rho_0 c} A_+ \cos\left(\omega t - \left(k_x x + k_y y\right) + \theta\right)$$

where:

$$n_x = \frac{k_x}{\sqrt{k_x^2 + k_y^2}}; \quad n_y = \frac{k_y}{\sqrt{k_x^2 + k_y^2}}$$

Euler's force equation: $\nabla p(x, y, t) = -\rho_o \frac{\partial}{\partial t} \vec{u}(x, y, t)$

Exercise 8.2: (Calculations of Wavenumber Vectors and Natural Frequencies)
The dimensions of a resonant cavity are $L_x = 1$ [m], $L_y = \frac{1}{3}$ [m] and $L_z = \frac{1}{2}$ [m].

(a) Calculate the wavenumber vectors $\vec{k}_{lmn}$ of the eigenmodes $(l,m,n) = (2,1,3)$.

(b) Calculate the natural frequencies f_{lmn} of the eigenmodes $(l,m,n) = (2,1,3)$.

Use 340 [m/s] for the speed of sound (c) in air. Show units in the Meter-Kilogram-Second (MKS) system.

(Answers): (a) $\vec{k}_{213} = 2\pi\left[\frac{1}{m}\right]\hat{e}_x + 3\pi\left[\frac{1}{m}\right]\hat{e}_y + 6\pi\left[\frac{1}{m}\right]\hat{e}_z$; (b) $f_{213} = \frac{7}{2}\left[\frac{1}{m}\right]c$

Exercise 8.3: (3D Mode Shapes in a Rectangular Cavity)
The dimensions of a rectangular resonant cavity are given as $(L_x, L_y, L_z) = (3,9,24)$ [m]:

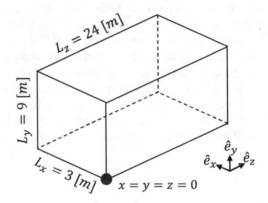

(a) Plot *nodal lines* and indicate *peaks and valleys* of eigenmode $(l,m,n) = (0,3,6)$.

(b) Plot *peak lines and valley lines*, and draw *four wavenumber vectors* (without values) of eigenmode $(l,m,n) = (0,3,6)$.

(c) Calculate the wavenumber vector $\vec{k}_{036}$ and natural frequency f_{036} of eigenmode $(l,m,n) = (0,3,6)$.

(d) Sketch a *wavenumber grid*, and draw the wavenumber vector $\vec{k}_{036}$ of eigenmode $(l,m,n) = (0,3,6)$ in the wavenumber grid.

Use 340 [m/s] for the speed of sound (c) in air. Show units in the Meter-Kilogram-Second (MKS) system.

Exercise 8.4: (Wavenumber Vectors in a Waveguide)
A waveguide has a rectangular cross-section of $L_x = 0.75$ [m] and $L_y = 1.5$ [m].

(a) Calculate all the possible transverse eigenmodes that can carry a 283.33 [Hz] pure tone in the waveguide. State the wavenumber vectors as $\vec{k}_{lm*} = k_{xl}\hat{e}_x + k_{ym}\hat{e}_y + k_{z*}\hat{e}_z$.

(b) Calculate the cutoff frequency of the waveguide.

Use 340 [m/s] for the speed of sound (c) in air. Show units in the Meter-Kilogram-Second (MKS) system.

(Answers): (a) $\vec{k}_{01*} = \frac{2}{3}\pi\hat{e}_y + \sqrt{\frac{21}{9}}\pi\hat{e}_z$; $\vec{k}_{02*} = \frac{4}{3}\pi\hat{e}_y + \pi\hat{e}_z$; $\vec{k}_{10*} = \frac{4}{3}\pi\hat{e}_x + \pi\hat{e}_z$;
$\vec{k}_{11*} = \frac{4}{3}\pi\hat{e}_x + \frac{2}{3}\pi\hat{e}_y + \sqrt{\frac{5}{9}}\pi\hat{e}_z$; (b) 113.33 [Hz]

Exercise 8.5: (Cutoff Frequency in Ventilation Fans)

A ventilation fan is installed on a long pipeline with a rectangular cross-section, generating a single-frequency noise that can be adjusted by modifying the speed of the fan motor. If the cross-section dimensions of the pipeline are $L_x = 0.75$ [m], $L_y = 1.5$[m]:

(a) Which transverse mode of the pipeline will carry the acoustic energy if the fan produces a noise signal at 115 [Hz]? State the mode number.
(b) What should the maximum frequency of the fan be so that the noise is not carried to long distances with any one of the transverse modes of the pipeline?

Use 340 [m/s] for the speed of sound (c) in air. Show units in the Meter-Kilogram-Second (MKS) system.

(Answers): (a) $(l, m) = (0, 1)$ mode; (b) 113.3 [Hz]

Exercise 8.6: (Echo Effect in Waveguides)

A 170 Hz pure tone travels in a cross-section waveguide with the dimensions of $(L_x, L_y) = (3, 4)$ [m].

Assuming that the length of the waveguide is 170 [m], calculate the:

(a) Traveling time of a direct wave $(l, m) = (0, 0)$ of the pure tone to arrive at the other end of the waveguide along the x-axis.
(b) Traveling time of the pure tone carried by the first transverse eigenmode in the x-direction $(l, m) = (1, 0)$ to arrive at the other end of the waveguide.
(c) Traveling time of the pure tone carried by the first transverse eigenmode in the y-direction $(l, m) = (0, 1)$ to arrive at the other end of the waveguide.
(d) What is the cutoff frequency of this waveguide?

Use 340 [m/s] for the speed of sound (c) in air. Show units in the Meter-Kilogram-Second (MKS) system.

(Answers): (a) 0.5 [s]; (b) 0.530 [s]; (c) 0.516 [s]; (d) 42.5 [Hz]

Exercise 8.7: (Echo Effect in Waveguide)

An 85 Hz pure tone travels in a cross-section waveguide with the dimensions of $(L_x, L_y) = (4, 5)$ [m].

Assume that the length of the waveguide is 240 [m], calculate the:

(a) Traveling time of a direct wave ($m = n = 0$ mode) of the pure tone to arrive at the other end of the waveguide along the x-axis.
(b) Traveling time of the pure tone carried by the first transverse eigenmode in the x-direction $(l, m) = (1, 0)$ to arrive at the other end of the waveguide.

(c) Traveling time of the pure tone carried by the first transverse eigenmode in the y-direction $(l, m) = (0, 1)$ to arrive at the other end of the waveguide.

(d) What is the cutoff frequency of this waveguide?

Use 340 [*m/s*] for the speed of sound (c) in air. Show units in the Meter-Kilogram-Second (MKS) system.

(Answers): (a) 0.706 [*s*]; (b) 0.815 [*s*]; (c) 0.770 [*s*]; (d) 34 [Hz]

Chapter 9
Sound Pressure Levels and Octave Bands

In the field of noise control, noise is quantified by sound pressure level (SPL). SPL can be either unweighted or weighted. In this chapter, you will learn how to calculate both unweighted SPL and weighted SPL. This chapter is divided into four sections and is summarized as follows:

Section 9.1 introduces and defines the decibel scale. The decibel scale is used for sound pressure level (SPL) and other power-like physical quantities in vibrations. In this section, you will learn how to calculate the combined level from separated levels of n incoherent radiating source using the formula below:

$$L_{tot} = 10 \log_{10} \left(10^{\frac{L_1 \, [dB]}{10}} + 10^{\frac{L_2 \, [dB]}{10}} + \ldots + 10^{\frac{L_n \, [dB]}{10}} \right) [dB]$$

Section 9.2 introduces and defines the sound pressure level (SPL). Parseval's theorem will be used to derive the formula for calculating SPL in the frequency domain. Parseval's theorem is important because it allows us to formulate RMS pressure in the frequency domain as shown in the flowchart below:

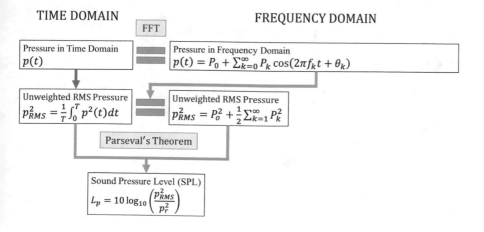

TIME DOMAIN | FREQUENCY DOMAIN

FFT

Pressure in Time Domain
$p(t)$

Pressure in Frequency Domain
$p(t) = P_0 + \sum_{k=0}^{\infty} P_k \cos(2\pi f_k t + \theta_k)$

Unweighted RMS Pressure
$p_{RMS}^2 = \frac{1}{T} \int_0^T p^2(t) \, dt$

Unweighted RMS Pressure
$p_{RMS}^2 = P_0^2 + \frac{1}{2} \sum_{k=1}^{\infty} P_k^2$

Parseval's Theorem

Sound Pressure Level (SPL)
$L_p = 10 \log_{10} \left(\frac{p_{RMS}^2}{p_r^2} \right)$

© The Author(s), under exclusive license to Springer Nature Switzerland AG 2021
H. Lin et al., *Lecture Notes on Acoustics and Noise Control*,
https://doi.org/10.1007/978-3-030-88213-6_9

Section 9.3 introduces the octave band and the 1/3 octave band, which are the most used bands. Bands are needed for formulations of weighted sound pressure level (SPL) because the weighting is related to frequencies and frequencies are grouped into bands. The calculation of either the octave band or the 1/3 octave band requires the frequency contents at discretized frequencies. The frequency contents are given in the form of the Fourier series as a result of the Fourier transform. Fourier transform is not covered in this course.

Section 9.4 introduces and formulates the weighted sound pressure level (SPL). Three commonly used weighting systems will be introduced. A-weighting is used in examples and homework exercises. This section will explain why weighted SPL can only be calculated in the frequency domain but not the time domain.

The goal of this chapter is to equip students with the understanding and technique to calculate a weighted sound pressure for measured sound pressure data.

9.1   Decibel Scale

9.1.1   Review of Logarithm Rules

Some logarithm rules related to levels and decibel scale are shown below for review:

A) **Logarithm Definition**

When b is raised to the power of y, it is equal to x, and y can then be calculated using the logarithm of x with base b:

$$b^y = x \rightarrow \log_b(x) = y$$

B) **Product Rule**

$$\log_b(x \cdot y) = \log_b(x) + \log_b(y)$$

C) **Quotient Rule**

$$\log_b(x/y) = \log_b(x) - \log_b(y)$$

D) **Power Rule**

$$\log_b(x^y) = y \cdot \log_b(x)$$

9.1.2 Levels and Decibel Scale

Sound pressure and power and vibration amplitudes are commonly expressed in the decibel scale instead of the linear scale. Since the quantities involved in the sound and vibration field cover an extremely large range of numbers, using a logarithmic scale to do calculations simplifies the comparison of values involved, and a smaller range of numbers can thus be used.

Decibel is defined as "10 times the logarithm base 10 of the ratio of 'power-like' quantities." The argument of the logarithm has no unit. Hence, the designations in decibels are referred to as levels. The name "decibel" is one-tenth of one "Bel," named in honor of Alexander Graham Bell, and is written in short as dB.

The mathematical expression for a level is:

$$L[dB] = 10 \log_{10} \left(\frac{U}{U_r} \right) \leftrightarrow 10^{\frac{L[dB]}{10}} = \frac{U}{U_r}$$

where:

U is a power-like physical quantity.
U_r is a reference value of the same physical quantity.
L is the level in units of dB (reference to U_r units).

If the level is known, the power-like physical quantity can be determined from:

$$U = U_r 10^{\frac{L\ [dB]}{10}}$$

Example 9.1
Given:
   $U = 100$ (units of power)
   $U_r = 0.1$ (units of power)
   Find [dB].

Example 9.1: Solution

$$L = 10 \log_{10} \left(\frac{U}{U_r} \right) = 10 \log_{10} \left(\frac{100}{0.1} \right) = 30\ [dB]$$

Example 9.2
Given:

$$L = 30\ [dB]$$

   $U_r = 0.1$ (units of power)
   Find U (units of power).

Example 9.2: Solution

$$U = U_r 10^{\frac{L\ [dB]}{10}} = 0.1 \times 10^{\frac{30\ [dB]}{10}} = 0.1 \times 10^3 = 100 \text{ (units of power)}$$

9.1.3 Decibel Arithmetic

The total (combined) level is the addition of power-like quantities. The formula for the sum of the sound pressure levels of n incoherent radiating source is:

$$L_{tot} = 10 \log_{10}\left(10^{\frac{L_1\ [dB]}{10}} + 10^{\frac{L_2\ [dB]}{10}} + \ldots + 10^{\frac{L_n\ [dB]}{10}}\right) [dB]$$

Use the law of conservation of energy to show that the above formula for the total (combined) level is valid (see Exercise 9.1).

Example 9.3
Two sound pressure levels are given as L_1 and L_2.
Use the law of conservation of energy ($U_{tot} = U_1 + U_2$) to find the total (combined) sound pressure level L_{tot}.

Example 9.3: Solution
Based on the definition of sound pressure level:

$$L_1[dB] = 10 \log_{10}\left(\frac{U_1}{U_r}\right) \rightarrow \frac{U_1}{U_r} = 10^{\frac{L_1\ [dB]}{10}}$$

$$L_2[dB] = 10 \log_{10}\left(\frac{U_2}{U_r}\right) \rightarrow \frac{U_2}{U_r} = 10^{\frac{L_2\ [dB]}{10}}$$

$$L_{tot} = 10 \log_{10}\left(\frac{U_{tot}}{U_r}\right) [dB]$$

The law of conservation of energy:
$U_{tot} = U_1 + U_2$
Therefore:

$$L_{tot} = 10 \log_{10}\left(\frac{U_{sum}}{U_r}\right) [dB] = 10 \log_{10}\left(\frac{U_1 + U_2}{U_r}\right) [dB]$$

$$\rightarrow L_{tot} = 10 \log_{10}\left(10^{\frac{L_1\ [dB]}{10}} + 10^{\frac{L_2\ [dB]}{10}}\right) [dB]$$

9.2 Sound Pressure Levels

9.2.1 Power-Like Quantities

All power-like physical quantities can be formulated as levels or using the decibel scale.

The definition and the units of sound power w are shown below:

$$(POWER)\; w\left[\frac{energy}{time}\right] = (PRESSURE)\; p\left[\frac{force}{area}\right] \times (VELOCITY)\; v\left[\frac{distance}{time}\right]$$
$$\times (AREA)\; S\; [area]$$

The sound power w can be formulated in terms of pressure, velocity, and area:

$$w = P_{RMS} V_{RMS} S \left[\frac{J}{s}\right]$$

For plane waves moving at speed c, because $V_{RMS} = \frac{P_{RMS}}{\rho_o c}$, we arrive at:

$$w = \frac{P_{RMS}^2}{\rho_o c} S \left[\frac{J}{s}\right] \; [Plane\; Waves]$$

The following are some power-like physical quantities that can be formulated as levels:

- Sound power level (SWL):

$$L_w\, [dB] = 10\log_{10}\left(\frac{W}{W_r}\right)$$

- Sound pressure level (SPL):

$$L_p = 10\log_{10}\left(\frac{P_{RMS}^2}{P_r^2}\right)$$

- RMS Velocity Square:

$$L_V\, [dB] = 10\log_{10}\left(\frac{V_{RMS}^2}{V_r^2}\right)$$

- Peak-to-peak velocity square:

$$L_V \, [dB] = 10 \log_{10} \left(\frac{V^2}{V_r^2} \right)$$

- Peak-to-peak acceleration square:

$$L_A \, [dB] = 10 \log_{10} \left(\frac{A^2}{A_r^2} \right)$$

- Peak-to-peak displacement square:

$$L_X \, [dB] = 10 \log_{10} \left(\frac{X^2}{X_r^2} \right)$$

Remarks
- Sound pressure level is a power-like quantity because $W = \mathrm{Re}\,(P)\,\mathrm{Re}\,(U)S = \mathrm{Re}\,(P)\,\mathrm{Re}\,\left(\frac{P}{Z}\right)$, where Z is the acoustic impedance that relates pressure P and velocity U.
- Only (1) sound pressure level and (2) sound power level have well-established reference values, but there is no established international reference for other power-like physical quantities. For levels other than the sound pressure level and power level, their reference values must be provided for calculating their associated power-like quantities.
- The definition of levels and the decibel scale requires a reference value. When a power-like physical quantity is presented as a level and decibel scale, this power-like quantity is compared to the reference value. Therefore, the benefit of using a level and decibel scale is to make comparing different physical quantities easier. For example, RMS pressure has an international reference value of 20 [μPa], and this reference value allows comparing RMS pressure using dB instead of μPa.

9.2.2   Sound Power Levels and Decibel Scale

A sound power level describes the acoustic power radiated by a source with respect to the international reference $w_r = 10^{-12} [W = \text{Watt}]$:

$$L_w = 10 \log \left(\frac{w}{10^{-12}} \right) = 10 \log (w) + 120$$

9.2.3   Sound Pressure Levels and Decibel Scale

Sound pressure is not a power-like quantity, but its squared value is proportional to sound power: $w \propto p^2$. The sound pressure level in decibels (SPL in dB) is thus:

$$L_p = 10 \log_{10}\left(\frac{P_{RMS}^2}{P_r^2}\right) = 10 \log_{10}\left(\frac{P_{RMS}}{P_r}\right)^2 = 20 \log_{10}\left(\frac{P_{RMS}}{P_r}\right)$$

where P_r is the international reference sound pressure based on an auditory threshold of human hearing at 1000 Hz:

$$P_r = 20 \cdot 10^{-6}\ [Pa] = 20\ [\mu Pa]\ \text{(RMS reference)}$$

In addition, the subscript RMS of the pressure P_{RMS} is often omitted for simplicity.

Remarks

- The number 20 before the log is the result of 10×2 where 2 is related to the power of $\left(\frac{P_{RMS}}{P_r}\right)$.

- Sound pressure level 0 [dB] means that the sound pressure is equal to the reference pressure:

$$L_p = 20 \log_{10}\left(\frac{20 \cdot 10^{-6}\ [Pa]}{P_r}\right)$$

$$= 20 \log_{10}\left(\frac{20 \cdot 10^{-6}\ [Pa]}{20 \cdot 10^{-6}\ [Pa]}\right) = 20 \cdot 0 = 0\ [dB]$$

9.2.4   Sound Pressure Levels Calculated in Time Domain

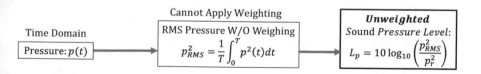

Example 9.4: Sound Pressure Level Calculated in Time Domain
Given:

The single-frequency pressure time domain function $p(t)$:

$$p(x, t) = P_1 \cos\left(2\pi f_1 t\right)$$

where:

$$P_1 = 3 \, [Pa]$$
$$f_1 = 500 \, [Hz]$$

Calculate:

(a) The RMS pressure in the time domain
(b) The unweighted sound pressure level

Example 9.4: Solution

(a) The RMS pressure calculated in the time domain is:

$$p_{RMS}^2 = \frac{1}{T}\int_0^T p^2(t)dt = \frac{1}{T}\int_0^T P_1^2 \cos^2(2\pi f_1 t)dt = \frac{P_1^2}{2} = \frac{3^2}{2} \, \left[(Pa)^2\right]$$

or:

$$p_{RMS} = \frac{P_1}{\sqrt{2}} = \frac{3}{\sqrt{2}} \, [Pa]$$

(b) The unweighted sound pressure level is:

$$L_p = 10\log_{10}\left(\frac{P_{RMS}^2}{P_r^2}\right) = 10\log_{10}\left(\frac{P_{RMS}^2}{P_r^2}\right) = 10\log_{10}\left(\frac{\left(\frac{3^2}{2}\right)}{\left(20\times10^{-6}\right)^2}\right) \, [dB]$$

or equivalently:

$$L_p = 10 \log_{10} \left(\frac{P_{RMS}}{P_r} \right)^2 = 20 \log_{10} \left(\frac{P_{RMS}}{P_r} \right) = 20 \log_{10} \left(\frac{\frac{3}{\sqrt{2}}}{20 \times 10^{-6}} \right) \, [dB]$$

9.2.5 Sound Pressure Level Calculated in Frequency Domain

In the previous section, sound pressure level (SPL) was defined and calculated in the time domain. However, there is a limitation to calculating SPL in the time domain: weighted SPL cannot be calculated in the time domain because the weighting is frequency-dependent. For this reason, SPL is commonly formulated and calculated in the frequency domain:

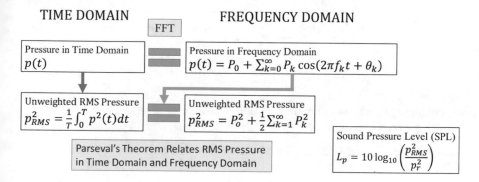

Parseval's theorem is used to formulate the RMS pressure in the frequency domain. Parseval's theorem is shown and proved as follows:

Parseval's Theorem

$$p_{RMS}^2 = P_o^2 + \frac{1}{2} \sum_{k=1}^{\infty} P_k^2$$

Proof of Parseval's Theorem
Based on the Fourier transform, a pressure function in time domain $p(t)$ can be formulated in the frequency domain with the frequency contents P_k as:

$$p(t) = P_o + \sum_{k=1}^{\infty} P_k \cos \left(2\pi f_k t + \theta_k \right)$$

The RMS pressure of this function can be formulated with Parseval's theorem as:

$$p_{RMS}^2 = \frac{1}{T} \int_0^T p^2(t) dt$$

$$= \frac{1}{T} \int_0^T \left\{ P_o + \sum_{k=1}^{\infty} P_k \cos\left(2\pi f_k t + \theta_k\right) \right\}^2 dt$$

Because Fourier series is an orthogonal basis which means that for $i \neq j$:

$$\int_0^T \cos\left(2\pi f_i t + \theta_k\right) \cos\left(2\pi f_j t + \theta_k\right) dt = 0$$

Apply the above orthogonal property for p_{RMS}^2 results:

$$p_{RMS}^2 = \frac{1}{T} \int_0^T \left\{ P_o + \sum_{k=1}^{\infty} P_k \cos\left(2\pi f_k t + \theta_k\right) \right\}^2 dt$$

$$= P_o^2 \frac{1}{T} \int_0^T dt + \sum_{k=1}^{\infty} P_k^2 \frac{1}{T} \int_1^T \cos^2\left(2\pi f_k t + \theta_k\right) dt$$

$$= P_o^2 + \frac{1}{2} \sum_{k=1}^{\infty} P_k^2$$

Frequency contents of amplitude P_k and phase θ_k can be treated as a sound source radiating a pure tone in frequency f_k with an amplitude P_k and phase θ_k.

Example 9.5: Sound Pressure Level Calculated in Frequency Domain
Given:

The single-frequency pressure time domain function $p(t)$:

$$p(x, t) = P_1 \cos\left(2\pi f_1 t\right)$$

where:

$$P_1 = 3 \ [Pa]$$
$$f_1 = 500 \ [Hz]$$

Calculate:

(a) The RMS pressure in the frequency domain
(b) The unweighted sound pressure level

Example 9.4: Solution

(a) The RMS pressure can be formulated in the frequency domain using Parseval's theorem as:

$$p_{RMS}^2 = P_o^2 + \frac{1}{2}\sum_{k=1}^{\infty} P_k^2$$

The RMS pressure can be calculated in the frequency domain by substituting the given frequency contents P_k into the equation above:

$$p_{RMS}^2 = P_o^2 + \frac{1}{2}\sum_{k=1}^{\infty} P_k^2 = \frac{P_1^2}{2} = \frac{3^2}{2}\left[(Pa)^2\right]$$

(b) The unweighted SPL calculated in the frequency domain is:

$$L_p = 20\log_{10}\left(\frac{P_{RMS}}{P_r}\right) = 20\log_{10}\left(\frac{\frac{3}{\sqrt{2}}}{20\times10^{-6}}\right) [dB]$$

Note that both RMS pressure and unweighted SPL calculated in the frequency domain (this example) are the same as if calculated in the time domain (the previous example). This is the expected result because of Parseval's theorem.

9.3 Octave Bands

Octave bands are commonly used in the fields of acoustics and vibration. Two major benefits of using octave bands are as follows: (1) it provides insight of power distribution vs. frequency, and (2) it allows to apply weight to the different frequencies.

9.3.1 Center Frequencies and Upper and Lower Bounds of Octave Bands

Spectra obtained by Fourier transform (FT) using frequency in linear scale are called the narrow band spectra. However, such detail in spectral resolution is not always needed. Hence, spectra can be obtained in wider frequency bands for easier analysis. The most commonly used frequency bands are octave bands (frequency in logarithmic scale). Each octave band has a center frequency and a band width, defined as:

$$f_o = 1000 * 2^{\left(\frac{k\ [octave]}{1\ [octave]}\right)} [Hz]; k = -5\tilde{4}; [\text{center frequency}]$$

$$f_{lower} = \frac{f_o}{\sqrt{2}} = f_o \cdot 2^{\frac{-1}{2}}; [\text{lower limit frequency}]$$

$$f_{upper} = \sqrt{2} f_o = f_o \cdot 2^{\frac{1}{2}}; [\text{upper limit frequency}]$$

$$\therefore f_{upper} = f_{lower} \cdot 2^1$$

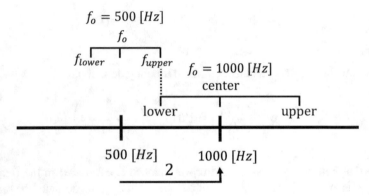

The center frequencies and lower and upper limit frequencies of octave bands are:

1 Octave Band	Center (Quoted)	Center (Quoted)	Center (Calculated)	Lower (Limit)	Upper (Limit)
[-]		[Hz]	[Hz]	[Hz]	[Hz]
			fo	fo*2^(-1/2)	fo*2^(1/2)
0	1000*2^(-5)	31.50	31.25	22.10	44.19
1	1000*2^(-4)	63.00	62.50	44.19	88.39
2	1000*2^(-3)	125.00	125.00	88.39	176.78
3	1000*2^(-2)	250.00	250.00	176.78	353.55
4	1000*2^(-1)	500.00	500.00	353.55	707.11
5	1000*2^0	1000.00	1000.00	707.11	1414.21
6	1000*2^1	2000.00	2000.00	1414.21	2828.43
7	1000*2^2	4000.00	4000.00	2828.43	5656.85
8	1000*2^3	8000.00	8000.00	5656.85	11313.71
9	1000*2^4	16000.00	16000.00	11313.71	22627.42

9.3.2   Lower and Upper Bounds of Octave Band and 1/3 Octave Band

There are other popular wide bands in the analysis that can use $\frac{1}{3}$ octave bands, $\frac{1}{10}$ octave bands, and $\frac{1}{12}$ octave bands. The formula for $\frac{1}{3}$ octave bands is shown below:

$$f_o = 1000 * 2^{\left(\frac{k \ [octave]}{3 \ [octave]}\right)} \ [Hz]; k = -19\tilde{1}3; [center \ frequency]$$

$$f_{lower} = \frac{f_o}{(\sqrt{2})^{1/3}} = f_o \cdot 2^{\frac{-1}{6}};$$

$$f_{upper} = \left(\sqrt{2}\right)^{1/3} f_o = f_o \cdot 2^{\frac{1}{6}}$$

$$\therefore f_{upper} = f_{lower} \cdot 2^{\frac{1}{3}}$$

In general, m-octave bands can be generated using the relationships below:

$$f_o = 1000 \cdot (2^m)^k, \text{where } m = 1, \frac{1}{2}, \frac{1}{3}, \frac{1}{4}$$

$$f_{lower} = 2^{-m/2} f_o; f_{upper} = 2^{m/2} f_o; f_{upper} = 2^m f_{lower}$$

Example 9.6
Given the time domain function $p(t)$ with frequency contents P_k as below:

$$p(t) = \sum_{k=1}^{4} P_k \cos(2\pi f_k t + \theta_k)$$

where:

$$f_k = 250 \times k \ [Hz]$$

And the magnitudes and phases are:

$P_1 = 0.023[Pa]; \quad \theta_1 = 0 \ [deg]$
$P_2 = 0.127[Pa]; \quad \theta_2 = 270[deg]$
$P_3 = 0.127[Pa]; \quad \theta_3 = 270[deg]$
$P_4 = 0.283[Pa]; \quad \theta_3 = 270[deg]$

Determine the combined sound pressure level (SPL) using the following three methods:

(a) Calculate the combined SPL from the summation of the square of the RMS pressures.
(b) Calculate the combined SPL from the SPL spectrum of all the existing frequencies.
(c) Calculate the combined SPL from the SPL spectrum of the octave bands.

Example 9.6: Solution
(a) Calculate the combined SPL from the summation of the square of the RMS pressures.

The RMS pressures of each frequency are:

$$p^2_{RMS,250Hz} = \frac{P^2_1}{2} = \frac{0.023^2}{2} \; [Pa^2]$$

$$p^2_{RMS,500Hz} = p^2_{RMS,750Hz} = \frac{P^2_2}{2} = \frac{0.127^2}{2} \; [Pa^2]$$

$$p^2_{RMS,1000Hz} = \frac{P^2_4}{2} = \frac{0.283^2}{2} \; [Pa^2]$$

The combined SPL from the summation of the square of the RMS pressure is:

$$L_p = 10 \log_{10} \left(\frac{p^2_{RMS,total}}{P^2_r} \right)$$

$$= 10 \log_{10} \left(\frac{\frac{0.023^2}{2} + \frac{0.127^2}{2} + \frac{0.127^2}{2} + \frac{0.283^2}{2}}{[20 \cdot 10^{-6}]^2} \right) [dB] = 81.495[dB]$$

(b) Calculate the combined SPL from the SPL spectrum of all the existing frequencies:

$$L_{p,250Hz} = 10 \log_{10} \left(\frac{p^2_{RMS,\,250Hz}}{P^2_r} \right) = 10 \log_{10} \left(\frac{\frac{0.023^2}{2}}{[20 \cdot 10^{-6}]^2} \right) [dB] = 58.2037[dB]$$

$$L_{p,500Hz} = 10 \log_{10} \left(\frac{p^2_{RMS,500Hz}}{P^2_r} \right) = 10 \log_{10} \left(\frac{\frac{0.127^2}{2}}{[20 \cdot 10^{-6}]^2} \right) [dB] = 73.0452[dB]$$

$$L_{p,750Hz} = L_{p,500Hz}$$

$$L_{p,1000Hz} = 10 \log_{10} \left(\frac{p^2_{RMS,1000Hz}}{P^2_r} \right) = 10 \log_{10} \left(\frac{\frac{0.283^2}{2}}{[20 \cdot 10^{-6}]^2} \right) [dB] = 80.0048[dB]$$

$$L_{p,tot} = 10 \log_{10} \left(10^{\frac{L_{p,250Hz}}{10}} + 10^{\frac{L_{p,500Hz}}{10}} + 10^{\frac{L_{p,750Hz}}{10}} + 10^{\frac{L_{p,1000Hz}}{10}} \right) [dB]$$

$$= 10 \log_{10} \left(10^{\frac{58.2037}{10}} + 10^{\frac{73.0452}{10}} + 10^{\frac{73.0452}{10}} + 10^{\frac{80.0048}{10}} \right) [dB]$$

$$= 81.495 [dB]$$

(c) Calculate the combined SPL from the SPL spectrum of the octave bands:

$$p^2_{RMS,b250Hz} = p^2_{RMS,250Hz} = \frac{0.023^2}{2} \ [Pa^2]$$

$$p^2_{RMS,b500Hz} = p^2_{RMS,500Hz} = \frac{0.127^2}{2} \ [Pa^2]$$

$$p^2_{RMS,b1000Hz} = p^2_{RMS,750Hz} + p^2_{RMS,1000Hz} = \frac{0.127^2}{2} + \frac{0.283^2}{2} \ [Pa^2]$$

$$L_{p,b250Hz} = 10 \log_{10} \left(\frac{p^2_{RMS,b250Hz}}{P^2_r} \right) = 10 \log_{10} \left(\frac{\frac{0.023^2}{2}}{[20 \cdot 10^{-6}]^2} \right) [dB]$$

$$= 58.2037 [dB]$$

$$L_{p,b500Hz} = 10 \log_{10} \left(\frac{p^2_{RMS,500Hz}}{P^2_r} \right) = 10 \log_{10} \left(\frac{\frac{0.127^2}{2}}{[20 \cdot 10^{-6}]^2} \right) [dB] = 73.0452 [dB]$$

$$L_{p,b1000Hz} = 10 \log_{10} \left(\frac{p^2_{RMS,750Hz} + P^2_{RMS,1000Hz}}{P^2_r} \right)$$

$$= 10 \log_{10} \left(\frac{\frac{0.127^2}{2} + \frac{0.283^2}{2}}{[20 \cdot 10^{-6}]^2} \right) [dB] = 80.8017 [dB]$$

$$L_{p,tot} = 10 \log_{10} \left(10^{\frac{L_{p,b250Hz}}{10}} + 10^{\frac{L_{p,b500Hz}}{10}} + 10^{\frac{L_{p,b1000Hz}}{10}} \right) [dB]$$

$$= 10 \log_{10} \left(10^{\frac{58.2037}{10}} + 10^{\frac{73.0452}{10}} + 10^{\frac{80.8017}{10}} \right) [dB]$$

$$= 81.495 [dB]$$

9.3.3 Preferred Speech Interference Level (PSIL)

Speech interference level (SIL) was established to determine the effect of steady background noise on speech communication in the work environment. With the acceptance of the new designation of the octave bands, the SIL was adapted to new octave bands and renamed preferred speech interference level (PSIL). PSIL is defined as the arithmetic average of SPL in the 500, 1000, and 2000 Hz octave bands [4,7]:

$$PSIL \equiv \frac{L_{p_{500}} + L_{p_{1000}} + L_{p_{2000}}}{3}$$

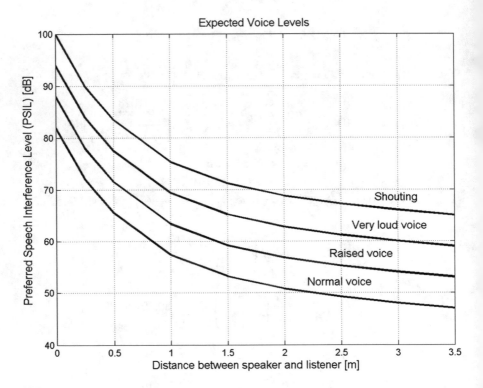

PSIL curves

9.4   Weighted Sound Pressure Level

Weightings were developed as a method to better subjectively evaluate the impact of noise on the human ear. The human ear is more sensitive to sounds at certain frequency ranges. Therefore, we apply weighting corresponding to sensitivity at different frequencies.

As mentioned in Sect. 9.2.5, weighting cannot be applied in SPL in the time domain. The weighted SPL can only be calculated in the frequency domain. The formula for calculating RMS pressure, which is used for calculating SPL, in the frequency domain is derived by Parseval's theorem as shown in the flowchart below. Parseval's theorem was proved in Sect. 9.2.5:

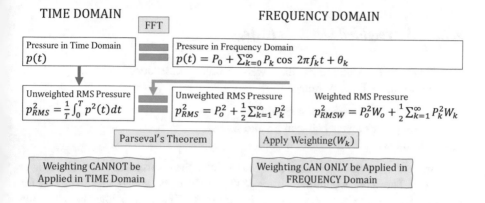

9.4.1   Logarithm of Weighting

A weighted RMS pressure is an RMS pressure multiplied by a weighting factor W as:

$$P_{RMS,w}^2 = P_{RMS}^2 \cdot W$$

Note that the weightings are applied to the square of RMS pressure.
Assume that an RMS pressure P_{RMS} has a sound pressure level of L_p as:

$$L_p = 10 \log_{10}\left(\frac{P_{RMS}^2}{P_r^2}\right)$$

The weighted sound pressure level $L_{p,\ w}$ of the weighted RMS pressure is:

$$L_{p,w} = 10 \log_{10}\left(\frac{P_{RMS}^2 \cdot W}{P_r^2}\right) = 10 \log_{10}\left(\frac{P_{RMS}^2}{P_r^2}\right) + 10 \log_{10}(W)$$
$$= L_p + 10 \log_{10}(W)$$

Based on the equations above, there are two ways to calculate the weighted sound pressure level $L_{p,\ w}$:

1. Applying the weighting W by multiplying $P_{RMS}^2 \cdot W$.
2. Applying the weighting W by adding $10\log_{10}(W)$

It is easier to calculate the weighted sound pressure level $L_{p,\ w}$ by adding $10\log_{10}(W)$ as a gain or loss of dB than by multiplying $P^2_{RMS} \cdot W$.

9.4.2 A-Weighted Decibels (dBA)

There are three weightings that were initially introduced for noise levels corresponding to different ranges.

A-weighting is for levels below 55 dB.
B-weighting is for levels between 55 and 85 dB.
C-weighting is for levels above 85 dB.

However, A-weighting is used today to evaluate the response of human hearing at all levels. The three weightings $10\log_{10}(W)$ in dB to be added as a gain/loss for different frequencies are shown in the figure below. The table below also gives the A-weighting for some frequencies.

When sound pressure levels are measured, their spectrum is applied by the weighting attenuation at different frequencies to generate weighted sound pressure levels. The resulting spectral levels can be added by the dB addition rule to find the total (combined) weighted sound pressure levels. Weighted SPLs have units in dB, with the weighting letter appended (i. e., dBA) to indicate the type of weighting. A-weighting can be applied to narrow or m-octave band spectrums and can be either an analog filter or, in the case of digital equipment, simply an attenuation applied to the calculated spectrum.

Example 9.7
Given the octave band spectrum below:

Center frequency $f_o[Hz]$	125	250	500	1000	2000
Sound pressure level $L_p[dB]$	50	58	76	80	46

(a) What is the PSIL and communication voice level at a distance of 1.2 m between the speaker and the listener?
(b) Determine the A-weighted spectrum.
(c) Calculate the A-weighted sound pressure levels.

Example 9.7: Solution
(a) Preferred speech interference level (PSIL) is calculated based on the three bands:

$$\text{PSIL} = \frac{L_{p_{500}} + L_{p_{1000}} + L_{p_{2000}}}{3} = \frac{76 + 80 + 46}{3} = 67.3\ [dB]$$

In PSIL curves of Sect. 9.3.3, based on PSIL=67.3 dB and Distance $= 1.2\ m.$, the voice level required is approximately "very loud voice"; see figure below:

Table 9.1 Sound level conversion chart from a flat response to A-weighting

1 Octave band	Frequency	A-weighting
[−]	[Hz]	[dB]
0	31.5	−39.5
1	63	−26.2
2	125	−16.2
3	250	−8.7
4	500	−3.2
5	1000	0
6	2000	1.2
7	4000	1.0
8	8000	−1.1
9	16,000	−6.7

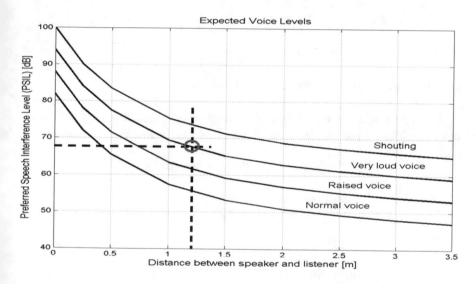

(b) A-weighted spectrum.

Get A-weighting from Table 9.1 (or Fig. 9.1) and calculate:

Center frequency $f_o[Hz]$	125	250	500	1000	2000	A-weighted
Sound pressure level $L_p[dB]$	50	58	76	80	46	
+ A-weight gain/lose [dB]	−16.2	−8.7	−3.2	0	1.2	
= A-weighted SPL $L_p[dBA]$	33.8	49.3	72.8	80	47.2	80.8
$10^{\frac{L_p}{10}} = \frac{p_{RMS}^2}{p_r^2}$	2399	85,114	1.9E7	10.0E7	52,481	11.9E7

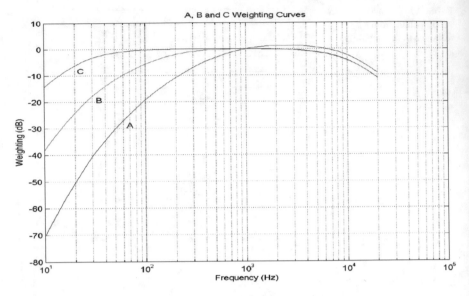

Fig. 9.1 Frequency response for the A-, B-, and C-weighting

A-weighted sound pressure level (spectrum) can be calculated by adding the unweighted sound pressure level and A-weighted gain or loss as shown below:

$$L_p[dBA] = L_p[dB] + \text{A} - \text{weighting}$$

Therefore, A-weighted spectrum is given by:
(33.8, 49.3, 72.8, 80, 47.2) *dB* at (125, 250, 500, 1000, 2000) *Hz*, respectively

(c) A-weighted SPL [*dBA*]:

$$L_{p_T}[dBA] = 10\log\left(10^{\frac{33.8}{10}} + 10^{\frac{49.3}{10}} + 10^{\frac{72.8}{10}} + 10^{\frac{80}{10}} + 10^{\frac{47.2}{10}}\right)$$
$$= 10\log\left(2399 + 85114 + 1.9\text{E}7 + 10.0\text{E}7 + 52481\right) = 10\log\left(11.9\text{E}7\right) = 80.8\ [dBA]$$

Therefore, the A-weighted SPL = 80.8 [*dBA*].

9.5   Homework Exercises

Exercise 9.1
The mathematical expression for a level is:

$$L\,[dB] = 10\log_{10}\left(\frac{U}{U_r}\right)$$

Use the law of conservation of energy to show that the following formula for the total (combined) level is valid:

$$L_{tot} = 10\log_{10}\left(10^{\frac{L_1\,[dB]}{10}} + 10^{\frac{L_2\,[dB]}{10}} + \ldots + 10^{\frac{L_n\,[dB]}{10}}\right)[dB]$$

Exercise 9.2

The sound pressure levels measured in the following five octave bands are:

Center frequency $f_o[Hz]$	250	500	1000	2000	4000
Sound pressure level $L_p[dB]$	90	72	82	65	50

(a) Determine the unweighted sound pressure level.
(b) Determine the A-weighted spectrum of the octave bands.
(c) Calculate the A-weighted sound pressure levels.
(d) What is the PSIL and communication voice level at 1.2 m between a speaker and a listener?

(Answers): (a) 90.7 [dB]; (b) (81.3, 68.8, 82.0, 66.2, 51.0) [dBA]; (c) 84.9 [dBA]; (d) 73 [dB], shouting.

Exercise 9.3

The sound pressure levels measured in five of the octave bands are given below:

Center frequency $f_o[Hz]$	125	250	500	1000	2000
Sound pressure level $L_p[dB]$	35	45	65	70	60

(a) Determine the unweighted sound pressure level.
(b) Determine the A-weighted spectrum of the octave bands.
(c) Calculate the A-weighted sound pressure levels.
(d) What is the PSIL and communication voice level at 1.5 m from a listener?

(Answers): (a) 71.5 [dB]; (b) (18.8, 36.3, 61.8, 70.0, 61.2) [dBA]; (b) 71.1 [dBA]; (c) 65 [dB], very loud voice.

Exercise 9.4

Given the time domain function $p(t)$ below:
$p(t) = 0.02\cos(2\pi \times 500 \times t) + 0.04\sin(2\pi \times 1000 \times t) +$
$0.06\cos(2\pi \times 1500 \times t) + 0.08\sin(2\pi \times 2000 \times t)[Pa]$

(a) Determine the total (combined) sound pressure level (SPL) using:

    (i) The summation of the squares of the RMS pressures.
    (ii) The SPLs of all individual frequencies.
    (iii) The SPL spectrum of octave bands.

(b) Determine the A-weighted sound pressure level (SPL) based on the sound level conversion table.

(Answers): (a) i-iii 71.76 [dB]; (b) 72.72 [dBA].

Exercise 9.5
Given the sound pressure function $p(t)$:

$$p(t) = 0.03 \cos (2\pi \times 400 \times t) + 0.04 \sin (2\pi \times 400 \times t)$$
$$+0.06 \cos (2\pi \times 1200 \times t) + 0.08 \sin (2\pi \times 1200 \times t)$$
$$+0.09 \cos (2\pi \times 1600 \times t) + 0.12 \sin (2\pi \times 1600 \times t)[Pa]$$

(a) Calculate the unweighted sound pressure level.
(b) Calculate the A-weighted sound pressure level.
(c) Calculate the preferred speech interference level (PSIL).
(d) Determine the communication voice level at 0.6 m from a listener.

    Hint: Combine the harmonic waves of the same frequencies into one harmonic wave function using the four equivalent forms.

(Answers): (a) 76.41[dB]; (b) 77.08[dBA]; (c) 70.14 [dB]; (d) raised voice.

Exercise 9.6
Given the sound pressure function $p(t)$:

$$p(t) = 0.05 \cos (800\pi t) + 0.10 \cos (2400\pi t) + 0.15 \cos (3200\pi t)[Pa]$$

(a) Calculate the unweighted sound pressure level.
(b) Calculate the A-weighted sound pressure level based on the sound level conversion table.
(c) Calculate the preferred speech interference level (PSIL).
(d) Determine the communication voice level at 0.6 m from a listener.

Answers): (a) 76.41[dB]; (b) 77.08[dBA]; (c) 70.14 [dB]; (d) raised voice.

Exercise 9.7

The sound pressure levels measured in five of the octave bands are given below:

Center frequency f_o[Hz]	125	250	500	1000	2000
Sound pressure level L_p[dB]	35	45	65	70	60

(a) Determine the unweighted total sound pressure level.
(b) Calculate the A-weighted sound pressure level.
(c) Calculate the preferred speech interference level (PSIL).
(d) Determine the communication voice level at 1 *m* from a listener.

(Answer): (a) 71.52 [*dB*]; (b) 71.08; (c) 65 [*dB*]; (d) raised voice.

Chapter 10
Room Acoustics and Acoustical Partitions

This chapter introduces room acoustics and acoustical partitions. Room acoustics studies the change of energy density at different locations due to the conditions of direct and reflected sound. Acoustical partitions are used to reduce noise from a source.

Formulas for evaluating transmission loss (TL) and noise reduction (NR) in a room due to partition walls are formulated in this chapter.

This chapter is organized into five sections as described below:

Section 10.1 introduces and defines three acoustic quantities: sound power (Sect. 10.1.1), acoustic intensity (Sect. 10.1.2), and energy density (Sect. 10.1.3). These acoustic quantities are used for deriving the formulas of transmission loss (TL) and noise reduction (NR). These three acoustic quantities are also commonly used in the analysis of room acoustics.

Section 10.2 introduces and defines three room quantities: absorption coefficient of surfaces (Sect. 10.2.1), room constant (Sect. 10.2.2), and reverberation time (Sect. 10.2.3). These room quantities are used in the analysis of room acoustics in the next section (Sect. 10.3).

Section 10.3 focuses on the analysis of room acoustics. A formula for calculating sound pressure level due to both direct and reverberant sound waves from a sound source is summarized below:

$$L_P = L_w + 10 \, \log \left(\frac{Q}{4\pi r^2} + \frac{4}{R} \right)$$

where L_P is the sound pressure level at a distance r from the acoustic source considering both the direct and reverberant sound waves, L_w is the sound power of the acoustic source, Q is the directivity factor of the acoustic, and R is room constant. The equation above will be derived in this section.

Section 10.4 introduces transmission loss (TL) due to acoustical partitions. A formula that relates the transmission loss (TL) to the incident sound pressure level (L_{wi}) and transmitted sound pressure level (L_{wt}) will be defined and is summarized below. Also, another formula for calculating the same transmission loss (TL) through a partition wall will be derived and is summarized as follows:

$$\text{TL} \equiv L_{wi} - L_{wt} = 10 \, \log \left(\frac{1}{\tau} \right)$$

$$\tau \equiv \frac{w_t}{w_i}$$

where τ is the transmission coefficient, w_i is the incident sound power, and w_t is the transmitted sound power.

Section 10.5 introduces noise reduction (NR) due to acoustical partitions. A formula that relates the noise reduction (NR) to the sound pressure level near the wall in Room 1 (L_{p_1}) and sound pressure level near the wall in Room 2 (L_{p_2}) will be defined and is summarized below. Also, a formula for calculating sound pressure level due to a sound source through a partition wall will be derived and is summarized as follows:

$$\text{NR} = L_{p_1} - L_{p_2}$$

$$L_P = L_{p,wall} + 10 \, \log \left(\frac{1}{4} + \frac{S_{wall}}{R} \right)$$

where $L_{p, wall}$ is the sound pressure level (SPL) near the wall without considering the reflection wave and L_P is the SPL near the wall considering both the direct and reflected waves:

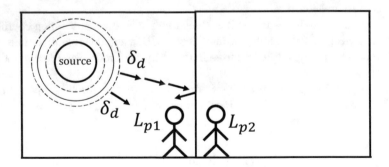

10.1   Sound Power, Acoustic Intensity, and Energy Density

10.1.1   Definition of Sound Power

The sound power w is defined as the energy per unit time and can be formulated in terms of the RMS pressure P_{RMS}, the RMS velocity V_{RMS}, and an area S as:

$$w = P_{RMS} V_{RMS} S$$

And the units in the MKS system are:

$$w \left[\frac{J}{s} \right] = P \left[\frac{N}{m^2} \right] \times V \left[\frac{m}{s} \right] \times S \left[m^2 \right]$$

For plane waves moving at the speed of sound c, because $V_{RMS} = \frac{P_{RMS}}{\rho_o c}$, the sound power is:

$$w = \frac{P_{RMS}^2}{\rho_o c} S \ (plane \ waves)$$

Remarks

The sound power of an acoustic source is the energy of an enclosed surface of the acoustic source.

10.1.2   Definition of Acoustic Intensity

The acoustic intensity I is defined as the energy per unit time per unit area or (by the definition of sound power) the sound power w per unit area S as:

$$I \equiv \frac{w}{S}$$

Based on the definition of sound power w, acoustic intensity I can be formulated in terms of the RMS pressure P_{RMS} and the RMS velocity V_{RMS} as:

$$I = \frac{w}{S} = \frac{P_{RMS} V_{RMS} S}{S} = P_{RMS} V_{RMS}$$

And the units in the MKS system are:

$$I \left[\frac{J}{sm^2} \right] = P \left[\frac{N}{m^2} \right] \times V \left[\frac{m}{s} \right]$$

For plane waves moving at the speed of sound c, because $V_{RMS} = \frac{P_{RMS}}{\rho_o c}$ as shown in Chap. 4, the acoustic intensity is:

$$I = \frac{P_{RMS}^2}{\rho_o c} \ (plane \ waves)$$

10.1.3   Definition of Energy Density

The energy density δ is defined as the energy per unit volume and can be formulated in terms of sound power w (energy per unit time) as:

$$\delta \equiv \frac{w}{cS}$$

Based on the definition of sound power w, energy density δ can be formulated in terms of the RMS pressure P_{RMS} and the RMS velocity V_{RMS} as:

$$\delta = \frac{w}{cS} = \frac{P_{RMS} V_{RMS} S}{cS} = P_{RMS} V_{RMS} \frac{1}{c}$$

And the units in the MKS system are:

$$\delta \left[\frac{J}{m^3}\right] = P\left[\frac{N}{m^2}\right] \times V\left[\frac{m}{s}\right] \times \frac{1}{c}\left[\frac{s}{m}\right]$$

For plane waves moving at the speed of sound c, because $V_{RMS} = \frac{P_{RMS}}{\rho_o c}$ as shown in Chap. 4, the acoustic intensity is:

$$\delta = \frac{P^2_{RMS}}{\rho_o c^2} \ (plane \ waves)$$

Note that the acoustic intensity is useful for deriving formulas for room acoustics because it is defined as the energy per unit volume and can be formulated based on the conservation of energy.

10.2   Absorption Coefficients, Room Constant, and Reverberation Time

10.2.1   Absorption Coefficient of Surface

The statistical absorption coefficient or random incidence sound absorption coefficient, α, is frequency-dependent and is defined as the ratio of acoustic energy absorbed by a surface to the acoustical energy incident upon the surface when the incident sound field is perfectly diffused, $\alpha \equiv \frac{w_i - w_r}{w_i}$.

The absorption coefficient of a material is a number between 0 and 1 that indicates the proportion of sound which is absorbed by the surface compared to the incident sound. A large, fully open window would offer no reflection, as any sound reaching it would pass straight out and no sound would be reflected. This would have an absorption coefficient of 1. Conversely, a thick, smooth painted concrete ceiling would be the acoustic equivalent of a mirror and have an absorption coefficient very close to 0.

Sound absorption coefficients of common materials used in buildings are presented in the table below.

Table: Absorption coefficients of general building materials

Materials	Descriptions	Octave band center frequency [Hz]					
		125	250	500	1000	2000	4000
Brick	Smooth, painted or glazed	0.01	0.01	0.02	0.02	0.02	0.03
Brick	Course surface	0.03	0.03	0.03	0.04	0.05	0.07
Concrete	Smooth, painted, or glazed	0.01	0.01	0.02	0.02	0.02	0.02
Concrete	Smooth, unpainted	0.01	0.01	0.02	0.02	0.03	0.05
Concrete	Coarse	0.01	0.02	0.04	0.06	0.08	0.10
Marble	Floor	0.01	0.01	0.01	0.02	0.02	0.02
Wood	Hardwood floor	0.15	0.10	0.06	0.08	0.10	0.10
Wood	Thin plywood	0.40	0.20	0.10	0.08	0.06	0.06
Carpet	Thin carpet, on concrete	0.10	0.15	0.25	0.30	0.33	0.30
Carpet	Thin carpet, on foam rubber	0.20	0.25	0.30	0.30	0.33	0.30
Carpet	Heavy carpet, on concrete	0.02	0.06	0.14	0.37	0.60	0.65
Carpet	Heavy carpet, on foam rubber	0.08	0.24	0.57	0.69	0.71	0.73
Glass	Normal window	0.35	0.25	0.18	0.12	0.07	0.04
Glass	Large window	0.30	0.10	0.05	0.12	0.07	0.04
Water	Swimming pool surface	0.01	0.01	0.01	0.01	0.02	0.03

Since different surfaces of a room will, in general, have different absorption coefficients, an average sound absorption coefficient $\bar{\alpha}$ need be calculated for a given room:

$$\bar{\alpha} = \frac{\sum_{i=1}^{n} S_i \alpha_i}{\sum_{i=1}^{n} S_i}$$

where:

S_i = area of the i-th surface.
α_i = absorption coefficient at the i-th surface.

Excess Absorption Coefficient Due to Air

For large rooms and frequencies above 2000 Hz, air absorption (not including surface absorption) may become significant. The absorption due to air in the room is represented by the excess absorption coefficient and is given by:

$$\bar{\alpha}_{ex} = kD = k\frac{4V}{S} = k\left[\frac{1}{Length}\right]\frac{4V}{S}\left[\frac{Volume}{Area}\right]$$

$k=$ experimentally determined coefficient of air (see table below)
$D = \frac{4V}{S}$ (distance between ceiling and floor)
$V=$ volume of the room
$S=$ total surface area

where D is the average traveling distance between ceiling and floor. The formula D is based on a half cube of 10 [m] by 10 [m] by 5 [m] as shown below:

The distance between ceiling and floor   Half Cube

$$D = 5 = \frac{4 \times 5 \times 10 \times 10}{2 \times 10 \times 10 + 4 \times 5 \times 10} = \frac{4V}{S}$$

where
$V = 10 \times 10 \times 5 = 500$
$S = 10 \times 10 \times 2 + 10 \times 5 \times 4$
$\quad = 400$

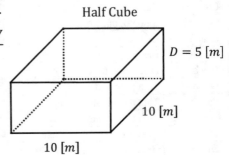

The total average absorption coefficient then becomes:

$$\bar{\alpha}' = \bar{\alpha}_{tot} = \bar{\alpha} + \bar{\alpha}_{ex}$$

Approximate values for the coefficient k in metric units [3]

		Frequency [Hz]		
		2000	4000	8000
Relative humidity [%]	Temperature [C]	k [1/m]	k [1/m]	k [1/m]
30	20	0.0030	0.0095	0.0340
30	25	0.0029	0.0078	
30	30	0.0028	0.0070	
50	20	0.0024	0.0061	0.0215
50	25	0.0024	0.0059	
50	30	0.0023	0.0058	
70	20	0.0021	0.0053	0.0150
70	25	0.0021	0.0053	
70	30	0.0021	0.0052	

10.2.2   *Room Constant*

The absorption coefficient is defined as the ratio of acoustic energy absorbed by a surface to the acoustical energy incident upon the surface when the incident sound field is perfectly diffused:

$$\alpha \equiv \frac{w_i - w_r}{w_i}$$

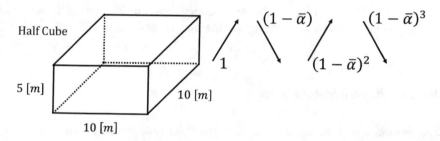

The summation of all the reflected ratios of a sound, except the direct wave, is equal to $\frac{(1-\bar{\alpha})}{\bar{\alpha}}$ as:

$$(1-\bar{\alpha}) + (1-\bar{\alpha})^2 + \ldots + (1-\bar{\alpha})^n = \frac{(1-\bar{\alpha})}{\bar{\alpha}}; n = \infty$$

The following is the proof of the formula above of the summation of the infinite series.

Let:

$$\frac{S}{R} = (1-\bar{\alpha}) + (1-\bar{\alpha})^2 + \ldots + (1-\bar{\alpha})^n$$

Multiplying both sides of the above equation by $(1-\bar{\alpha})$ yields:

$$\frac{S}{R}(1-\bar{\alpha}) = (1-\bar{\alpha})^2 + (1-\bar{\alpha})^3 + \ldots + (1-\bar{\alpha})^{n+1}$$

Subtract these two equations to arrive at:

$$\frac{S}{R}\bar{\alpha} = (1-\bar{\alpha}) - (1-\bar{\alpha})^{n+1}$$

For $n = \infty$ and $(1-\bar{\alpha}) < 1$, $(1-\bar{\alpha})^{n+1} = 0$, we obtain:

$$\frac{S}{R} = \frac{(1 - \overline{\alpha})}{\overline{\alpha}} = (1 - \overline{\alpha}) + (1 - \overline{\alpha})^2 + \ldots + (1 - \overline{\alpha})^n$$

The variable R in the equation above is known as the room constant. The room constant R is related to the total area of a room S and the total reverberant ratio and is defined as:

$$R = \frac{\overline{\alpha}}{1 - \overline{\alpha}} S \left[m^2 \right]$$

where the factor $\frac{S}{R} = \frac{(1 - \overline{\alpha})}{\overline{\alpha}}$ is the summation of the infinite series: $(1 - \overline{\alpha}), (1 - \overline{\alpha})^2, (1 - \overline{\alpha})^3$, etc. The series components are the ratios of the reflected energies to the initial incident energy for the first, second, third, etc. reflections.

10.2.3   Reverberation Time

The reverberation time T_{60} is the time required for reflections of a direct sound to decay to 60 dB:

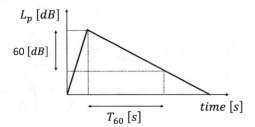

When the sound pressure level L_P reaches the peak:

$$L_{P,T0} = 10 \, \log_{10} \left(\frac{w_{T0}}{w_r} \right)$$

After the sound pressure level L_P decreases 60 dB from the peak:

$$L_{P,T60} = 10 \, \log_{10} \left(\frac{w_{T60}}{w_r} \right)$$

Therefore:

$$L_{P,T0} - L_{P,T60} = 60 \, [dB]$$

$$\rightarrow 60 = 10 \, \log_{10} \left(\frac{w_{T0}}{w_r} \right) - 10 \, \log_{10} \left(\frac{w_{T60}}{w_r} \right)$$

$$\rightarrow \log_{10} \left(\frac{w_{T0}}{w_{T60}} \right) = 6$$

$$\rightarrow \frac{w_{T60}}{w_{T0}} = 10^{-6}$$

For each reflection from the boundary surface, the ratio of the reflected energy to the incident energy is $(1 - \bar{\alpha})$, and the average travel time between reflections is approximately $\frac{4V}{Sc}$ (base on a half cube):

$$D = 5 = \frac{4 \times 5 \times 10 \times 10}{2 \times 10 \times 10 + 4 \times 5 \times 10} = \frac{4V}{S}$$

Half Cube

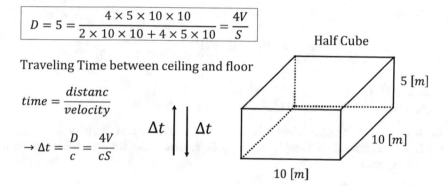

Traveling Time between ceiling and floor

$$time = \frac{distanc}{velocity}$$

$$\rightarrow \Delta t = \frac{D}{c} = \frac{4V}{cS}$$

$\Delta t \quad \Delta t$

Assume that the number of reflections required for the reflected energy to decay to 60 dB is $T_{60} / \left(\frac{4V}{Sc} \right)$.

Therefore:

$$(1 - \bar{\alpha})^{\frac{Sc}{4V} T_{60}} = 10^{-6}$$

Taking the natural logarithm of the above equation yields:

$$\frac{Sc}{4V} T_{60} \ln (1 - \bar{\alpha}) = -6 \ln (10)$$

or (use $c = 343 \, m/s$):

$$T_{60} = \frac{-24V \, \ln (10)}{Sc \, \ln (1 - \bar{\alpha})} = \frac{-0.161V}{S \, \ln (1 - \bar{\alpha})} \quad [\text{arbitary } \bar{\alpha} < 1]$$

The equation above is the general formula for calculating T_{60} for an arbitrary absorption coefficient $\bar{\alpha}$.

Special Case: Small Absorption Coefficient

For a small $\bar{\alpha}$ ($\bar{\alpha} < 0.2$), the formula above can be simplified as:

$$(1 - \bar{\alpha})^{\frac{-1}{\bar{\alpha}}} \cong e$$

Hence:

$$(1 - \bar{\alpha})^{\frac{Sc}{4V}T_{60}} = \left((1 - \bar{\alpha})^{\frac{-1}{\bar{\alpha}}}\right)^{-\bar{\alpha}\frac{Sc}{4V}T_{60}} \cong e^{-\bar{\alpha}\frac{Sc}{4V}T_{60}} = 10^{-6}$$

Again, taking the natural logarithm of the above equation yields:

$$-\bar{\alpha}\frac{Sc}{4V}T_{60} = -6 \ln(10)$$

Hence:

$$T_{60} = \frac{24V \ln(10)}{Sc\,\bar{\alpha}} = \frac{0.161V}{S\,\bar{\alpha}} \text{ [for } \bar{\alpha} < 0.2 \text{ only]}$$

Example 10.1

A (half cube) room has the dimensions $10 \times 10 \times 5\ m$.

(a) If the absorption coefficient is $\bar{\alpha} = 0.5$, determine the reverberation time T_{60}.
(b) If the reverberation time is $T_{60} = 0.32\ s$, determine the average absorption coefficient $\bar{\alpha}$.

Example 10.1: Solution

(a) For the reverberation time:

$$V = 10^3/2 = 500\ m^3$$

$$S = 10^2 \times 4 = 400\ m^2\ (4V/S = 10/2 = 5\ m)$$

$$T_{60} = \frac{-0.161V}{S \ln(1 - \bar{\alpha})} = \frac{-0.161 \times 500}{400 \ln(1 - 0.5)} = 0.29\ s$$

(b) For the average absorption coefficient:

$$V = 10^3/2 = 500\ m^3$$

$$S = 10^2 \times 4 = 400\ m^2\ (4V/S = 10/2 = 5\ m)$$

$$\bar{\alpha} = 1 - e^{\frac{-0.161V}{ST_{60}}} = 1 - e^{\frac{-0.161 \times 500}{400 \times 0.32}} = 0.467$$

10.3 Room Acoustics

10.3.1 *Energy Density due to an Acoustic Source*

The total energy density δ is the sum of the direct energy δ_d and the reverberant energy density δ_r as:

$$\delta = \delta_d + \delta_r$$

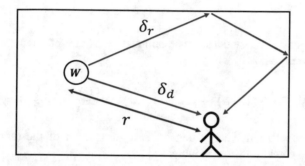

The energy density δ is related to the sound power w by an area S as shown in Sect. 10.1.3 as:

$$\delta = \frac{w}{cS}$$

Direct Energy Density

The direct energy density due to a sound source of acoustic power w can be derived from the equation above by replacing S with $4\pi r^2$ as:

$$\delta_d = \frac{w}{Sc} = \frac{1}{4\pi r^2} \frac{Q_d w}{c} \left[\frac{J}{m^3}\right]$$

[*spherical source*]

where S is the area of the sphere at a distance r from the center of the spherical source as:

$$S = 4\pi r^2$$

and Q_d is the directivity factor defined as:

$$Q_d = \left.\frac{P_d^2}{\overline{P}^2}\right|_r \ [unitless]$$

where:

P_d = RMS pressure in a particular direction at a distance r from the source

$\overline{P}$ = average RMS pressure over a sphere at a distance r from the source

The directivity factor Q_d for an isotropic acoustic source is:

$Q_d = 1$

[isotropic source]

In general, acoustic sources are not isotropic, and the direct energy density must be modified according to the directivity of the source.

Reverberant Energy Density

The energy density δ is related to the sound power w by an area S as:

$$\delta = \frac{w}{cS}$$

The reverberant energy density is a summation of the reflected energy between two parallel walls as:

$$(1 - \overline{\alpha}) + (1 - \overline{\alpha})^2 + \ldots + (1 - \overline{\alpha})^n = \frac{(1 - \overline{\alpha})}{\overline{\alpha}} = \frac{S}{R}; n = \infty$$

Also, the reverberant energy density is a summation of the reflected energy. Therefore, the sound power w is increased by:

$$\delta_r = \frac{w}{cS} = \frac{Q_r w}{cS} \times \left[\frac{(1 - \overline{\alpha})}{\overline{\alpha}}\right] = \frac{4w}{cS} \times \left(\frac{S}{R}\right) = \frac{4w}{cR}$$

where S is the total area of the room and R is the room constant considering the reverberant effect and is related to the area S as shown in Sect. 10.2.2 as:

$$R = \frac{\overline{\alpha}}{1 - \overline{\alpha}} S$$

And Q_r is the reflectivity factor equals:

$$Q_r = 4$$

Total Energy Density

The total energy density δ is the sum of the direct and reverberant energy densities:

$$\delta = \delta_d + \delta_r$$

$$= \frac{wQ}{4\pi r^2 c} + \frac{4w}{cR}$$

10.3.2 Sound Pressure Level due to an Acoustic Source

Equating the energy density expression above to the energy density from the sound pressure $\delta = \frac{P_{RMS}^2}{\rho_o c^2}$ (see Sect. 12.1.3) yields a relationship between the RMS pressure in a room and the acoustic source, as shown below:

$$\delta = \delta_d + \delta_r$$

$$\rightarrow \frac{P_{RMS}^2}{\rho_o c^2} = \frac{wQ}{4\pi r^2 c} + \frac{4w}{cR}$$

$$\rightarrow P_{RMS}^2 = w\rho_o c\left(\frac{Q}{4\pi r^2} + \frac{4}{R}\right)$$

If we divide both sides of the square RMS pressure expression by the square of the international pressure reference $P_r = 20 \times 10^{-6}$ [Pa], we obtain:

$$\frac{P_{RMS}^2}{(20 \times 10^{-6})^2} = \frac{w\rho_o c}{(20 \times 10^{-6})^2}\left(\frac{Q}{4\pi r^2} + \frac{4}{R}\right)$$

$$\rightarrow \frac{P_{RMS}^2}{(20 \times 10^{-6})^2} = \frac{w}{10^{-12}} \frac{\rho_o c}{400}\left(\frac{Q}{4\pi r^2} + \frac{4}{R}\right)$$

Since $\rho_o c \cong 400$:

$$\frac{P_{RMS}^2}{(20 \times 10^{-6})^2} \cong \frac{w}{10^{-12}}\left(\frac{Q}{4\pi r^2} + \frac{4}{R}\right)$$

Take log base 10 of both sides of the above equation to arrive at:

$$10 \log\left(\frac{P_{RMS}^2}{(20 \times 10^{-6})^2}\right) = 10 \log\left(\frac{w}{10^{-12}}\right) + 10 \log\left(\frac{Q}{4\pi r^2} + \frac{4}{R}\right)$$

Because the $10^{-12}\left[\frac{J}{s}\right]$ in the equation above is the international sound power reference, therefore, the equation above can be formulated in the form of levels as:

$$L_P = L_w + 10 \log\left(\frac{Q}{4\pi r^2} + \frac{4}{R}\right)$$

where L_P is the sound pressure level at a distance r from the acoustic source considering both the direct and reverberant sound waves, L_w is the sound power of the acoustic source, Q is the directivity factor of the acoustic source and is equal to one for isotropic acoustic sources, and R is room constant that is related to the area and the absorption coefficient of the surface.

10.4   Transmission Loss due to Acoustical Partitions

10.4.1   Transmission Coefficient

The transmission coefficient τ is defined as the ratio of the transmitted acoustic power from a wall to the incident acoustic power on the other side of the wall:

$$\tau \equiv \frac{w_t}{w_i}$$

where:
    w_i = incident sound power
    w_t = transmitted sound power

$$w_i = \text{incident sound power}$$

$$w_t = \text{transmitted sound power}$$

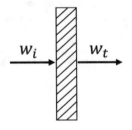

Average Transmission Coefficient of Mixed Wall
For a mixed wall containing different elements (i.e., doors, windows, etc.), the average transmission coefficient ($\bar{\tau}$) is calculated as:

$$\bar{\tau} = \frac{\sum_{i=1}^{n} S_i \tau_i}{\sum_{i=1}^{n} S_i} = \frac{1}{S} \sum_{i=1}^{n} S_i \tau_i$$

where:
S_i = surface area of each element
τ_i = transmission coefficient of each element
S = total surface area of the wall

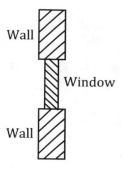

10.4.2 Transmission Loss (TL)

Transmission loss is defined as the difference between the incident sound power level L_{wi} and the transmitted sound power level L_{wt} as:

$$TL \equiv L_{wi} - L_{wt}$$

Based on the definition of sound power level, the transmission loss becomes:

$$TL \equiv L_{wi} - L_{wt} = 10 \log\left(\frac{w_i}{w_r}\right) - 10 \log\left(\frac{w_t}{w_r}\right)$$

$$= 10 \log\left(\frac{\frac{w_i}{w_r}}{\frac{w_t}{w_r}}\right) = 10 \log\left(\frac{w_i}{w_t}\right)$$

where w_r is the international power reference.

Because the transmission coefficient τ is defined as the ratio of the transmitted acoustic power to the incident acoustic power as:

$$\tau \equiv \frac{w_t}{w_i}$$

Therefore, the transmission loss can be formulated as:

$$TL = 10 \log\left(\frac{w_i}{w_t}\right) = 10 \log\left(\frac{1}{\tau}\right)$$

On the other hand, because the sound power w is linearly related to the square of the RMS pressure P_{RMS} as discussed in Sect. 10.1.3 for plane waves as:

$$w = \frac{P_{RMS}^2}{\rho_o c} S \ [plane \ waves]$$

The transmission loss can also be formulated in terms of the RMS pressure P_{RMS} as:

$$TL = 10 \ \log \left(\frac{w_i}{w_t}\right) = 10 \ \log \left(\frac{p_{RMS,i}^2}{p_{RMS,t}^2}\right) = 10 \ \log \left(\frac{\frac{p_{RMS,i}^2}{P_r^2}}{\frac{p_{RMS,t}^2}{P_r^2}}\right)$$

$$= 10 \ \log \left(\frac{p_{RMS,i}^2}{P_r^2}\right) - 10 \ \log \left(\frac{p_{RMS,t}^2}{P_r^2}\right) = L_{pi} - L_{pt}$$

The definition and formulas of the transmission loss (TL) can be summarized as:

$$TL \equiv L_{wi} - L_{wt} = L_{pi} - L_{pt} = 10 \ \log \left(\frac{1}{\tau}\right)$$

10.5   Noise Reduction due to Acoustical Partitions

10.5.1   Energy Density due to a Partition Wall

The total energy density δ is the sum of the direct energy δ_d and the reverberant energy density δ_r as:

$$\delta = \delta_d + \delta_r$$

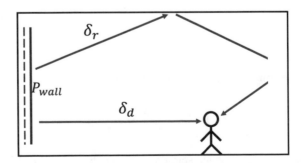

The energy density δ is related to the sound power w by an area S as shown in Sect. 10.1.3 as:

$$\delta = \frac{w}{cS}$$

Direct Energy Density

The formula for direct energy density due to plane waves was derived in Sect. 10.1.3 and is shown below for reference:

$$\delta_d = \frac{w}{Sc} = \frac{P_{RMS}^2}{\rho_o c^2} \ [plane\ waves]$$

where P_{RMS} is the RMS pressure in the room radiated from the four walls.

Because P_{wall} (RMS pressure on the wall) contributes to only part of P_{RMS} (RMS pressure on the room), P_{RMS} can be related to P_{wall} with a ratio of $\frac{S_{wall}}{S}$ as:

$$P_{RMS}^2 = P_{wall}^2 \frac{S_{wall}}{S}$$

Therefore, the direct energy density δ_d can be formulated in terms of P_{wall} as:

$$\delta_d = \frac{P_{wall}^2}{\rho_o c^2} \frac{S_{wall}}{S}$$

Reverberant Energy Density

The reverberant energy density is a summation of all the reflected energy from the direct energy density:

$$\delta_r = \delta_d \left[(1 - \bar{\alpha}) + (1 - \bar{\alpha})^2 + \ldots + (1 - \bar{\alpha})^n \right] = \delta_d \frac{S}{R}$$

where $\frac{S}{R}$ was derived in the previous section as:

$$\frac{S}{R} = \frac{(1 - \bar{\alpha})}{\bar{\alpha}} = (1 - \bar{\alpha}) + (1 - \bar{\alpha})^2 + \ldots + (1 - \bar{\alpha})^n$$

Therefore:

$$\delta_r = \delta_d \frac{S}{R}$$

Total Energy Density

The total energy density δ is the sum of the direct and reverberant energy densities:

$$\delta = \delta_d + \delta_r$$

$$= \delta_d + \delta_d \frac{S}{R}$$

$$= \delta_d \left(1 + \frac{S}{R}\right)$$

Because the direct energy density δ_d can be formulated in terms of P_{wall} as:

$$\delta_d = \frac{P_{wall}^2}{\rho_o c^2} \frac{S_{wall}}{S}$$

therefore, the total energy density δ can be formulated in terms of P_{wall} as:

$$\delta = \frac{P_{wall}^2}{\rho_o c^2} \frac{S_{wall}}{S} \left(1 + \frac{S}{R}\right)$$

10.5.2   Sound Pressure Level due to a Partition Wall

Equating the energy density expression above to the energy density from the sound pressure $\delta = \frac{P_{RMS}^2}{\rho_o c^2}$ (see Sect. 10.1.3) yields a relationship between the RMS pressure in a room and the acoustic source, as shown below:

$$\delta = \delta_d \left(1 + \frac{S}{R}\right)$$

$$\rightarrow \frac{P_{RMS}^2}{\rho_o c^2} = \frac{P_{wall}^2}{\rho_o c^2} \frac{S_{wall}}{S} \left(1 + \frac{S}{R}\right)$$

$$\rightarrow \frac{P_{RMS}^2}{\rho_o c^2} = \frac{P_{wall}^2}{\rho_o c^2} \left(\frac{S_{wall}}{S} + \frac{S_{wall}}{R}\right)$$

$$\rightarrow P_{RMS}^2 = P_{wall}^2 \left(\frac{S_{wall}}{S} + \frac{S_{wall}}{R}\right)$$

The ratio between the wall area S_{wall} and the room surface S can be approximated as:

$$\frac{S_{wall}}{S} = \frac{1}{4}$$

Then:

$$P_{RMS}^2 = P_{wall}^2 \left(\frac{1}{4} + \frac{S_{wall}}{R} \right)$$

Divide both sides by the square of the international pressure reference $P_r = 20 \times 10^{-6}$ [Pa], and take log base 10 of both sides of the above equation as:

$$10 \log \left(\frac{P_{RMS}^2}{P_r^2} \right) = 10 \log \left[\frac{P_{wall}^2}{P_r^2} \left(\frac{1}{4} + \frac{S_{wall}}{R} \right) \right]$$

$$\rightarrow L_P = L_{p,wall} + 10 \log \left(\frac{1}{4} + \frac{S_{wall}}{R} \right)$$

where $L_{p,\,wall}$ is the sound pressure level (SPL) near the wall without considering the reflection wave and L_P is the SPL near the wall considering both the direct and reflected waves. Note that the first term $\frac{1}{4}$ in the logarithmic function above is the result of the direct wave and the second term $\frac{S_{wall}}{R}$ is the result of the reflection wave.

Sound Pressure Level Away from the Wall
The sound pressure level (SPL) located far from the wall is formulated based on the above formulas for SPL located near the wall. At locations far from the partition wall, considering only the reflection wave from the above equation, we obtain:

$$P_{RMS}^2 = P_{wall}^2 \left(0 + \frac{S_{wall}}{R} \right)$$

Divide both sides by the square of the international pressure reference $P_r = 20 \times 10^{-6}$ [Pa], and take log base 10 of both sides of the above equation to arrive at

$$L_P = L_{p,wall} + 10 \log \left(\frac{S_{wall}}{R} \right)$$

10.5.3 Noise Reduction (NR)

The noise reduction of the enclosure is equal to the TL of the walls of the enclosure:

$$NR = L_{p_1} - L_{p_2}$$

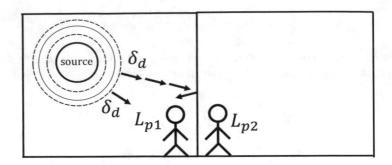

Example 10.2: Common Wall

Two rooms are separated by a common wall that has the dimensions $8 \times 5\ m^2$ with a TL of 30 dB. Room 1 contains a noise source that produces an SPL of 108 dB near the wall. The average absorption coefficient of room 2 is 0.4, and the total surface area of room 2 is 225 m^2.

(a) What is the SPL near the wall in the second room?
(b) What is the SPL far from the wall in the second room?

Example 10.2: Solution

(a) The SPL near the wall in the second room is:

$$\bar{\alpha}_2 = 0.4$$

$$S_w = 8 \times 5 = 40\ m^2$$

$$R_2 = \frac{\bar{\alpha}_2 \times S_2}{1 - \bar{\alpha}_2} = \frac{0.4 \times 225}{0.6} = 150$$

$$\text{TL} = 30\ dB$$

$$L_{p_1} = 108\ dB$$

$$L_{p_2} = L_{p,wall\ 1} + 10\ \log\left(\frac{1}{4} + \frac{S_w}{R_2}\right) = L_{p_1} - \text{TL} + 10\log\left(\frac{1}{4} + \frac{S_w}{R_2}\right)$$

$$L_{p_2} = 108 - 30 + 10\log\left(\frac{1}{4} + \frac{40}{150}\right) = 108 - 30 - 2.9 = 75.1\ dB$$

(a) The SPL far from the wall in the second room is:

$$L'_{p_2} = L_{p1} - \text{TL} + 10\log\left(\frac{S_w}{R_2}\right)$$

$$L'_{p_2} = 108 - 30 - 5.7 = 72.3 \ dB$$

Example 10.3

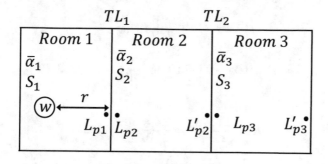

Given:

$$Q = 1 \qquad w = 2 \ watts \qquad r = 2 \ m$$
$$\bar{\alpha}_1 = 0.2 \qquad S_1 = 100 \ m^2 \qquad S_{w1} = S_{w2} = 10 \ m^2$$
$$TL_1 = 25 \ dB \qquad TL_2 = 20 \ dB \qquad \bar{\alpha}_2 = 0.2$$
$$S_2 = 200 \ m^2 \qquad \bar{\alpha}_3 = 0.4 \qquad S_3 = 400 \ m^2$$

Determine:

(a) L_{p_1}, (b) L_{p_2}, (c) L'_{p_2}, (d) L_{p_3}, (e) L'_{p_3}, and (f) the noise reduction $(NR = L_{p_1} - L'_{p_3})$ through the rooms

Example 10.3: Solution

$$L_w = 10 \log w + 120 = 10 \log 2 + 120 = 123 \ dB$$

$$R_1 = \frac{\bar{\alpha}_1 \times S_1}{1 - \bar{\alpha}_1} = \frac{0.2 \times 100}{1 - 0.2} = 25$$

$$L_{p_1} = L_w + 10 \log \left(\frac{Q}{4\pi r^2} + \frac{4}{R_1} \right)$$

$$= 123 + 10 \log \left(\frac{1}{4\pi 2^2} + \frac{4}{25} \right) = 115.5605 \ dB$$

$$R_2 = \frac{\bar{\alpha}_2 \times S_2}{1 - \bar{\alpha}_2} = \frac{0.2 \times 200}{1 - 0.2} = 50$$

$$L_{p_2} = L_{p,wall\ 1} + 10\ \log\left(\frac{1}{4} + \frac{S_{w1}}{R_2}\right) = L_{p1} - TL_1 + 10\log\left(\frac{1}{4} + \frac{S_{w1}}{R_2}\right)$$

$$= 115.5605 - 25 + 10\log\left(\frac{1}{4} + \frac{10}{50}\right) = 87.0926\ dB$$

$$L'_{p_2} = L_{p1} - TL_1 + 10\log\left(\frac{S_{w1}}{R_2}\right)$$

$$= 115.5605 - 25 + 10\log\left(\frac{10}{50}\right) = 83.5708\ dB$$

$$R_3 = \frac{\bar{\alpha}_3 \times S_3}{1 - \bar{\alpha}_3} = \frac{0.4 \times 400}{0.6} = 266.7$$

$$L_{p_3} = L_{p,wall\ 2} + 10\ \log\left(\frac{1}{4} + \frac{S_{w2}}{R_3}\right) = L'_{p_2} - TL_2 + 10\log\left(\frac{1}{4} + \frac{S_{w2}}{R_3}\right)$$

$$= 83.5708 - 20 + 10\log\left(\frac{1}{4} + \frac{10}{266.7}\right) = 58.1572\ dB$$

$$L'_{p_3} = L'_{p_2} - TL_2 + 10\log\left(\frac{S_{w2}}{R_3}\right)$$

$$= 83.5708 - 20 + 10\ \log\left(\frac{10}{266.6}\right) = 49.3111\ dB$$

$$NR = L_{p_1} - L'_{p_3} = 115.5605 - 49.3111 = 66.2494\ dB > 25 + 20\ dB$$

Example 10.4
Given:

$$L_{p_1} = 100\ dB$$

$$\tau_1 = 0.01, \tau_2 = 0.01$$

$$\bar{\alpha}_{1,2} = 0.6, \bar{\alpha}_2 = 0.4$$

$$S_w = 10\ m^2, S_2 = 200\ m^2$$

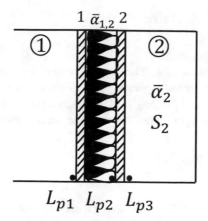

Determine the noise reduction of a double-leaf wall with sound absorption materials between the leaves.

Example 10.4: Solution

$$L_{p_1} = 100 \; dB$$

$$\tau_1 = 0.01$$

$$\bar{\alpha}_{1,2} = 0.6$$

$$S_w = 10 \; m^2$$

$$R_{1,2} = \frac{2S_w\bar{\alpha}_{1,2}}{1 - \bar{\alpha}_{1,2}} = \frac{20 \times 0.6}{0.4} = 30 \; m^2$$

$$L_{p_2} = 100 - 10\log\frac{1}{0.01} + 10\log\left(\frac{1}{4} + \frac{10}{30}\right) = 100 - 20 - 2.3 = 77.7 \; d\dot{B}$$

$$\tau_2 = 0.01$$

$$\bar{\alpha}_2 = 0.4$$

$$S_2 = 200 \; m^2$$

$$R_2 = \frac{S_2\bar{\alpha}_2}{1 - \bar{\alpha}_2} = \frac{200 \times 0.4}{0.6} = 133.3 \; m^2$$

$$L_{p_3} = 77.7 - 20 + 10\log\left(\frac{1}{4} + \frac{10}{133.3}\right) = 77.7 - 20 - 4.9 = 52.8 \; dB$$

10.6   Homework Exercises

Exercise 10.1: Partition Wall
Two rooms are separated by a wall having a transmission coefficient equal to 0.063.
If the SPL in Room 1 near the separating wall is 102 *dB*, what would the SPL be in
the second room near the separating wall and away from it? The wall surface area is
15 m^2, the average absorption coefficient of the second room is 0.4, and the total
surface area is 200 m^2.

(Answers): 85.6 [*dB*]; 80.5 [*dB*]

Exercise 10.2
Two rooms are separated with a wall having a transmission coefficient equal to 0.06.
The separating wall surface area is 25 m^2. The total surface area of the second room
is 250 m^2, and the average absorption coefficient of the second room is 0.3. If the
SPL in Room 1 near the separating wall is 100 *dB*, determine:

(a) The SPL in the second room near the separating wall
(b) The SPL in the second room away from the separating wall

(Answers): (a) 84.8 [*dB*]; (b) 81.7 [*dB*]

Exercise 10.3
The total surface areas of Room 1 and Room 2 are 150 m^2 and 200 m^2, respectively.
The two rooms are separated by a wall with a surface area of 15 m^2 and having a
transmission coefficient of 0.01. The average absorption coefficients of Room 1 and
Room 2 are 0.4 and 0.3, respectively. If Room 1 contains a 0.1-watt isotropic noise
source 2 meters away from the wall:

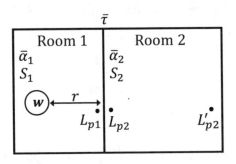

(a) What is the sound pressure level (L_{p_1}) in Room 1 near the separating wall?
(b) What is the sound pressure level (L_{p_2}') in Room 2 away from the separating wall?

(Answers): (a) 97.7739 *dB*; (b) 70.2042 *dB*

Exercise 10.4

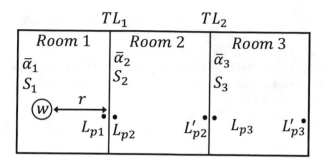

Given:

$Q = 0.9$ $w = 0.01\ watts$ $r = 3\ m$
$\bar{\alpha}_1 = 0.2$ $S_1 = 100\ m^2$ $S_{w1} = S_{w2} = 10\ m^2$
$\bar{\tau}_1 = 0.001$ $\bar{\tau}_2 = 0.010$ $\bar{\alpha}_2 = 0.2$
$S_2 = 200\ m^2$ $\bar{\alpha}_3 = 0.4$ $S_3 = 400\ m^2$

Determine:

(a) L_{p_1}
(b) L'_{p_2}
(c) L'_{p_3}
(d) The noise reduction (NR $= L_{p_1} - L'_{p_3}$) through the rooms

(Answers): (a) 92.6452 [dB]; (b) 55.6555 [dB]; (c) 21.3958 [dB]; (d) 71.2494 [dB]

Exercise 10.5: Mixed Partition Wall
A large room is divided into two rooms by a mixed partition wall. Room 1 contains a 2-watt isotropic noise source 2 meters away from the mixed wall:

	Room 1		Room 2		Mixed wall	
	Walls and ceiling	Floor	Walls and ceiling	Floor	Wall	Window
Avg. absorption coefficient	0.25	0.1	0.45	0.3	0.25	0.1
Surface area (m^2)	200	50	300	75	10	3
Transmission loss (dB)	–		–		20	10

(a) Calculate the room constant R of Rooms 1 and 2. The mixed wall is also a part of these rooms.
(b) Calculate the transmission loss of the mixed partition wall.
(c) What is the sound pressure level in the second room near the partition and away from it?

(Answers): (a) 74.2 $[m^2]$, 269.6 $[m^2]$; (b) 15.2 $[dB]$; (c) 91.1 $[dB]$, 83.2 $[dB]$

Exercise 10.6: Double-Leaf Partition

A double-leaf partition divides a room into two smaller rooms. Room 1 contains a 2-watt isotropic noise source 2 meters away from the partition. Using the data provided in the table below:

	Room 1	Room 2	Partitionleaf 1	Partition leaf 2	Inside partition
Avg. absorption coefficient	0.15	0.45	–	–	0.6
Surface area (m^2)	200	300	10	10	20
Transmission loss (dB)	–	–	20	15	–

(a) Calculate the noise reduction of the double-leaf partition.
(b) Calculate the insertion loss of the double-leaf partition.

Assume that:

1. When the dividing partition is not present, the average absorption coefficient of the surfaces, which previously belonged to Rooms 1 and 2, remains the same as listed in the table.
2. The total surface area of the new larger room is equal to the sum of the surface areas of Rooms 1 and 2.
3. The location for the sound pressure estimate necessary for insertion loss calculations, without the partition in place, is also 2 m away from the noise source (i.e., double-leaf partition thickness is negligible).
4. The acoustic pressure inside the partition cavity is uniform (i.e., same everywhere):

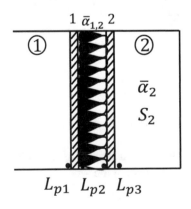

(Answers): (a) $NR = 42.7$ $[dB]$; (b) $IL = 37.1$ $[dB]$.

Chapter 11
Power Transmission in Pipelines

This chapter and the next chapter introduce both analytical and numerical methods for filter designs. This chapter will focus on developing formulas of pressure and acoustic impedance in pipelines. These formulas are the foundation for analyzing sound waves in pipes. The next chapter will extend pipelines to pipes with side branches which can be used to design any complex pipe system.

Pipelines are pipes connected in series without any side branches. Pipes with side branches will be discussed in the next chapter. The complex amplitudes, acoustic impedance, balancing equations, and power transmission in pipelines will be covered in this chapter and are organized into the following sections:

Section 11.1 defines complex amplitudes of pressure. Four formulas (Formulas 2A–2D) for transferring complex amplitudes of pressure from one end of a pipe to the other end are introduced.

Section 11.2 defines acoustic impedance in a pipe considering both the forward and backward waves. The acoustic impedance is defined by the complex amplitude of pressure introduced in Section 11.1. Two formulas (Formulas 3A–3B) for calculating acoustic impedance from complex amplitudes of pressure are introduced.

Section 11.3 introduces two equations for (1) balancing pressure and (2) conservation of mass at the intersection between two pipes. Four formulas (Formulas 4A–4D) for transferring complex amplitudes of pressure between two adjacent pipes are introduced.

Section 11.4 introduces a formula (Formula 5) for transferring acoustic impedances between two adjacent pipes.

Section 11.5 demonstrates two techniques for the frequency domain analysis of pipelines. The first technique is for calculating the equivalent acoustic impedance of pipelines (based on Formula 5). The second technique is for transferring the complex amplitudes of pressure in pipelines (based on Formulas 4A–4D). Example 11.4 shows the procedures for calculating power transmission coefficient of pipelines using a numerical approach. This example uses the computer program in Sect. 11.9.

Section 11.6 demonstrates a numerical method for modeling pipelines with three pipes in series. This method uses three MATLAB functions provided in the

Computer Program Section in Sect. 11.7. This method will plot the power transmission coefficient vs. the frequency of the pipeline.

Section 11.7 shows three MATLAB functions that can be used for the numerical analysis of pipelines described in the previous section (Sect. 11.6). These functions are based on Formulas 1, 4, 5, and 6 derived in this chapter.

Section 11.8 includes a numerical project to calculate and plot the power transmission coefficient vs. the frequency of a pipeline with an arbitrary number of pipes in series. The procedures for this project are similar to the procedures described in the Numerical Method Section (Sect. 11.6). The solutions are not provided.

11.1 Complex Amplitude of Pressure and Acoustic Impedance

We will define the complex amplitude of pressure in this section.

11.1.1 Definition of Complex Amplitude of Pressure

Complex amplitudes of sound pressure are used for frequency domain analysis of pipes. For frequency domain analysis of pipes, all time-related parts must be eliminated. Therefore, complex amplitudes must have the following two properties:

- Complex amplitudes of pressure are time-independent.
- Complex amplitudes of pressure are location-dependent.

Based on these two properties, the complex amplitude P_{fx_o} of pressure $p_f(x, t)$ of a forward wave can be demonstrated as:

$$
\begin{aligned}
p_f(x, t) &= \frac{1}{2}\left(A_+ e^{j(\omega t - kx)} + cc\right) \\
&= \frac{1}{2}\left(A_+ e^{-jkx_o} e^{-jk(x-x_o)} e^{j\omega t} + cc\right) \\
&= \frac{1}{2}\left(P_{fx_o} e^{-jk(x-x_o)} e^{j\omega t} + cc\right)
\end{aligned}
$$

where complex amplitude P_{fx_o} of pressure $p_f(x, t)$ at $x = x_o$ is defined as:

$$
P_{fx_o} = A_+ e^{-jkx_o}
$$

Note that:

- The time-dependent term $e^{j\omega t}$ is not included in the complex amplitude of pressure P_{fx_o}.

- The complex amplitude of pressure P_{fx_o} is the amplitude of pressure at location $x = x_o$ where $e^{-jk(x-x_o)} = 1$.

We can reverse the procedures above to obtain the pressure $p_f(x, t)$ for a given complex amplitude P_{fx_o} at $x = x_o$. It is important to indicate the location of the complex amplitude of pressure when handling any complex amplitudes.

11.1.2 Definition of Acoustic Impedance

Three types of impedances are defined below and will be explained in the following sections:

Three Types of Impedances

Three types of impedances are defined below and will be explained in the following sections:

Specific acoustic impedance:

$$z_s \equiv \frac{\text{pressure}}{\text{velocity}} = \frac{P}{V} (\equiv z)$$

Acoustic impedance:

$$Z_a \equiv \frac{\text{pressure}}{\text{volume flow rate}} = \frac{P}{V}\frac{1}{S} = \frac{P}{U} (\equiv Z)$$

Mechanical impedance:

$$Z_m \equiv \frac{\text{force}}{\text{velocity}} = \frac{P}{V}\frac{S}{1}$$

In the impedances above, P is the complex amplitude of pressure, V is the complex amplitude of velocity, S is the cross-section area of a pipe, and U is defined as the complex amplitude of velocity V multiplied by the cross-section area S.

Complex Acoustic Impedance

Based on the definition of acoustic impedance shown above, acoustic impedance is a complex number because both pressure and velocity are formulated in complex.

Plane waves in a pipe usually result from the addition of incident waves (forward) and reflected waves (backward). The combined acoustic impedance Z_c of the combined (forward and backward waves) is a complex number and can be formulated as shown below:

$$Z_c \equiv \frac{P_f + P_b}{U_f + U_b} \qquad (f \text{ forward wave}, b \text{ backward wave})$$

where P_f, P_b, U_f, and U_b are the complex amplitudes of pressures and volume flow rates.

Therefore, the complex acoustic impedance has the same two properties of the complex amplitude of pressure as shown in Sect. 11.1 as:

- Complex acoustic impedance are time-independent.

  - That is, the time-dependent term is not included in the complex acoustic impedance.

- Complex acoustic impedance are location-dependent.

  - That is, the complex acoustic impedance is the acoustic impedance at the evaluated location.

Acoustic Impedances of One-Way Plane Waves (Real Numbers)

In general, combined acoustic impedance Z_c of forward and backward waves is a complex number.

For one-way plane waves such as forward plane waves and backward plane waves, the acoustic impedances Z_f and Z_b are always real numbers. The acoustic impedance Z_c of the combined (forward and backward waves) is a complex number.

Acoustic impedances for both traveling waves and combined waves can be formulated according to the pressure-velocity relationship as shown in Chapter 4 as:

$$u_{\pm} = \pm \frac{1}{\rho_o c} p_{\pm}(x, t)$$

Formulas 1A–1E

Acoustic impedances for traveling plane waves and combined plane waves can be formulated according to the pressure-velocity relationship above. Formulas 1A to 1C of acoustic impedances of forward plane traveling waves, backward plane traveling waves, and the combined plane waves are listed below:

$$Z_f \equiv \frac{P_f}{U_f} = \frac{P_f}{SV_f} = \frac{\rho_o c}{S} \qquad \text{(Formula 1A)}$$

$$Z_b \equiv \frac{P_b}{U_b} \equiv \frac{P_b}{SV_b} = \frac{-\rho_o c}{S} = -Z_f \qquad \text{(Formula 1B)}$$

$$Z_c \equiv \frac{P_c}{U_c} \equiv \frac{P_f + P_b}{U_f + U_b} = \frac{P_f + P_b}{\frac{P_f}{Z_f} + \frac{P_b}{Z_b}} = \frac{P_f + P_b}{\frac{P_f}{Z_f} - \frac{P_b}{Z_f}} = Z_f \frac{P_f + P_b}{P_f - P_b} \qquad \text{(Formula 1C)}$$

where:

$$P_c \equiv P_f + P_b \qquad \text{(Formula 1D)}$$

$$U_c \equiv U_f + U_b \qquad \text{(Formula 1E)}$$

Proof of Formula 1C

Based on Formulas 1D and 1E, the acoustic impedance Z_{c2R} at the RHS of Pipe 2 (as an example) is defined as:

$$Z_{c2R} \equiv \frac{P_{c2R}}{U_{c2R}} = \frac{P_{f2R} + P_{b2R}}{U_{f2R} + U_{b2R}}$$

Based on Formulas 1A and 1B, the equation above becomes:

$$Z_{c2R} = \frac{P_{f2R} + P_{b2R}}{U_{f2R} + U_{b2R}} = \frac{P_{f2R} + P_{b2R}}{\frac{P_{f2R}}{Z_{f2}} + \frac{P_{b2R}}{Z_{b2}}} = \frac{P_{f2R} + P_{b2R}}{\frac{P_{f2R}}{Z_{f2}} - \frac{P_{b2R}}{Z_{f2}}} = Z_{f2} \frac{P_{f2R} + P_{b2R}}{P_{f2R} - P_{b2R}}$$

11.1.3   Transfer Pressure

Formulas 2A–2G

The complex amplitude of the pressure P_{f2L} of the forward wave $p_{f2}(x, t)$ at $x = x_1$ (LHS of Pipe 2) is related to the complex amplitude of the pressure P_{f2R} of the forward wave $p_{f2}(x, t)$ at $x = x_2$ (RHS of Pipe 2) in Formula 2A below. The complex amplitude of the pressure P_{b2L} of the backward wave $p_{b2}(x, t)$ at $x = x_1$ (LHS of Pipe 2) is related to the complex amplitude of the pressure P_{b2R} of the backward wave $p_{b2}(x, t)$ at $x = x_2$ (RHS of Pipe 2) in Formula 2B:

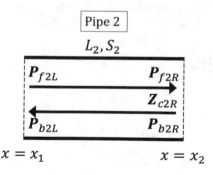

$$P_{f2L} = P_{f2R} e^{jkL_2} \tag{Formula 2A}$$

$$P_{b2L} = P_{b2R} e^{-jkL_2} \tag{Formula 2B}$$

$$P_{f2R} = P_{f2L} e^{-jkL_2} \tag{Formula 2C}$$

$$P_{b2R} = P_{b2L} e^{jkL_2} \tag{Formula 2D}$$

$$\begin{Bmatrix} \boldsymbol{P}_{f2L} \\ \boldsymbol{P}_{b2L} \end{Bmatrix} = \frac{1}{e^{jkL_2} - e^{-jkL_2}} \begin{bmatrix} e^{jkL_2} & -1 \\ -e^{-jkL_2} & 1 \end{bmatrix} \begin{Bmatrix} \boldsymbol{P}_{c2L} \\ \boldsymbol{P}_{c2R} \end{Bmatrix} \qquad \text{(Formula 2E)}$$

$$\begin{Bmatrix} \boldsymbol{P}_{f2R} \\ \boldsymbol{P}_{b2R} \end{Bmatrix} = \frac{1}{e^{jkL_2} - e^{-jkL_2}} \begin{bmatrix} 1 & -e^{-jkL_2} \\ -1 & e^{jkL_2} \end{bmatrix} \begin{Bmatrix} \boldsymbol{P}_{c2L} \\ \boldsymbol{P}_{c2R} \end{Bmatrix} \qquad \text{(Formula 2F)}$$

$$\begin{Bmatrix} -\boldsymbol{U}_{c2L} \\ \boldsymbol{U}_{c2R} \end{Bmatrix} = \frac{j}{\boldsymbol{Z}_{f2}\sin(kL_2)} \begin{bmatrix} \cos(kL_2) & -1 \\ -1 & \cos(kL_2) \end{bmatrix} \begin{Bmatrix} \boldsymbol{P}_{c2L} \\ \boldsymbol{P}_{c2R} \end{Bmatrix} \qquad \text{(Formula 2G)}$$

Proof of Formulas 2A and 2C

In Pipe 2 as shown in the figure above, the pressure $p_{f2}(x, t)$ of a forward wave and the pressure $p_{b2}(x, t)$ of a backward wave are formulated in terms of complex amplitudes of the pressures $\boldsymbol{P}_{f2L}$ and $\boldsymbol{P}_{b2L}$ at $x = x_1$ (LHS of Pipe 2) as:

$$p_{f2}(x, t) = \frac{1}{2}\left\{ \boldsymbol{P}_{f2L}e^{-jk(x-x_1)}e^{j\omega t} + \boldsymbol{cc} \right\}$$

$$p_{b2}(x, t) = \frac{1}{2}\left\{ \boldsymbol{P}_{b2L}e^{jk(x-x_1)}e^{j\omega t} + \boldsymbol{cc} \right\}$$

The pressure $p_{f2}(x, t)$ of the forward wave as a function of space and time in terms of the complex amplitude of pressure $\boldsymbol{P}_{f2L}$ at $x = x_1$ (LHS of Pipe 2) is:

$$p_{f2}(x, t) = \frac{1}{2}\left\{ \boldsymbol{P}_{f2L}e^{-jk(x-x_1)}e^{j\omega t} + \boldsymbol{cc} \right\}$$

The pressure $p_{f2}(x, t)$ of the forward wave as a function of space and time in terms of the complex amplitude of pressure $\boldsymbol{P}_{f2R}$ at $x = x_2$ (RHS of Pipe 2) is:

$$p_{f2}(x, t) = \frac{1}{2}\left\{ \boldsymbol{P}_{f2R}e^{-jk(x-x_2)}e^{j\omega t} + \boldsymbol{cc} \right\}$$

Comparing the two identical equations above yields:

$$\boldsymbol{P}_{f2L}e^{-jk(x-x_1)} = \boldsymbol{P}_{f2R}e^{-jk(x-x_2)}$$

$$\rightarrow \boldsymbol{P}_{f2L} = \boldsymbol{P}_{f2R}e^{jk(x_2-x_1)} = \boldsymbol{P}_{f2R}e^{jkL_2}$$

$$\rightarrow \boldsymbol{P}_{f2R} = \boldsymbol{P}_{f2L}e^{-jk(x_2-x_1)} = \boldsymbol{P}_{f2L}e^{-jkL_2}$$

Proof of Formulas 2E and 2F

Based on Formula 1D, $\boldsymbol{P}_{c2L} = \boldsymbol{P}_{f2L} + \boldsymbol{P}_{b2L}$ and $\boldsymbol{P}_{c2R} = \boldsymbol{P}_{f2R} + \boldsymbol{P}_{b2R}$, and using Formula 2C–2D or 2A–2B gets:

$$\begin{Bmatrix} P_{c2L} \\ P_{c2R} \end{Bmatrix} = \begin{bmatrix} 1 & 1 \\ e^{-jkL_2} & e^{jkL_2} \end{bmatrix} \begin{Bmatrix} P_{f2L} \\ P_{b2L} \end{Bmatrix} \text{ and } \begin{Bmatrix} P_{c2L} \\ P_{c2R} \end{Bmatrix} = \begin{bmatrix} e^{jkL_2} & e^{-jkL_2} \\ 1 & 1 \end{bmatrix} \begin{Bmatrix} P_{f2R} \\ P_{b2R} \end{Bmatrix}$$

Inverse the 2x2 matrix to get:

$$\begin{Bmatrix} P_{f2L} \\ P_{b2L} \end{Bmatrix} = \frac{1}{e^{jkL_2} - e^{-jkL_2}} \begin{bmatrix} e^{jkL_2} & -1 \\ -e^{-jkL_2} & 1 \end{bmatrix} \begin{Bmatrix} P_{c2L} \\ P_{c2R} \end{Bmatrix}$$

$$\begin{Bmatrix} P_{f2R} \\ P_{b2R} \end{Bmatrix} = \frac{1}{e^{jkL_2} - e^{-jkL_2}} \begin{bmatrix} 1 & -e^{-jkL_2} \\ -1 & e^{jkL_2} \end{bmatrix} \begin{Bmatrix} P_{c2L} \\ P_{c2R} \end{Bmatrix}$$

Proof of Formula 2G

Based on Formulas 1E and 1A–1B, $U_{c2L} = U_{f2L} + U_{b2L} = (P_{f2L} - P_{b2L})/Z_{f2}$ and $U_{c2R} = U_{f2R} + U_{b2R} = (P_{f2R} - P_{b2R})/Z_{f2}$, and using Formula 2A–2B gets:

$$\begin{Bmatrix} -U_{c2L} \\ U_{c2R} \end{Bmatrix} = \frac{1}{Z_{f2}} \begin{bmatrix} -e^{jkL_2} & e^{-jkL_2} \\ 1 & -1 \end{bmatrix} \begin{Bmatrix} P_{f2R} \\ P_{b2R} \end{Bmatrix}$$

$$= \frac{1}{Z_{f2}} \frac{1}{e^{jkL_2} - e^{-jkL_2}} \begin{bmatrix} -e^{jkL_2} & e^{-jkL_2} \\ 1 & -1 \end{bmatrix} \begin{bmatrix} 1 & -e^{-jkL_2} \\ -1 & e^{jkL_2} \end{bmatrix} \begin{Bmatrix} P_{c2L} \\ P_{c2R} \end{Bmatrix}$$

$$= \frac{1}{Z_{f2}} \frac{-1}{e^{jkL_2} - e^{-jkL_2}} \begin{bmatrix} e^{jkL_2} + e^{-jkL_2} & -2 \\ -2 & e^{jkL_2} + e^{-jkL_2} \end{bmatrix} \begin{Bmatrix} P_{c2L} \\ P_{c2R} \end{Bmatrix}$$

Using Euler's formula, the equation above arrives at:

$$\begin{Bmatrix} -U_{c2L} \\ U_{c2R} \end{Bmatrix} = \frac{j}{Z_{f2}\sin(kL_2)} \begin{bmatrix} \cos(kL_2) & -1 \\ -1 & \cos(kL_2) \end{bmatrix} \begin{Bmatrix} P_{c2L} \\ P_{c2R} \end{Bmatrix}$$

Proof of Formulas 2B and 2D
(Homework Exercise 11.1 Part a)

11.2 Complex Acoustic Impedance

Formulas 3A–3B

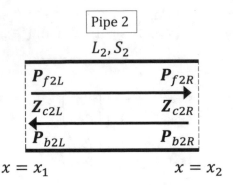

The acoustic impedance Z_{c2L} at LHS of Pipe 2 is related to the acoustic impedance Z_{f2} as shown in Formula 3A. The acoustic impedance Z_{c2R} at RHS of Pipe 2 is
related to the acoustic impedance Z_{f2} as shown in Formula 3B:

$$Z_{c2L} = Z_{f2}\frac{P_{f2L} + P_{b2L}}{P_{f2L} - P_{b2L}} = Z_{f2}\frac{P_{f2R}e^{jkL_2} + P_{b2R}e^{-jkL_2}}{P_{f2R}e^{jkL_2} - P_{b2R}e^{-jkL_2}} \qquad \text{(Formula 3A)}$$

$$Z_{c2R} = Z_{f2}\frac{P_{f2R} + P_{b2R}}{P_{f2R} - P_{b2R}} = Z_{f2}\frac{P_{f2L}e^{-jkL_2} + P_{b2L}e^{jkL_2}}{P_{f2L}e^{-jkL_2} - P_{b2L}e^{jkL_2}} \qquad \text{(Formula 3B)}$$

where $Z_{f2} = \frac{\rho_{\varrho}c}{S_2}$ is a real number.

Proof of Formula 3B
In Pipe 2 as shown in the figure above, the pressure $p_{f2}(x, t)$ of a forward wave and
the pressure $p_{b2}(x, t)$ of a backward wave are formulated in terms of complex
amplitudes of the pressures P_{f2L} and P_{b2L} at LHS of Pipe 2 ($x = x_1$) as:

$$p_{f2}(x, t) = \frac{1}{2}\left\{P_{f2L}e^{-jk(x-x_1)}e^{j\omega t} + cc\right\}$$

$$p_{b2}(x, t) = \frac{1}{2}\left\{P_{b2L}e^{jk(x-x_1)}e^{j\omega t} + cc\right\}$$

Based on Formula 3B, the acoustic impedance Z_{c2R} at the RHS of Pipe 2 is:

$$Z_{c2R} = Z_{f2}\frac{P_{f2R} + P_{b2R}}{P_{f2R} - P_{b2R}} \qquad \text{(Formula 3B)}$$

Based on Formulas 2C and 2D, the pressures P_{f2R} and P_{b2R} at RHS of Pipe 2 are
related to the pressures P_{f2L} and P_{b2L} at LHS of Pipe 2 as:

$$P_{f2R} = P_{f2L}e^{-jkL_2} \qquad \text{(Formula 2C)}$$

$$P_{b2R} = P_{b2L}e^{jkL_2} \qquad \text{(Formula 2D)}$$

Substituting Formulas 2C and 2D to Formula 3B yields:

$$Z_{c2R} = Z_{f2}\frac{P_{f2R} + P_{b2R}}{P_{f2R} - P_{b2R}} = Z_{f2}\frac{P_{f2L}e^{-jkL_2} + P_{b2L}e^{jkL_2}}{P_{f2L}e^{-jkL_2} - P_{b2L}e^{jkL_2}} \qquad \text{(Formula 3B)}$$

where:

$$Z_{f2} = \frac{\rho_o c}{S_2}$$

Proof of Formula 3A
(Homework Exercise 11.1 Part b)

11.3 Balancing Pressure and Conservation of Mass

The cross-sectional areas of Pipe 1 and Pipe 2 are S_1 and S_2, respectively, as shown in the figure below. The pressures of the forward wave and the backward wave in Pipe 1 at RHS of Pipe 1 ($x = x_2$) are P_{f1R} and P_{b1R}, respectively. The pressures of the forward wave and backward wave in Pipe 2 at LHS of Pipe 2 ($x = x_2$) are P_{f2L} and P_{b2L}, respectively:

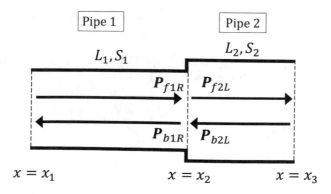

The following conditions must be satisfied at the intersection between the two pipes:

I. All pressures are balanced at the intersection according to Newton's third law of motion:

$$P_{f1R} + P_{b1R} = P_{f2L} + P_{b2L} \qquad \text{(state of equilibrium)}$$

II. The volume flow rate ($U = VS$) inward is equal to the volume flow rate outward according the conservation of mass, assuming that the air has a constant density:

$$U_{f1R} + U_{b1R} = U_{f2L} + U_{b2L} \qquad \text{(conservation of mass)}$$

11.4 Transformation of Pressures

Formulas 4A–4D
The cross-sectional areas of Pipe 1 and Pipe 2 are S_1 and S_2, respectively, as shown in the figure below. The pressures of the forward wave and the backward wave in Pipe 1 at RHS of Pipe 1 ($x = x_2$) are P_{f1R} and P_{b1R}, respectively. The pressures of the forward wave and the backward wave in Pipe 2 at LHS of Pipe 2 ($x = x_2$) are P_{f2L} and P_{b2L}, respectively:

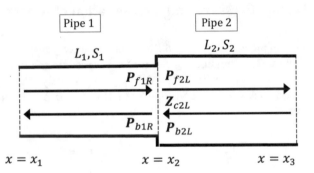

The pressures P_{f1R} and P_{b1R} (RHS of Pipe 1) can be formulated in terms of the pressures P_{f2L} and P_{b2L} (LHS in Pipe 2) as:

$$P_{f1R} = \frac{1}{2}\left(1 + \frac{Z_{f1}}{Z_{f2}}\right)P_{f2L} + \frac{1}{2}\left(1 - \frac{Z_{f1}}{Z_{f2}}\right)P_{b2L} \qquad \text{(Formula 4A')}$$

$$P_{b1R} = \frac{1}{2}\left(1 - \frac{Z_{f1}}{Z_{f2}}\right)P_{f2L} + \frac{1}{2}\left(1 + \frac{Z_{f1}}{Z_{f2}}\right)P_{b2L} \qquad \text{(Formula 4B')}$$

$$P_{f1R} = \frac{1}{2}\left(1 + \frac{Z_{f1}}{Z_{c2L}}\right)P_{c2L} \qquad \text{(Formula 4A)}$$

$$P_{b1R} = \frac{1}{2}\left(1 - \frac{Z_{f1}}{Z_{c2L}}\right)P_{c2L} \qquad \text{(Formula 4B)}$$

$$\frac{P_{b1R}}{P_{f1R}} = \frac{Z_{c2L} - Z_{f1}}{Z_{c2L} + Z_{f1}} \qquad \text{(Formula 4C)}$$

$$\frac{P_{c2L}}{P_{f1R}} = \frac{2Z_{c2L}}{Z_{c2L} + Z_{f1}} \qquad \text{(Formula 4D)}$$

where:

$$P_{c2L} = P_{f2L} + P_{b2L} \qquad \text{(Formula 1D)}$$

$$Z_{c2L} = Z_{f2} \frac{P_{f2L} + P_{b2L}}{P_{f2L} - P_{b2L}} \qquad \text{(Formula 3A)}$$

$$Z_{f2} = \frac{\rho_o c}{S_2}$$

Proof of Formulas 4A–4D

Based on Newton's third law of motion (state of equilibrium) and conservation of mass:

$$P_{f1R} + P_{b1R} = P_{f2L} + P_{b2L} \qquad \text{(state of equilibrium)} \qquad (1)$$

$$U_{f1R} + U_{b1R} = U_{f2L} + U_{f2L} \qquad \text{(conservation of mass)} \qquad (2)$$

Based on Formulas 1A and 1B, changing the volume flow rate U to P/Z and changing Z_{b1} and Z_{b2} to $-Z_{f1}$ and $-Z_{f2}$, respectively, yield:

$$\frac{P_{f1R}}{Z_{f1}} + \frac{P_{b1R}}{Z_{b1}} = \frac{P_{f2L}}{Z_{f2}} + \frac{P_{b2L}}{Z_{b2}}$$

$$\rightarrow \frac{P_{f1R}}{Z_{f1}} - \frac{P_{b1R}}{Z_{f1}} = \frac{P_{f2L}}{Z_{f2}} - \frac{P_{b2L}}{Z_{f2}}$$

$$\rightarrow P_{f1R} - P_{b1R} = \frac{Z_{f1}}{Z_{f2}} \left(P_{f2L} - P_{b2L} \right) \qquad (3)$$

Solve P_{f1R} and P_{b1R} from Eq.(1) and Eq.(3) to get:

$$P_{f1R} = \frac{1}{2} \left(1 + \frac{Z_{f1}}{Z_{f2}} \right) P_{f2L} + \frac{1}{2} \left(1 - \frac{Z_{f1}}{Z_{f2}} \right) P_{b2L} \qquad \text{(Formula 4A')}$$

$$P_{b1R} = \frac{1}{2} \left(1 - \frac{Z_{f1}}{Z_{f2}} \right) P_{f2L} + \frac{1}{2} \left(1 + \frac{Z_{f1}}{Z_{f2}} \right) P_{b2L} \qquad \text{(Formula 4B')}$$

Based on Formula 1D, $P_{c2L} = P_{f2L} + P_{b2L}$, and on Formula, 3A, $Z_{c2L} = Z_{f2} \frac{P_{f2L} + P_{b2L}}{P_{f2L} - P_{b2L}}$, rearrange Eq.(3) to get:

$$\rightarrow P_{f1R} - P_{b1R} = \frac{Z_{f1}}{Z_{f2} \frac{P_{f2L} + P_{b2L}}{P_{f2L} - P_{b2L}}} \left(P_{f2L} + P_{b2L} \right)$$

$$\rightarrow P_{f1R} - P_{b1R} = \frac{Z_{f1}}{Z_{c2L}} P_{c2L} \qquad (4)$$

Solve P_{f1R} and P_{b1R} from Eq.(1) and Eq.(4) to get:

$$P_{f1R} = \frac{1}{2}\left(1 + \frac{Z_{f1}}{Z_{c2L}}\right)P_{c2L} \qquad \text{(Formula 4A)}$$

$$P_{b1R} = \frac{1}{2}\left(1 - \frac{Z_{f1}}{Z_{c2L}}\right)P_{c2L} \qquad \text{(Formula 4B)}$$

$$\frac{P_{b1R}}{P_{f1R}} = \frac{Z_{c2L} - Z_{f1}}{Z_{c2L} + Z_{f1}} \qquad \text{(Formula 4C)}$$

$$\frac{P_{c2L}}{P_{f1R}} = \frac{2Z_{c2L}}{Z_{c2L} + Z_{f1}} \qquad \text{(Formula 4D)}$$

where:

$$P_{c2L} = P_{f2L} + P_{b2L} \qquad \text{(Formula 1D)}$$

$$Z_{c2L} = Z_{f2}\,\frac{P_{f2L} + P_{b2L}}{P_{f2L} - P_{b2L}} \qquad \text{(Formula 3A)}$$

$$Z_{f2} = \frac{\rho_o c}{S_2}$$

Note that when there is only an outward wave in Pipe 2 ($P_{b2L} = 0$), $P_{c2L} = P_{f2L}$ and $Z_{c2L} = Z_{f2}$.

11.5   Transformation of Acoustic Impedance

Formula 5
Acoustic Impedance in the Reflected Pipe
   The combined acoustic impedance Z_{c1L} at the LHS of Pipe 1 can be calculated from Z_{c2L} at the LHS of Pipe 2 as:

$$Z_{c1L} = Z_{f1}\,\frac{Z_{c2L} + jZ_{f1}\,tan\,(kL_1)}{jZ_{c2L}\,tan\,(kL_1) + Z_{f1}} \qquad \text{(Formula 5)}$$

where:

$$Z_{f1} = \frac{\rho_o c}{S_1}$$

Proof of Formula 5
Based on Formula 3A, the acoustic impedance Z_{c1L} at the LHS of Pipe 1 is given as:

$$Z_{c1L} = Z_{f1}\,\frac{P_{f1R}e^{jkL_1} + P_{b1R}e^{-jkL_1}}{P_{f1R}e^{jkL_1} - P_{b1R}e^{-jkL_1}} = Z_{f1}\,\frac{e^{jkL_1} + \frac{P_{b1R}}{P_{f1R}}e^{-jkL_1}}{e^{jkL_1} - \frac{P_{b1R}}{P_{f1R}}e^{-jkL_1}}$$

According to Formula 4C, $\frac{P_{b1R}}{P_{f1R}} = \frac{Z_{c2L}-Z_{f1}}{Z_{c2L}+Z_{f1}}$, the equation above can be rearranged as:

$$Z_{c1L} = Z_{f1}\frac{e^{jkL_1} + \left(\frac{Z_{c2L}-Z_{f1}}{Z_{c2L}+Z_{f1}}\right)e^{-jkL_1}}{e^{jkL_1} - \left(\frac{Z_{c2L}-Z_{f1}}{Z_{c2L}+Z_{f1}}\right)e^{-jkL_1}} = Z_{f1}\frac{\left(Z_{c2L}+Z_{f1}\right)e^{jkL_1} + \left(Z_{c2L}-Z_{f1}\right)e^{-jkL_1}}{\left(Z_{c2L}+Z_{f1}\right)e^{jkL_1} - \left(Z_{c2L}-Z_{f1}\right)e^{-jkL_1}}$$

Divide both the numerator and denominator by $\left(e^{jkL_1} + e^{-jkL_1}\right)$ as:

$$Z_{c1L} = Z_{f1}\frac{Z_{c2L}\left(e^{jkL_1} + e^{-jkL_1}\right) + Z_{f1}\left(e^{jkL_1} - e^{-jkL_1}\right)}{Z_{c2L}\left(e^{jkL_1} - e^{-jkL_1}\right) + Z_{f1}\left(e^{jkL_1} + e^{-jkL_1}\right)} = Z_{f1}\frac{Z_{c2L} + Z_{f1}\frac{\left(e^{jkL_1}-e^{-jkL_1}\right)}{\left(e^{jkL_1}+e^{-jkL_1}\right)}}{Z_{c2L}\frac{\left(e^{jkL_1}-e^{-jkL_1}\right)}{\left(e^{jkL_1}+e^{-jkL_1}\right)} + Z_{f1}}$$

because:

$$\tan(kL_1) = \frac{\sin(kL_1)}{\cos(kL_1)} = \frac{\left(\frac{e^{jkL_1}-e^{-jkL_1}}{2j}\right)}{\left(\frac{e^{jkL_1}+e^{-jkL_1}}{2}\right)} = \frac{1}{j}\frac{e^{jkL_1}-e^{-jkL_1}}{e^{jkL_1}+e^{-jkL_1}}$$

Therefore, the combined acoustic impedance Z_{c1L} at the LHS of Pipe 1 can be calculated from Z_{c2L} at the LHS of Pipe 2 as:

$$Z_{c1L} = Z_{f1}\frac{Z_{c2L} + jZ_{f1}\tan(kL_1)}{jZ_{c2L}\tan(kL_1) + Z_{f1}} \qquad \text{(Formula 5)}$$

where:

$$Z_{f1} = \frac{\rho_o c}{S_1}$$

This concludes the proof of the formula for combined acoustic impedance in the pipe.

Two Special far End Boundary Conditions
Case 1: Closed End Case $P_{f1R} = P_{b1R}$
The relationship above between the forward wave P_{f1R} and backward wave P_{b1R} can be obtained by Formula 4C for closed end ($Z_{coL} = \infty$) as:

$$\frac{P_{f1R}}{P_{b1R}} = \frac{Z_{coL} + Z_{f1}}{Z_{coL} - Z_{f1}} \qquad \text{(Formula 4C)}$$

Substitute $Z_{coL} = \infty$ into Formula 4C as:

$$\frac{P_{f1R}}{P_{b1R}} = \frac{Z_{coL} + Z_{f1}}{Z_{coL} - Z_{f1}} = \frac{\infty + Z_{f1}}{\infty - Z_{f1}} = \frac{\infty + \frac{\rho_o c}{S_1}}{\infty - \frac{\rho_o c}{S_1}} = 1$$

$$\rightarrow P_{f1R} = P_{b1R}$$

where $Z_{coL} = \infty$ due to the closed far end at $x = 0$ (special case of Z_{coL} for the closed end boundary condition).

The same relationship between the forward wave and backward wave can be obtained by the conservation of mass for the closed end boundary condition as:

$$U_{f1R} + U_{b1R} = 0 \qquad \text{(conservation of mass)}$$

$$\rightarrow \frac{P_{f1R}}{Z_{f1}} + \frac{P_{b1R}}{Z_{b1}} = 0 \quad \rightarrow \frac{P_{f1R}}{Z_{f1}} - \frac{P_{b1R}}{Z_{f1}} = 0 \quad \rightarrow P_{f1R} = P_{b1R}$$

Case 2: Open End Case : $P_{f1R} = -P_{b1R}$

The relationship above between the forward wave P_{f1R} and backward wave P_{b1R} can be obtained by Formula 4C for open end ($Z_{coL} = 0$) as:

$$\frac{P_{fR}}{P_{bR}} = \frac{Z_{coL} + Z_{f1}}{Z_{coL} - Z_{f1}} \qquad \text{(Formula 4C)}$$

Substitute $Z_{coL} = 0$ into Formula 4C as:

$$\frac{P_{fR}}{P_{bR}} = \frac{Z_{coL} + Z_{f1}}{Z_{coL} - Z_{f1}} = \frac{0 + Z_{f1}}{0 - Z_{f1}} = -1$$

$$\rightarrow P_{f1R} = -P_{b1R}$$

where $Z_{coL} = 0$ due to the far open end at $x = 0$ (special case of Z_{coL} for the open end boundary condition).

The same relationship between the forward wave P_{f1R} and backward wave P_{b1R} can be obtained by balancing pressure for the open end boundary condition ($P_{f0L} = P_{b0L} = 0$ at the outside of an open end pipe) as:

$$P_{f1R} + P_{b1R} = P_{f0L} + P_{b0L} \qquad \text{(state of equilibrium)}$$

$$\rightarrow P_{f1R} + P_{b1R} = 0$$

$$\rightarrow P_{f1R} = -P_{b1R}$$

Four Combinations of the Complex Acoustic Impedances for the Special far End Conditions Stated Above and the Special Near end Conditions Below:

- General case of Z_{c2L}:

$$Z_{c2L} \equiv \frac{P_{c2L}}{U_{c2L}} \equiv \frac{P_{f2L} + P_{b2L}}{U_{f2L} + U_{b2L}} \equiv Z_{f2}\frac{P_{f2L} + P_{b2L}}{P_{f2L} - P_{b2L}} \qquad \text{(general case)}$$

- Special case of Z_{c2L} for closed end boundary condition ($U_{c2L} = 0$):

$$Z_{c2L} = \frac{P_{c2L}}{U_{c2L}} = \frac{P_{c2L}}{0} = \infty \qquad \text{(closed end)}$$

- Special case of Z_{c2L} for open end boundary condition ($P_{c2L} = 0$):

$$Z_{c2L} = \frac{P_{c2L}}{U_{c2L}} = \frac{0}{U_{c2L}} = 0 \qquad \text{(open end)}$$

Consider a single pipe with a finite length L. Assume the acoustic impedance at the near end is Z_{c1L} and the acoustic impedance at the far end is Z_{c2L} as shown in the figure above.

The four combinations of the complex acoustic impedances Z_{c1L} and Z_{c2L} of a single pipe for special end conditions are shown below:

	Near end: ($x = -L$)	Far end: ($x = 0$)
Case 1	CLOSED: $Z_{c1L} = \infty$	CLOSED: $Z_{c2L} = \infty$
Case 2	OPEN: $Z_{c1L} = 0$	OPEN: $Z_{c2L} = 0$
Case 3	OPEN: $Z_{c1L} = 0$	CLOSED: $Z_{c2L} = \infty$
Case 4	CLOSED: $Z_{c1L} = \infty$	OPEN: $Z_{c2L} = 0$

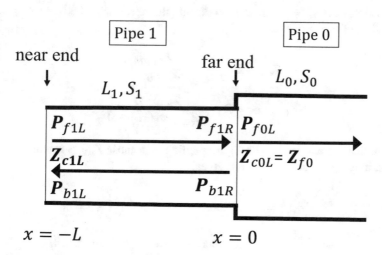

For case 1 and case 2, the detail calculations based on the combined acoustic impedance are given in Example 11.1 and 11.2. Case 3 and case 4 will be left as two exercises. Note that the results are identical to the solutions in Examples 7.1 and 7.2.

Example 11.1: (Case 1: Closed-Closed)
For Case 1: a pipe with a CLOSED near end and a CLOSED far end

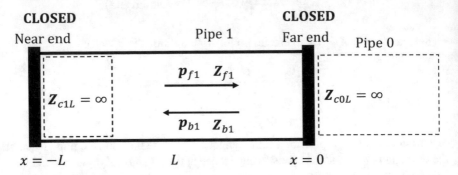

The relationship between Z_{c1L} of Pipe 1 and Z_{c0L} of Pipe 0 is:

$$Z_{c1L} = Z_{f1} \frac{Z_{c0L} + jZ_{f1} \tan(kL)}{jZ_{c0L} \tan(kL) + Z_{f1}} \qquad \text{(Formula 5)}$$

The mode shapes are related to the wavenumber k that satisfies the given two boundary conditions:

Near end: CLOSED: $Z_{c1L} = \infty$ at $x = -L$
Far end: CLOSED: $Z_{c0L} = \infty$ at $x = 0$

Substituting $Z_{c1L} = \infty$ and $Z_{c0L} = \infty$ into the relationship between Z_{c1L} of Pipe 1 and Z_{c0L} of Pipe 0 yields:

$$Z_{c1L} = Z_{f1} \frac{Z_{c0L} + jZ_{f1} \tan(kL)}{jZ_{c0L} \tan(kL) + Z_{f1}}$$

$$\rightarrow \infty = Z_{f1} \frac{Z_{c0L} + 0}{jZ_{c0L} \tan(kL) + 0}$$

Note that the two terms with Z_{f1} are negligible because Z_{c0L} is infinite. Rearranging the above equation gives:

$$\frac{-jZ_{f1}}{\tan(kL)} = \infty \qquad\qquad \rightarrow jZ_{f1} \frac{\cos(kL)}{\sin(kL)} = \infty$$

$$\rightarrow \sin(k_n L) = 0 \qquad \text{where } k_n = n\frac{\pi}{L}, \text{ and } n = 1, 2, 3, \cdots$$

The resonant frequencies ($\omega_n = 2\pi f_n = ck_n$) are:

$$f_n = \frac{ck_n}{2\pi} = \frac{c}{2\pi} n\frac{\pi}{L} = \frac{cn}{2L}, \text{ where } n = 1, 2, 3, \cdots$$

And at these resonant frequencies $\omega_n = ck_n$, the mode shapes of the pressure with eigenfrequencies obtained above are:

$$p = P_{f1R}e^{j(-kx)} + P_{b1R}e^{j(kx)}$$

$$= P_{f1R}e^{j(-kx)} + P_{f1R}e^{j(kx)}$$

$$= P_{f1R}\left(e^{j(-kx)} + e^{j(kx)}\right)$$

$$= 2P_{f1R}\cos(k_nx)$$

$$= P\cos\left(\frac{n\pi}{L}x\right), \text{where } n = 1, 2, 3, \cdots$$

where the complex amplitude P_{b1R} is equal to the complex amplitude P_{f1R} as:

$$P_{f1R} = P_{b1R}$$

The relationship above between the forward wave P_{f1R} and backward wave P_{b1R} can be obtained by Formula 4C for closed end ($Z_{coL} = \infty$) as shown in Example 11.2 as:

$$\frac{P_{f1R}}{P_{b1R}} = \frac{Z_{coL} + Z_{f1}}{Z_{coL} - Z_{f1}} \qquad \text{(Formula 4C)}$$

Substitute $Z_{coL} = \infty$ into Formula 4C as:

$$\frac{P_{f1R}}{P_{b1R}} = \frac{Z_{coL} + Z_{f1}}{Z_{coL} - Z_{f1}} = \frac{\infty + Z_{f1}}{\infty - Z_{f1}} = \frac{\infty + \frac{\rho_o c}{S_1}}{\infty - \frac{\rho_o c}{S_1}} = 1$$

$$\rightarrow P_{f1R} = P_{b1R}$$

Hence, the first three modes of pressure in the pipe of the closed-closed ends are:
Mode 1:

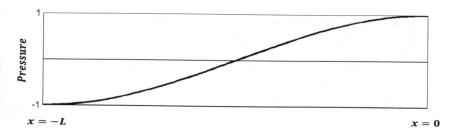

$$\omega_1 = \frac{\pi c}{L}; p_1(x) = P\cos\left(\frac{\pi}{L}x\right); k_1 = \frac{\pi}{L} = \frac{2\pi}{\lambda_1} \rightarrow \lambda_1 = 2L$$

Mode 2:

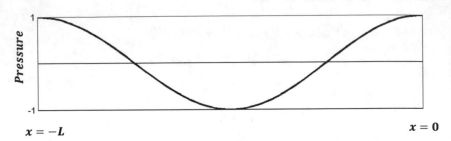

$$\omega_2 = \frac{2\pi c}{L}; p_2(x) = P\cos\left(\frac{2\pi}{L}x\right); k_2 = \frac{2\pi}{L} = \frac{2\pi}{\lambda_2} \rightarrow \lambda_2 = L$$

Mode 3:

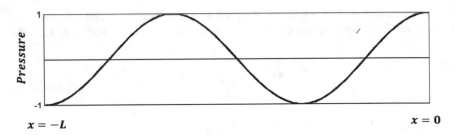

$$\omega_3 = \frac{3\pi c}{L}; p_3(x) = P\cos\left(\frac{3\pi}{L}x\right); k_3 = \frac{3\pi}{L} = \frac{2\pi}{\lambda_3} \rightarrow \lambda_3 = \frac{2}{3}L$$

Example 11.2 (Case 2: Open-Open)
For Case 2: a pipe with an OPEN near end and an OPEN far end

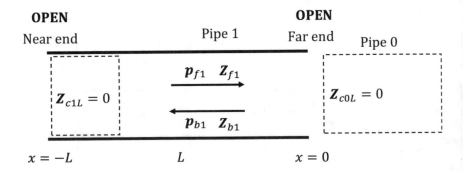

The relationship between Z_{c1L} of Pipe 1 and Z_{c0L} of Pipe 0 is:

$$Z_{c1L} = Z_{f1} \frac{Z_{c0L} + jZ_{f1} \tan(kL)}{jZ_{c0L} \tan(kL) + Z_{f1}} \qquad \text{(Formula 5)}$$

The mode shapes are related to the wavenumber k that satisfies the given two boundary conditions:

Near end: OPEN: $Z_{c1L} = 0$ at $x = -L$
Far end: OPEN: $Z_{c0L} = 0$ at $x = 0$

Substituting $Z_{c1L} = 0$ and $Z_{c0L} = 0$ into the relationship between Z_{c1L} of Pipe 1 and Z_{c0L} of Pipe 0 yields:

$$Z_{c1L} = Z_{f1} \frac{Z_{c0L} + jZ_{f1} \tan(kL)}{jZ_{c0L} \tan(kL) + Z_{f1}}$$

$$\rightarrow 0 = Z_{f1} \frac{0 + jZ_{f1} \tan(kL)}{0 + Z_{f1}}$$

$$\rightarrow jZ_{f1} \tan(kL) = 0$$

The eigenfrequencies that satisfy the above equation are:

$$\sin(k_n L) = 0 \qquad \text{where } k_n = n\frac{\pi}{L} \text{ and } n = 1, 2, 3, \cdots$$

The resonant frequencies ($\omega_n = 2\pi f_n = ck_n$) are:

$$f_n = \frac{ck_n}{2\pi} = \frac{c}{2\pi} n\frac{\pi}{L} = \frac{cn}{2L}, \text{ where } n = 1, 2, 3, \cdots$$

And at these resonant frequencies $\omega_n = ck_n$, the mode shapes of the pressure with eigenfrequencies obtained above are:

$$p = P_{f1R} e^{j(-kx)} + P_{b1R} e^{j(kx)}$$

$$= P_{f1R} e^{j(-kx)} - P_{f1R} e^{j(kx)}$$

$$= P_{f1R} \left(e^{j(-kx)} - e^{j(kx)} \right)$$

$$= -2jP_{f1R} \sin\left(\frac{n\pi}{L}x\right), \text{ where } n = 1, 2, 3,$$

where the complex amplitude P_{b1R} is equal to the complex amplitude P_{f1R} as:

$$P_{f1R} = -P_{b1R}$$

The relationship above between the forward wave P_{f1R} and backward wave P_{b1R} can be obtained by Formula 4C for open end ($Z_{c0L} = 0$, based on Formula 4C) as:

$$\frac{P_{f1R}}{P_{b1R}} = \frac{Z_{coL} + Z_{f1}}{Z_{coL} - Z_{f1}} \qquad \text{(Formula 4C)}$$

Substitute $Z_{coL} = 0$ into Formula 4C as:

$$\frac{P_{f1R}}{P_{b1R}} = \frac{Z_{coL} + Z_{f1}}{Z_{coL} - Z_{f1}} = \frac{0 + Z_{f1}}{0 - Z_{f1}} = -1$$

$$\rightarrow P_{f1R} = - P_{b1R}$$

Hence, the first three modes of the pipe for the open-open ends are:
Mode 1:

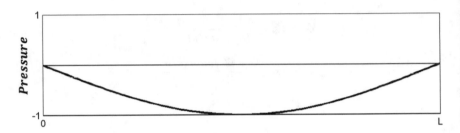

$$\omega_1 = \frac{\pi c}{L}; p_1(x) = P \sin\left(\frac{\pi}{L}x\right); k_1 = \frac{\pi}{L} = \frac{2\pi}{\lambda_1} \rightarrow \lambda_1 = 2L$$

Mode 2:

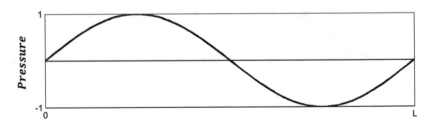

$$\omega_2 = \frac{2\pi c}{L}; p_2(x) = P \sin\left(\frac{2\pi}{L}x\right); k_2 = \frac{2\pi}{L} = \frac{2\pi}{\lambda_2} \rightarrow \lambda_2 = L$$

Mode 3:

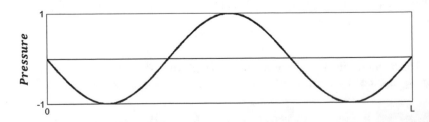

$$\omega_3 = \frac{3\pi c}{L} ; p_3(x) = P \sin\left(\frac{3\pi}{L}x\right); k_3 = \frac{3\pi}{L} = \frac{2\pi}{\lambda_3} \rightarrow \lambda_3 = \frac{2}{3}L$$

11.6 Power Reflection and Transmission

11.6.1 Definition of Power of Acoustic Waves

The power of an acoustic wave is calculated from the pressure and velocity as the following:

$$(\text{POWER}) \left[\tfrac{work}{time}\right] = (\text{PRESSURE}) \left[\tfrac{force}{area}\right] \times (\text{VELOCITY}) \left[\tfrac{distance}{ime}\right] \times (\text{AREA}) \, [area]$$

Note that the work in the above equation is the force multiplied by the distance ($work = force \times distance$):

Based on the above definition of power, the power of the acoustic wave is:

$$w = Re\,(P)\,Re\,(V)S = Re\,(P)\,Re\,(U)$$
$$= Re\,(P)\,Re\left(\frac{P}{Z}\right) = \frac{1}{4}\left(P + P^*\right)\left[\frac{P}{Z} + \left(\frac{P}{Z}\right)^*\right]$$

11.6.2 Power Reflection and Transmission Coefficients of One-to-One Pipes

Formulas 6A–6B

The power reflection and transmission coefficients in one-to-one pipes are defined as:

$$R_w \equiv \frac{w_r}{w_i} = \frac{Re\,(P_r)\,Re\left(\frac{P_r}{Z_r}\right)}{Re\,(P_i)\,Re\left(\frac{P_i}{Z_i}\right)} \qquad \text{(Formula 6A)}$$

$$T_w \equiv \frac{w_o}{w_i} = \frac{Re\,(P_o)\,Re\left(\frac{P_o}{Z_o}\right)}{Re\,(P_i)\,Re\left(\frac{P_i}{Z_i}\right)} \qquad \text{(Formula 6B)}$$

11.6.3 Simplified Cases of Power Reflection and Transmission in One-to-One Pipes

Based on the definition of power transmission:

$$T_w \equiv \frac{w_o}{w_i} = \frac{Re\,(P_o)\,Re\left(\frac{P_o}{Z_o}\right)}{Re\,(P_i)\,Re\left(\frac{P_i}{Z_i}\right)}$$

The following two simplified formulas can be derived for the relevant conditions:
If Z_i and Z_o are real:

Then, $T_w = \frac{[Re\,(P_o)]^2}{[Re\,(P_i)]^2}\frac{Z_i}{Z_o}.$

If Z_i, Z_o and P_i are real:

Then, $T_w = \left[Re\left(\frac{P_o}{P_i}\right)\right]^2\frac{Z_i}{Z_o}.$

11.6.4 Special Case of Power Reflection and Transmission: One-to-One Pipes

Formulas 7A–7E
Power of Traveling Plane Waves

Power reflection and transmission in one-to-one pipes can be formulated with real acoustic impedance Z and real magnitude P as:

$$w_i = P_i^2\,\frac{1}{Z_i} \qquad\qquad \text{(Formula 7A)}$$

$$w_r = P_r^2 \frac{1}{-Z_i} \qquad \text{(Formula 7B)}$$

$$R_w = \frac{w_r}{w_i} = -\left(\frac{P_r}{P_i}\right)^2 = -\left(\frac{Z_o - Z_i}{Z_o + Z_i}\right)^2 \qquad \text{(Formula 7C)}$$

$$T_w = \frac{w_o}{w_i} = \left(\frac{P_o}{P_i}\right)^2 \frac{Z_i}{Z_o} = \left(\frac{2Z_o}{Z_o + Z_i}\right)^2 \frac{Z_i}{Z_o} \qquad \text{(Formula 7D)}$$

$$w_i + w_r = w_o \qquad \text{(Formula 7E)}$$

where, for traveling plane waves, the following quantities are real numbers:

$$Z_i, Z_o, P_i, P_o \text{ real}$$

$\boldsymbol{Z_i, Z_o, P_i, P_o}$ real

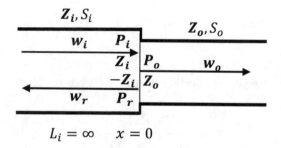

$$L_i = \infty \qquad x = 0$$

Proof of Formulas 7A and 7B

For traveling plane waves (forward and backward) in pipes:

- The acoustic impedance Z is a real number.
- The complex amplitude of pressure P is also a real number.

Therefore, Formula 7A for the power of acoustic waves can be formulated with real acoustic impedance Z and real magnitude P as:

$$w = Re\,(P)\,Re\left(\frac{P}{Z}\right) = P^2 \frac{1}{Z}$$

Because $-Z_r = -Z_i$ (Formula 1B), when the power is formulated with Z_i, the power carried by the forward wave P_i and backward wave P_r are formulated as:

$$w_i = P_i^2 \frac{1}{Z_i}$$

$$w_r = P_r^2 \frac{1}{Z_r} = P_r^2 \frac{1}{-Z_i}$$

Proof of Formulas 7C and 7D

The power reflection and transmission coefficients are defined as:

$$R_w \equiv \frac{w_r}{w_i} = \frac{Re\,(P_r)\,Re\left(\frac{P_r}{Z_r}\right)}{Re\,(P_i)\,Re\left(\frac{P_r}{Z_i}\right)}$$

$$T_w \equiv \frac{w_o}{w_i} = \frac{Re\,(P_o)\,Re\left(\frac{P_o}{Z_o}\right)}{Re\,(P_i)\,Re\left(\frac{P_r}{Z_i}\right)}$$

Because the following quantities for traveling plane waves in one-to-one pipes are real numbers:

$$Z_i, Z_r, Z_o, P_i, P_o \text{ real numbers}$$

and:

$$Z_r = -Z_i \qquad\qquad\qquad\qquad \text{(Formula 2B)}$$

Therefore, the power reflection and transmission coefficients can be simplified as:

$$R_w = \frac{Re\,(P_r)\,Re\left(\frac{P_r}{Z_r}\right)}{Re\,(P_i)\,Re\left(\frac{P_r}{Z_i}\right)} = \frac{\frac{P_r^2}{Z_r}}{\frac{P_i^2}{Z_i}} = -\left(\frac{P_r}{P_i}\right)^2$$

$$T_w = \frac{Re\,(P_o)\,Re\left(\frac{P_o}{Z_o}\right)}{Re\,(P_i)\,Re\left(\frac{P_r}{Z_i}\right)} = \frac{\frac{P_o^2}{Z_o}}{\frac{P_i^2}{Z_i}} = \left(\frac{P_o}{P_i}\right)^2 \frac{Z_i}{Z_o}$$

Based on Formulas 4C–4D:

$$\frac{P_{b1L}}{P_{f1R}} = \frac{Z_{foL} - Z_{f1}}{Z_{foL} + Z_{f1}} \qquad\qquad \text{(Formula 4C)}$$

$$\frac{P_{foL}}{P_{f1R}} = \frac{2Z_{foL}}{Z_{foL} + Z_{f1}} \qquad\qquad \text{(Formula 4D)}$$

The power reflection and transmission coefficients can be formulated as:

$$R_w = \frac{w_r}{w_i} = -\left(\frac{P_r}{P_i}\right)^2 = -\left(\frac{Z_o - Z_i}{Z_o + Z_i}\right)^2 \qquad \text{(Formula 7C)}$$

$$T_w = \frac{w_o}{w_i} = \left(\frac{P_o}{P_i}\right)^2 \frac{Z_i}{Z_o} = \left(\frac{2Z_o}{Z_o + Z_i}\right)^2 \frac{Z_i}{Z_o} \qquad \text{(Formula 7D)}$$

Proof of Formula 7E

Based on the law of conservation of energy, the power that goes into a system has to equal the power that comes out of it:

$$(\text{POWER IN}) = (\text{POWER OUT})$$

Therefore, the summation of power on the left-hand side pipe equals the summation of power on the right-hand side pipe:

$$w_i + w_r = w_o \qquad\qquad \text{(Formula 7E)}$$

Divide the equation above by w_i to obtain the power ratio:

$$1 + \frac{w_r}{w_i} = \frac{w_o}{w_i}$$

$$\rightarrow \quad 1 + R_w = T_w$$

$$\rightarrow \quad -R_w + T_w = 1$$

Note that w_i and w_o does positive work but w_r does negative work due to the negative impedance ($180°$ phase difference between the pressure and the velocity).

Example 11.3: Power Transmission Coefficient

The cross-section area of Pipe i and Pipe o as shown below are $S_i = 0.0830$ [m^2] and $S_o = 0.0415$ [m^2], respectively:

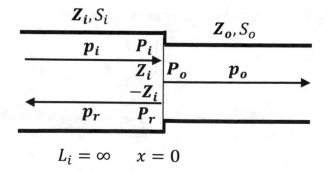

Use 415 rayls for the characteristic impedance ($\rho_o c$) of air.

(a) Determine the power reflection coefficient R_w.
(b) Determine the power transmission coefficient T_w.
(c) Validate your answers in parts a and b using the conservation of energy.

Solution of Example 11.3

The acoustic impedances in Pipe i and Pipe o can be calculated based on Formulas 1A and 1B as:

$$Z_i = \frac{\rho_o c}{S_i} = \frac{415}{0.083} = 50 \qquad \text{(Formula 1A)}$$

$$Z_r = \frac{-\rho_o c}{S_i} = \frac{-415}{0.083} = -50 \qquad \text{(Formula 1B)}$$

$$Z_o = \frac{\rho_o c}{S_o} = \frac{415}{0.0415} = 100 \qquad \text{(Formula 1A)}$$

Note that Z_i, Z_r, and Z_o are real numbers.

(a) The power reflection coefficients can be calculated based on Formula 7C as:

$$R_w = -\left(\frac{Z_o - Z_i}{Z_o + Z_i}\right)^2 = -\left(\frac{100 - 50}{100 + 50}\right)^2 = \frac{-1}{9} \qquad \text{(Formula 7C)}$$

(b) The power transmission coefficients can be calculated based on Formula 7D as:

$$T_w = \left(\frac{2Z_o}{Z_o + Z_i}\right)^2 \frac{Z_i}{Z_o} = \left(\frac{2 \times 100}{100 + 50}\right)^2 \frac{50}{100} = \frac{8}{9} \qquad \text{(Formula 7D)}$$

(c) Based on the conservation of energy (Formula 7E):

$$w_i + w_r = w_o$$

Divide the equation above by w_i to obtain the power ratio:

$$1 + \frac{w_r}{w_i} = \frac{w_o}{w_i}$$

$$\rightarrow \quad 1 + R_w = T_w$$

Substituting the power reflection ($R_w = \frac{-1}{9}$) and power transmission ($T_w = \frac{8}{9}$) calculated in Parts a and b into the conservation of energy ($1 + R_w = T_w$) gives:

$$1 + R_w = T_w$$

$$\rightarrow 1 - \frac{1}{9} = \frac{8}{9}$$

The equation above validates the power reflection (R_w) and power transmission (T_w) calculated in Parts a and b.

11.7 Numerical Method for Molding of Pipelines

Objective

Numerical methods can be used to calculate power transmission coefficients of pipelines. A pipeline is constructed with three pipes as shown below. Acoustic waves enter the left-hand side of the pipeline and exit the right-hand side:

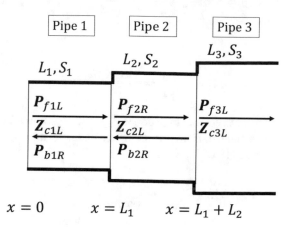

Given:

Pipe 1: $L_1 = 0.1\ [m], S_1 = 0.5\ [m^2]$
Pipe 2: $L_2 = 0.1\ [m], S_2 = 0.5\ [m^2]$
Pipe 3: $L_3 = 0.1\ [m], S_3 = 0.5\ [m^2]$

$$P_{f3L} = 1; P_{b3L} = 0$$

At 1000 Hz, calculate:

(a) The complex acoustic impedance Z_{c1L} of the combined wave at the LHS of Pipe 1
(b) The complex amplitude of pressure of both the forward wave (P_{f1L}) and backward wave (P_{b1L}) at the LHS of Pipe 1
(c) The power transmission coefficient T_w

Hint: Doing this example by hand can be difficult because the complex acoustic impedances are complex numbers. We will use the three MATLAB functions provided in the MATLAB program section (Sect. 11.8) for this example. The same results can be obtained with any suitable programming languages

Solution

The method is demonstrated using three functions described in the Computer Program Section (Sect. 11.8). The following is the logic for this numerical method:

Part a: Get the Complex Acoustic Impedance Z_{c1L}

According to Formula 5, the complex acoustic impedance Z_{c1L} of the combined wave at the LHS of Pipe 1 can be calculated from the complex acoustic impedance Z_{c2L} of the combined wave at the LHS of Pipes 2 as:

$$Z_{c1L} = Z_{f1} \frac{Z_{c2L} + jZ_{f1} \tan(kL)}{jZ_{c2L} \tan(kL) + Z_{f1}} \qquad \text{(Formula 5) (5)}$$

where:

$$Z_{f1} = \frac{\rho_o c}{S_1} = \frac{415 \, [rayles]}{S_1};$$

Similarly, the complex acoustic impedance Z_{c2L} of the combined wave at the LHS of Pipe 2 can be calculated from the complex acoustic impedance Z_{c3L} of the combined wave at the LHS of Pipe 3 as:

$$Z_{c2L} = Z_{f2} \frac{Z_{c3L} + jZ_{f2} \tan(kL)}{jZ_{c3L} \tan(kL) + Z_{f2}} \qquad \text{(Formula 5) (6)}$$

where:

$$Z_{f2} = \frac{\rho_o c}{S_2} = \frac{415 \, [rayles]}{S_1}$$

Because there is only an outward wave in Pipe 3 ($P_{f3L} = 1$; $P_{b3L} = 0$), based on Formula 1C, the combined pressure P_{c3L} is equal to P_{f3L} as:

$$P_{c3L} = P_{f3L} + P_{b3L} = P_{f3L} = 1$$

Substituting P_{c3L} into Eq. (5) and Eq. (6) above in the reverse order gets Z_{c1L}.

Part b: Get the Complex Amplitudes of Pressure P_{c1L} **and** P_{b1L}

Substituting P_{f1R} and P_{b1R} of Formulas 4A and 4B into Formulas 2A and 2B gets the formulas for P_{f1L} and P_{b1L} in Pipe 1 as shown below. This formula can be used to transfer transforming pressures from $P_{c2L} P_{c2L}$ to P_{f1L} and P_{b1L} from Pipe 2 to Pipe 1 as:

$$P_{f1L} = \left[\frac{1}{2}\left(1 + \frac{Z_{f1}}{Z_{f2}}\right) P_{f2L} + \frac{1}{2}\left(1 - \frac{Z_{f1}}{Z_{f2}}\right) P_{b2L} \right] e^{jkL_1} \qquad \text{(Formula 4A}' + \text{2A)}$$

$$P_{b1L} = \left[\frac{1}{2}\left(1 - \frac{Z_{f1}}{Z_{f2}} \right) P_{f2L} + \frac{1}{2}\left(1 + \frac{Z_{f1}}{Z_{f2}} \right) P_{b2L} \right] e^{-jkL_1} \qquad \text{(Formula 4B}' + \text{2B)}$$

$$P_{f1L} = \left[\frac{1}{2}\left(1 + \frac{Z_{f1}}{Z_{c2L}} \right) P_{c2L} \right] e^{jkL_1} \qquad \text{(Formula 4A} + \text{2A)}$$

$$P_{b1L} = \left[\frac{1}{2}\left(1 - \frac{Z_{f1}}{Z_{c2L}} \right) P_{c2L} \right] e^{jkL_1} \qquad \text{(Formula 4B} + \text{2B)}$$

where:

$$P_{c2L} = P_{f2L} + P_{b2L} \qquad \text{(Formula 1D)}$$

Similarly, the formulas for transforming pressures from $P_{c3L} P_{c3L}$ to P_{f2L} and P_{b2L} from Pipe 3 to Pipe 2 are as follows:

$$P_{f2L} = \left[\frac{1}{2}\left(1 + \frac{Z_{f2}}{Z_{c3L}} \right) P_{c3L} \right] e^{jkL_2}$$

$$P_{b2L} = \left[\frac{1}{2}\left(1 - \frac{Z_{f2}}{Z_{c3L}} \right) P_{c3L} \right] e^{-jkL_2}$$

Because there is only an outward wave in Pipe 3 ($P_{f3L} = 1$; $P_{b3L} = 0$), based on Formula 1D, the combined pressure P_{c3L} is equal to P_{f3L} as:

$$P_{c3L} = P_{f3L} + P_{b3L} = P_{f3L} = 1 \qquad \text{(Formula 1D)}$$

Therefore, P_{f1L} and P_{b1L} can be calculated by solving the four equations above in the reverse order.

Part c: Get the Power Transmission Coefficient T_w

The power transmission coefficient T_w can be calculated using Formula 6B:

$$T_w \equiv \frac{w_o}{w_i} = \frac{Re\left(P_o\right) Re\left(\frac{P_o}{Z_o}\right)}{Re\left(P_i\right) Re\left(\frac{P_i}{Z_i}\right)} = \frac{Re\left(P_{f3L}\right) Re\left(\frac{P_{f3L}}{Z_{f3}}\right)}{Re\left(P_{f1L}\right) Re\left(\frac{P_{f1L}}{Z_{f1}}\right)} \qquad \text{(Formula 6B)}$$

11.8   Computer Program

Function 1: TRANSFER_IMPEDANCE

The function TRANSFER_IMPEDANCE calculates the complex acoustic impedance (Z_{c1L}) of the combined wave at the LHS of Pipe 1 based on the complex acoustic impedance (Z_{c2L}) of the combined wave at the LHS of Pipe 2. The input and output arguments of the function are as follows:

Input Arguments:

```
Zc2L: Complex acoustic impedance of the combined wave at the LHS of Pipe 2
L1: Length of Pipe 1
S1: Cross-section area of Pipe 1
Zc2L: Complex acoustic impedance of the combined wave at the LHS of Pipe 2
k: Wavenumber
LoC: Characteristic impedance
```

Output Arguments:

```
Zc1L: Complex acoustic impedance of the combined wave at the LHS of Pipe 1
```

Formulas:

This function is based on Formula 5 as:

$$Z_{c1L} = Z_{f1} \frac{Z_{c2L} + jZ_{f1} \tan(kL_1)}{jZ_{c2L} \tan(kL_1) + Z_{f1}} \qquad \text{(Formula 5)}$$

where:

$$Z_{f1} = \frac{\rho_o c}{S_1}$$

Function 2: TRANSFER_PRESSURE

The function TRANSFER_PRESSURE calculates the complex amplitude of pressure of both the forward wave (P_{f1L}) and backward wave (P_{b1L}) at the LHS of Pipe 1 based on the acoustic impedance at the LHS of Pipe 2. The input and output arguments of the function are as follows:

Input Arguments:

```
Pc2L: Complex amplitude of pressure of the combined wave at the LHS of
Pipe 2
L1: Length of Pipe 1
S1: Cross-section area of Pipe 1
Zc2L: Complex acoustic impedance of the combined wave at the LHS of Pipe 2
k: Wavenumber
LoC: Characteristic impedance
```

Output Arguments:

```
Pf1L: Complex amplitude of pressure of the forward wave at the LHS of Pipe
1
Pb1L: Complex amplitude of pressure of the backward wave at the LHS of
Pipe 1
```

Formulas:

This function is based on the two formulas below, which are the combined results of Formulas 2A–2B and 4A–4B as shown in Example 11.2:

$$P_{f1L} = \left[\frac{1}{2} \left(1 + \frac{Z_{f1}}{Z_{c2L}} \right) P_{c2L} \right] e^{jkL_1} \qquad \text{(Formulas 4A + 2A)}$$

$$P_{b1L} = \left[\frac{1}{2} \left(1 - \frac{Z_{f1}}{Z_{c2L}} \right) P_{c2L} \right] e^{-jkL_1} \qquad \text{(Formulas 4B + 2B)}$$

where:

$$Z_{f1} = \frac{\rho_o c}{S_1} \qquad \text{(Formula 1A)}$$

Function 3: Power Transmission Coefficient
Input arguments:
Pi: Complex amplitude of the pressure of the forward (incident) wave in the inlet pipe
Po: Complex amplitude of the pressure of the forward (outward) wave in the outlet pipe
Pr: Complex amplitude of the pressure of the backward (reflected) wave in the inlet pipe
Zi: acoustic impedance (real number) of the forward (incident) wave in the inlet pipe
Zo: the acoustic impedance of the forward (outward) wave in the outlet pipe
Zr: the acoustic impedance of the backward (reflected) wave in the inlet pipe

Output arguments:
Rw: Power reflection coefficient
Tw: Power transmission coefficient

Formulas:

$$R_w \equiv \frac{w_r}{w_i} = \frac{Re\,(P_r)\,Re\,\left(\frac{P_r}{Z_r} \right)}{Re\,(P_i)\,Re\,\left(\frac{P_i}{Z_i} \right)} \qquad \text{(Formula 6A)}$$

$$T_w \equiv \frac{w_o}{w_i} = \frac{Re\,(P_o)\,Re\,\left(\frac{P_o}{Z_o} \right)}{Re\,(P_i)\,Re\,\left(\frac{P_i}{Z_i} \right)} \qquad \text{(Formula 6B)}$$

MATLAB Code (Function)

```
function  [Pf1L,Pb1L]=TransferPressure_fb1Lfb2L(Zf1,Zf2,Pf2L,Pb2L,
L1,k)
% Purpose: Transfer comoplex pressure from Pipe 2 to Pipe 1
% Input variables:
% Zf1: acoustic impedance of forward wave of Pipe 1
% Zf2: acoustic impedance of forward wave of Pipe 2
% Pf2L: forward pressure at LHS of Pipe 2
```

```
% Pb2L: backward pressure at LHS of Pipe 2
% L1: length of Pipe 1
% k: wave number
% Output variable:
% Pf1L: forward pressrue at LHS of Pipe 1
% Pb1L: backward pressrue at LHS of Pipe 1
% Formula 4A'-B' + 2A-B
% Pf1L=0.5*((1+Zf1/Zf2)*Pf2L + (1-Zf1/Zf2)*Pb2L)*exp( j*k*L1);
% Pb1L=0.5*((1-Zf1/Zf2)*Pf2L + (1+Zf1/Zf2)*Pb2L)*exp(-j*k*L1);
% Formula 4A'-B' + 2A-B
[Pf1R,Pb1R]=TransferPressure_fb1Rfb2L(Zf1,Zf2,Pf2L,Pb2L);
[Pf1L,Pb1L]=TransferPressure_fb1Lfb1R(Pf1R,Pb1R,L1,k);
end
function [Pf1R,Pb1R]=TransferPressure_fb1Rfb2L(Zf1,Zf2,Pf2L,Pb2L)
% Purpose: Transfer comoplex pressure from Pipe 2 to Pipe 1
% Input variables:
% Zf1: acoustic impedance of forward wave of Pipe 1
% Zf2: acoustic impedance of forward wave of Pipe 2
% Pf2L: forward pressure at LHS of Pipe 2
% Pb2L: backward pressure at LHS of Pipe 2
% Output variable:
% Pf1R: forward pressrue at RHS of Pipe 1
% Pb1R: backward pressrue at RHS of Pipe 1
% Formula 4A'-B'
Pf1R=0.5*((1+Zf1/Zf2)*Pf2L + (1-Zf1/Zf2)*Pb2L);
Pb1R=0.5*((1-Zf1/Zf2)*Pf2L + (1+Zf1/Zf2)*Pb2L);
end
function [Pf1L,Pb1L]=TransferPressure_fb1Lfb1R(Pf1R,Pb1R,L1,k)
% Purpose: Transfer comoplex pressure from RHS to LHS of Pipe 1
% Input variables:
% Pf1R: forward pressure at RHS of Pipe 1
% Pb1R: backward pressure at RHS of Pipe 1
% L1: length of Pipe 1
% k: wave number
% Output variable:
% Pf1L: forward pressrue at LHS of Pipe 1
% Pb1L: backward pressrue at LHS of Pipe 1
% Formula 2A-B
Pf1L=Pf1R*exp( j*k*L1);
Pb1L=Pb1R*exp(-j*k*L1);
end
function [Pf1R,Pb1R]=TransferPressure_fb1Rfb1L(Pf1L,Pb1L,L1,k)
% Purpose: Transfer comoplex pressure from LHS to RHS of Pipe 1
% Input variables:
% Pf1L: forward pressrue at LHS of Pipe 1
% Pb1L: backward pressrue at LHS of Pipe 1
% L1: length of Pipe 1
% k: wave number
% Output variable:
% Pf1R: forward pressure at RHS of Pipe 1
% Pb1R: backward pressure at RHS of Pipe 1
% Formula 2C-D
Pf1R=Pf1L*exp(-j*k*L1);
Pb1R=Pb1L*exp( j*k*L1);
end
```

```
function [Pf1R,Pb1R]=TransferPressure_fb1Rfb2R(Zf1,Zf2,Pf2R,Pb2R,
L2,k)
% Purpose: Transfer comoplex pressure from Pipe 2 to Pipe 1
% Note: This function does not work for Pipe 2 is the end pipe
% Input variables:
% Zf1: acoustic impedance of forward wave of Pipe 1
% Zf2: acoustic impedance of forward wave of Pipe 2
% Pf2R: forward pressure at RHS of Pipe 2
% Pb2R: backward pressure at RHS of Pipe 2
% L2: length of Pipe 2
% k: wave number
% Output variable:
% Pf1R: forward pressrue at RHS of Pipe 1
% Pb1R: backward pressrue at RHS of Pipe 1
% Formula 2A-B + 4A'-B'
% Pf1R=0.5*((1+Zf1/Zf2)*Pf2R*exp(j*k*L2)+...
% (1-Zf1/Zf2)*Pb2R*exp(-j*k*L2));
% Pb1R=0.5*((1-Zf1/Zf2)*Pf2R*exp(j*k*L2)+
% (1+Zf1/Zf2)*Pb2R*exp(-j*k*L2));
% Formula 2A-B + 4A'-B'
[Pf2L,Pb2L]=TransferPressure_fb1Lfb1R(Pf2R,Pb2R,L2,k);
[Pf1R,Pb1R]=TransferPressure_fb1Rfb2L(Zf1,Zf2,Pf2L,Pb2L);
end
function[Pf1R,Pb1R]=TransferPressure_EndPipe(Zf1,Zf2,Pf2L)
% Purpose: Transfer comoplex pressure from Pipe 2 (end pipe) to Pipe 1
% Input variables:
% Zf1: acoustic impedance of forward wave of Pipe 1
% Zf2: acoustic impedance of forward wave of Pipe 2 (The end pipe)
% Pf2L: forward pressure at LHS of Pipe 2 (Pb2L=0) (The end pipe)
% Output variable:
% Pf1R: forward pressrue at RHS of Pipe 1
% Pb1R: backward pressrue at RHS of Pipe 1
% Formula 2A-B (with Pb2L=0)
Pf1R=0.5*(1+Zf1/Zf2)*Pf2L;
Pb1R=0.5*(1-Zf1/Zf2)*Pf2L;
End


function [Pf1R,Pb1R]=TransferPressure_fb1Rc2L(Zf1,Zc2L,Pc2L)
% Purpose: Transfer comoplex pressure from Pipe 2 to Pipe 1
% Input variables:
% Zf1: acoustic impedance of forward wave of Pipe 1
% Zc2L: acoustic impedance of forward wave of Pipe 2
% Pc2L: combined pressure at LHS of Pipe 2
% Output variable:
% Pf1R: forward pressrue at RHS of Pipe 1
% Pb1R: backward pressrue at RHS of Pipe 1
% Formula 4A-B
Pf1R=0.5*(1+Zf1/Zc2L)*Pc2L;
Pb1R=0.5*(1-Zf1/Zc2L)*Pc2L;
end
function [Zc1L]=TransferImpedance(Zf1,Zc2L,L1,k)
% Purpose: Transfer acoustic impedacne from Pike 2 to Pipe 1
% Input variables:
% Zf1: complex acoustic impedance of forward pressure of Pipe 1
```

```
% Zc1L: complex acoustic impedance of combine pressure at LHS of Pipe 1
% L1: length of Pipe 1
% k: wave number
% Output variable:
% Zc2L: complex acoustic impedance of combine pressure at LHS of Pipe 2
% Formula 5
Zc1L=Zf1*(Zc2L+j*Zf1*tan(k*L1))/(j*Zc2L*tan(k*L1)+Zf1);
end
function [Tw]=get_Tw_PipeLine(P_in,Z_in,P_out,Z_out)
% calculate the Power Transmission Coeffient Tw
% P_in: Complex pressure inward
% P_out: Complex pressure outward
% Z_in: Complex acoustic impedance inward
% Z_out: Complex acoustic impedance outward
% Formula 6B
Tw=real(P_out)*real(P_out/Z_out)/(real(P_in)*real(P_in/Z_in));
end
```

MATLAB Code (Main Code)

```
function ANC_PRJ051_FilterAndResonator_D191116B
clear all % remove all variables
clf % remove all figures
%--------------------------------------------------
--------------------%
% Section 1: Define Variables and Parameters
%--------------------------------------------------
------------------%
% define the parameter of the wave function
% define the air properties
c=340; % [m/s]
loC=415; % [Rayl]
% define the dimensions of the pipe line
iCase=3;
if iCase==1 % Test Case (Horn Shaped Pipeline)
% sPipe=[0.002; 0.004; 0.006; 0.008]; % area [m^2]
% lPipe=[0.025; 0.025; 0.025; 0.025]; % length [m]
sPipe=[0.2; 0.6; 0.9; 1.2]; % area [m^2]
lPipe=[0.025; 0.025; 0.025; 0.025]; % length [m]
elseif iCase==2 % Low-Pass Filter
sPipe=[0.002; 0.008; 0.002]; % area [m^2]
lPipe=[0.025; 0.015; 0.025]; % length [m]
elseif iCase==3 % Band Stop Filter
% sPipe=[0.002; 0.00001]; % area [m^2]
% lPipe=[0.05; 0.00001]; % length [m]
sPipe=[0.002; 0.004; 0.00001]; % area [m^2]
lPipe=[0.01; 0.04; 0.00001]; % length [m]
elseif iCase==4 % High-Pass Filter
sPipe=[0.001; 0.10]; % area [m^2]
lPipe=[0.020; 0.015]; % length [m]
end
% define the dimensions of the main pipe
SiMain=0.002; % m2 % area of the main line inlet pipe
SoMain=0.002; % m2 % area of the main line outlet pipe
% define the frequency sweep
nFreq=8192;
```

```
df=0.5; % [Hz]
freq=[0:1:nFreq-1]*df; % 0:0.5:4095.5 [Hz]
%----------------------------------------------------------------
------------%
% Section 2: Setup
%----------------------------------------------------------------
------------%
% define the dimensions
Tw=zeros(nFreq,1);
nPipe=max(size(lPipe)); % Number of pipes in pipe line side branch
ZiPipe=zeros(nPipe,1); % Incident Impedance of the pipe line side
branch
ZcPipe=zeros(nPipe,1); % combined Impedance of the pipe line side
branch
for iPipe=1:1:nPipe
 ZiPipe(iPipe)=loC/sPipe(iPipe);
end
%----------------------------------------------------------------
------------%
% Section 3: Calculation
%----------------------------------------------------------------
------------%
% calculate Tw for each frequency
for iw=1:1:nFreq
k=2*pi*freq(iw)/c;
% define the dimensions of the matrices
PfR=zeros(nPipe,1); % forward pressure at the right end
PbR=zeros(nPipe,1); % backward pressure at the right end
% pressure for the nth pipe (the left one)
PfR(nPipe)=1.0; % the pon is given
PbR(nPipe)=0.0;
% calculate the combined impedance for the nth pipe (the left one)
ZcPipe(nPipe)=ZiPipe(nPipe); % the last pipe (has no return wave)
% get PfR and PbR of pipe n-1
[PfR_iPipe_temp,PbR_iPipe_temp]=TransferPressure_EndPipe(...
ZiPipe(nPipe-1),ZiPipe(nPipe),PfR(nPipe));
PfR(nPipe-1)=PfR_iPipe_temp;
PbR(nPipe-1)=PbR_iPipe_temp;
% transfer Zc from (nPipe) to (nPipe-1)
ZcPipe(nPipe-1)=TransferImpedance(ZiPipe(nPipe-1),ZcPipe(nPipe),
lPipe(nPipe-1),k);
% transfer Zc from (nPipe-1) to (iPipe=1)
for iPipe=nPipe-2:-1:1
% pressure for the iPipe pipe
[PfR_temp,PbR_temp]=TransferPressure_fb1Rfb2R(ZiPipe(iPipe),
ZiPipe(iPipe+1),PfR(iPipe+1),PbR(iPipe+1),lPipe(iPipe+1),k);
PfR(iPipe)=PfR_temp;
PbR(iPipe)=PbR_temp;
% transfer Zc from (iPipe+1) to (iPipe)
ZcPipe(iPipe)=TransferImpedance(ZiPipe(iPipe),ZcPipe(iPipe+1),
lPipe(iPipe),k);
end
% calculate Tw for Low-Pass filternPipe=size(PfR,1);
PfR_PfR1=PfR/PfR(1); % make the input pressure 'REAL' by making it 1
Tw(iw,1)=get_Tw_PipeLine(PfR_PfR1(1),ZiPipe(1),PfR_PfR1(nPipe),
```

```
ZiPipe(nPipe));
% calculate Tw of the Main Pipe with a side-pipe-line
ZiMain=loC/SiMain;
ZoMain=loC/SoMain;
ZoPara=1.0/(1.0/ZcPipe(1)+1.0/ZoMain);
PoPipe_PiMain=2*ZoPara/(ZoPara+ZiMain);
PoMain_PiMain=PoPipe_PiMain;
PiMain=1;
Tw(iw,2)=get_Tw_PipeLine(PiMain,ZiMain,PoMain_PiMain,ZoMain);
end
%-------------------------------------------------------------------
%
% Section 4: Plotting
%-------------------------------------------------------------------
%
% plot the Tw of the pipe line
figure(1);
plot(freq,Tw(:,1));
xlabel('frequency [Hz]')
ylabel('Tw')
title('Tw of a Pipe line')
saveas(gcf,'Figure_Tw_of_a_Pipe_line','emf')
% plot the Tw of main pipe with a side branch of the pipe line
figure(2);
plot(freq,Tw(:,2));
xlabel('frequency [Hz]')
ylabel('Tw')
title('Tw of a Main Pipe with a Side Branch of a Pipe Line')
% saveas(gcf,'Figure_Tw_of_a_Main_Pipe_with_a_Side_Branch','emf')
end
```

11.9   Project

Any pipeline can be constructed by connecting n pipes as shown below. Acoustic waves enter the left-hand side of the pipeline and exit the right-hand side:

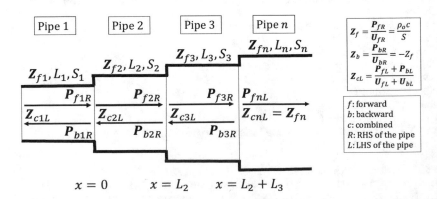

For given dimensions of the length L and the cross-sectional area S of each pipe, write a computer program using the provided MATLAB function (in Sect. 11.8) to calculate and plot power transmission coefficient T_w of the pipeline as a function of frequency.

Hint: Doing this example by hand can be difficult because (1) formulas of acoustic impedance are in complex number format and (2) power transmission coefficients need to be calculated at multiple frequencies. You can use the MATLAB functions provided in Computer Code Section (Sect. 11.7) or any suitable programming language for this project.

11.10 Homework Exercises

Exercise 11.1

In Pipe 2 as shown in the figure below, a pressure $p_{f2}(x, t)$ of a forward wave and a pressure $p_{b2}(x, t)$ of a backward ware are formulated in terms of the complex amplitudes of the pressures P_{f2R} and P_{b2R} at $x = x_2$ (RHS of Pipe 2) as:

$$p_{f2}(x, t) = \frac{1}{2}\left\{P_{f2R}e^{-jk(x-x_2)}e^{j\omega t} + cc\right\}$$

$$p_{b2}(x, t) = \frac{1}{2}\left\{P_{b2R}e^{jk(x-x_2)}e^{j\omega t} + cc\right\}$$

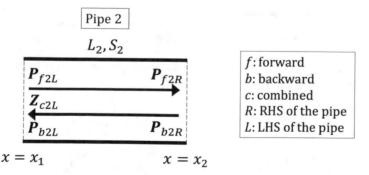

Show that:

Part a (Backward Plane Waves)

The complex amplitude of the pressure P_{b2L} of the forward wave $p_{b2}(x, t)$ at $x = x_2$ (RHS of Pipe 2) can be formulated in terms of the complex amplitude of the pressure P_{b2R} and length L_2 as:

$$P_{b2L} = P_{b2R}e^{-jkL_2} \qquad \text{((Formula 2B)}$$

Part b (LHS of a Pipe)

The acoustic impedance Z_{c2L} at $x = x_1$ (LHS of Pipe 2) can be formulated in terms of the complex amplitudes of the pressures P_{f2R}, P_{b2R}, the length L_2, and the cross-sectional area S_2 as:

$$Z_{c2L} = Z_{f2} \frac{P_{f2R} e^{jkL_2} + P_{b2R} e^{-jkL_2}}{P_{f2R} e^{jkL_2} - P_{b2R} e^{-jkL_2}} \qquad \text{(Formula 3A)}$$

where $Z_{f2} = \frac{\rho_o c}{S_2}$.

Exercise 11.2

The cross-sectional areas of Pipe 1 and Pipe 2 are S_1 and S_2, respectively, as shown in the figure below. The pressures of the forward wave and backward wave in Pipe 1 at $x = x_2$ (RHS of Pipe 1) are P_{f1R} and P_{b1R}, respectively. The pressures of the forward wave and backward wave in Pipe 2 at $x = x_3$ (RHS of Pipe 2) are P_{f2R} and P_{b2R}, respectively:

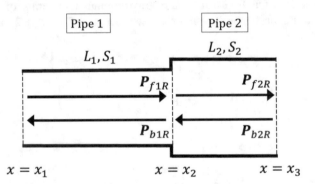

Show that the pressures P_{f1R} and P_{b1R} (RHS of Pipe 1) can be formulated in terms of the pressure P_{f2R} and P_{b2R} (RHS in Pipe 2) as:

$$P_{f1R} = \frac{1}{2}\left(1 + \frac{Z_{f1}}{Z_{f2}}\right) P_{f2R} e^{jkL_2} + \frac{1}{2}\left(1 - \frac{Z_{f1}}{Z_{f2}}\right) P_{b2R} e^{-jkL_2}$$

$$P_{b1R} = \frac{1}{2}\left(1 - \frac{Z_{f1}}{Z_{f2}}\right) P_{f2R} e^{jkL_2} + \frac{1}{2}\left(1 + \frac{Z_{f1}}{Z_{f2}}\right) P_{b2R} e^{-jkL_2}$$

Exercise 11.3 (Closed-Open Pipe)

A pipe has a length L and a cross-section area S_i. The far end is *open* ($Z_o = 0$) and the near end is *closed* by a rigid cap ($Z_1 = \infty$):

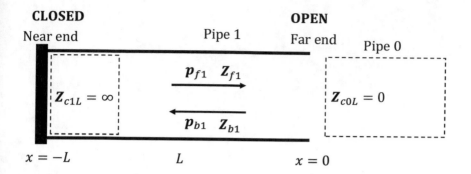

(a) Calculate the eigenfrequencies of the pipe in terms of the pipe length L.
(b) Calculate the eigenmodes of the pressure and plot the first three modes of the
   pipe using the equation below:

$$Z_{c1L} = Z_{f1} \frac{Z_{c0L} + jZ_{f1} \tan(kL)}{jZ_{c0L} \tan(kL) + Z_{f1}} \qquad \text{(Formula 5)}$$

where:

$$Z_{f1} = \frac{\rho_o c}{S_1}$$

(Answers): (a) $\omega_n = c \frac{(2n+1)\pi}{2L}$, $n = 0,\ 1,\ 2\ldots$; (b) $p_n = -2jP \sin\left[\frac{(2n+1)\pi}{2L}x\right]$, $n = 0, 1, 2\ldots$

Chapter 12
Filters and Resonators

In the previous chapter, we defined and derived formulas of complex amplitudes, acoustic impedance, and power transmission in the pipeline. In this chapter, we will extend the analysis of pipelines (pipes without side branches) to pipeline with side branches. Most of the formulas derived in the previous chapter will be reused in this chapter. In fact, a pipeline with a side branch can be treated as a one-to-two pipeline. The acoustic analysis of a one-to-two pipeline can be treated as a one-to-one pipeline by an equivalent acoustic impedance for the two pipes.

Three basic types of filters will be introduced in the second half of this chapter: low-pass filters (Sect. 12.3), high-pass filters (Sect. 12.4), and band-stop filters (Sect. 12.5) based on the principle of the Helmholtz resonator. This chapter will introduce the fundamental principles, the formulas, and the power transmission coefficient of each type of filter.

A numerical method for modeling pipelines with side branches is included in this chapter. A numerical project to model filters is provided without a solution. This chapter is organized into the following sections:

Section 12.1 introduces one-to-two pipes. Formulas for calculating the equivalent acoustic impedance of a one-to-two pipe are derived (Formulas 8A–8B) in this section.

Section 12.2 introduces the power transmission of one-to-two pipes. Formulas for calculating power reflection and transmission will be derived (Formulas 13A–13B) in this section.

Section 12.3 introduces low-pass filters. A formula for the power transmission coefficient of an open side branch as a low-pass filter will be derived in this section.

Section 12.4 introduces high-pass filters. A formula for the power transmission coefficient of a three pipeline as a high-pass filter will be derived in this section.

Section 12.5 introduces band-stop filters. A formula for the power transmission coefficient of the Helmholtz resonator as a band-stop filter is derived in this section.

Section 12.6 introduces a numerical method for calculating the power transmission coefficients for filter designs. This method can be used to model all three basic types of filters: high-pass filters, low-pass filters, and band-stop filters.

© The Author(s), under exclusive license to Springer Nature Switzerland AG 2021   317
H. Lin et al., *Lecture Notes on Acoustics and Noise Control*,
https://doi.org/10.1007/978-3-030-88213-6_12

12.1   Pressure in a One-to-Two Pipe

The power transmission coefficients of a one-to-two pipe as shown below will be calculated and analyzed in this section.

The dimensions and acoustic impedance of a one-to-two pipe are given as:

- Cross-section areas of the pipes are S_i, S_1, and S_2.
- Acoustic impedances of the pipes are Z_i, Z_r, Z_1, and Z_2:

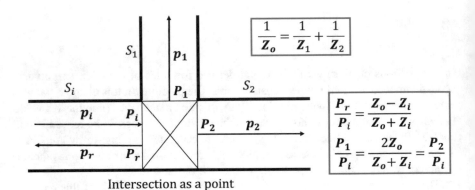

Intersection as a point

where:

S_i: cross-section area of the inlet pipe
S_1: cross-section area of outlet Pipe 1 (side branch)
S_2: cross-section area of outlet Pipe 2 (main pipe)
p_i: pressure of the incident wave in the inlet pipe
p_1: pressure of the transmitted wave in outlet Pipe 1 (side branch)
p_2: pressure of transmitted wave to outlet Pipe 2 (main branch)
$P_i = P_{iR}$: complex pressure of the incident wave at the RHS of the inlet pipe
P_1: complex pressure of the transmitted wave in the side branch pipe at the intersection
$P_2 = P_{2L}$: complex pressure of the transmitted wave at the LHS of the outlet pipe
Z_i: acoustic impedance of the incident wave at the RHS of the inlet pipe
Z_1: acoustic impedance of the transmitted wave inside the branch pipe at the intersection
Z_2: acoustic impedance of the transmitted wave at the LHS of the outlet pipe

Pressures and Velocities in a One-to-Two Pipe

The incident and transmission pressures are forward plane waves (P_i, P_1, and P_2) and have the same wave function format. The pressure and velocity functions of an incident wave are shown below for reference:

$$p_i(x,t) = \frac{1}{2}\left(P_i\, e^{\,j(\omega t - kx)} + P_i^*\, e^{-j(\omega t - kx)}\right) \qquad \text{(forward plane wave)}$$

$$v_i(x,t) = \frac{1}{2}\left(\frac{1}{\rho_o c}\, P_i\, e^{\,j(\omega t - kx)} + \frac{1}{\rho_o c}\, P_i^*\, e^{-j(\omega t - kx)}\right) \qquad \text{(forward plane wave)}$$

The reflection pressure is a backward plane wave (P_r) and is shown below for reference:

$$p_r(x,t) = \frac{1}{2}\left(P_r\, e^{\,j(\omega t + kx)} + P_r^*\, e^{-j(\omega t + kx)}\right) \qquad \text{(backward plane wave)}$$

The velocity function is calculated by Euler's force equation:

$$v_r(x,t) = \frac{1}{2}\left(\frac{1}{-\rho_o c}\, P_r\, e^{\,j(\omega t + kx)} + \frac{1}{-\rho_o c}\, P_r^*\, e^{-j(\omega t + kx)}\right) \qquad \text{(backward plane wave)}$$

- Note that pressure and velocity are functions of space and time. Also, the real pressure, $p_i(x,t)$, is the addition of a complex conjugate pair of the pressures $\frac{1}{2}\left[p_i(x,t) + p_i^*(x,t)\right]$.

Remarks
- The incident acoustic impedance and reflected backward acoustic impedance have a simple relationship:

$$Z_r = -Z_i$$

- Because transmission pressures P_1 and P_2 are outward only (no return), acoustic impedance $Z_1(=\rho_o c/S_1)$ and $Z_2(=\rho_o c/S_2)$ are real numbers.
- Because the incident and reflected pressures P_i and P_r are individual plane waves (forward and backward), acoustic impedance $Z_i(=\rho_o c/S_i)$ and $Z_r\left(=-\frac{\rho_o c}{S_i} = -Z_i\right)$ are real numbers.

12.1.1   Equivalent Acoustic Impedance of a One-to-Two Pipe

The formulas for a one-to-two pipe can be compared to the formulas for a one-to-one pipe as listed below:

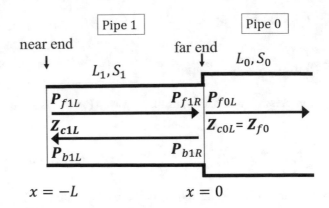

$$P_{f1R} = \frac{1}{2}\left(1 + \frac{Z_{f1}}{Z_{foL}}\right)P_{foL} \qquad \text{(Formula 8A)}$$

$$P_{b1R} = \frac{1}{2}\left(1 - \frac{Z_{f1}}{Z_{foL}}\right)P_{foL} \qquad \text{(Formula 8B)}$$

Formulas 8A–8B (Equivalent Acoustic Impedance of a One-to-Two Pipe)
The dimensions and acoustic impedance of a one-to-two pipe are given as follows:

- Cross-section areas of the pipes are S_i, S_1, and S_2.
- Acoustic impedances of the pipe are Z_i, Z_r, Z_1, and Z_2:

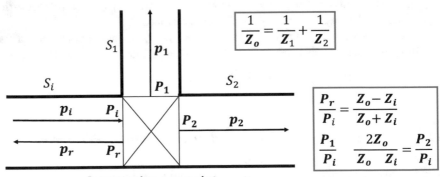

$$\frac{1}{Z_o} = \frac{1}{Z_1} + \frac{1}{Z_2}$$

$$\frac{P_r}{P_i} = \frac{Z_o - Z_i}{Z_o + Z_i}$$

$$\frac{P_1}{P_i} = \frac{2Z_o}{Z_o \; Z_i} = \frac{P_2}{P_i}$$

Intersection as a point

$$P_i = \frac{Z_o + Z_i}{2Z_o}P_2 \qquad \text{(Formula 8A)}$$

$$P_r = \frac{Z_o - Z_i}{2Z_o} P_2 \qquad \text{(Formula 8B)}$$

where Z_o is the equivalent acoustic impedance. This Z_o is the combined impedance of Z_1 and Z_2 and is formulated as:

$$\frac{1}{Z_o} = \frac{1}{Z_1} + \frac{1}{Z_2} \qquad \text{(Formula 12)}$$

Proof 12: Solution

The pressure and velocity functions in the previous section are general solutions for plane wave equations. The exact reflection and transmission pressures can be solved by the boundary conditions at the intersection.

The boundary of the intersection of three pipes can be considered as a point, and the following conditions must be satisfied:

(a) All pressures are equal in amplitude at the intersection
    (related to Newton's third law):

$$P_i + P_r = P_1 = P_2 \qquad (1)$$

(b) The total flow rate inward is equal to the total flow rate outward
    (based on the conservation of mass):

$$S(V_i + V_r) = S_1 V_1 + S_2 V_2$$
$$\rightarrow U_i + U_r = U_1 + U_2 \qquad (2)$$

where the relationship between V and U is:

$$v(x, t) = \frac{1}{2}\left(V e^{j(\omega t - kx)} + V^* e^{-j(\omega t - kx)} \right)$$

$$u(x, t) \equiv Sv(x, t) = \frac{1}{2}S\left(V e^{j(\omega t - kx)} + V^* e^{-j(\omega t - kx)} \right)$$

$$= \frac{1}{2}\left(U e^{j(\omega t - kx)} + U^* e^{-j(\omega t - kx)} \right)$$

The acoustic impedance (Z) gives the relationship between pressure and velocity as shown below:

$$Z = \frac{P}{U} = \frac{P}{SV} = \frac{1}{S}z \quad \leftarrow \quad z = \frac{P}{V}$$

Use the above acoustic impedance to replace U with P/Z in Eq. (2) to obtain:

$$\frac{P_i}{Z_i} - \frac{P_r}{Z_i} = \frac{P_1}{Z_1} + \frac{P_2}{Z_2}$$

Now, P_r, P_1, and P_2 can be solved from Eq. (1) and Eq. (3) as follows:

$$P_i + P_r = P_1 = P_2 \tag{1'}$$

$$P_i - P_r = \frac{Z_i}{Z_1} P_1 + \frac{Z_i}{Z_2} P_2 \tag{3}$$

Since $P_1 = P_2$, P_1 can be replaced with P_2 in Eq. (3) to reduce one unknown variable (P_2):

$$P_i - P_r = \frac{Z_i}{Z_1} P_1 + \frac{Z_i}{Z_2} P_2 = \left(\frac{1}{Z_1} + \frac{1}{Z_2}\right) \frac{Z_i}{1} P_2 \tag{4}$$

Define the equivalent acoustic impedance Z_o as the combined impedance of Z_1 and Z_2 and computed as:

$$\frac{1}{Z_o} = \frac{1}{Z_1} + \frac{1}{Z_2} \tag{Formula 12}$$

Based on the equivalent acoustic impedance Z_o defined above, Eq. (4) becomes:

$$P_i - P_r = \frac{Z_i}{Z_o} P_2 \tag{5}$$

Combining Eq. (1) and Eq. (5) yields Formula 8:

$$P_i = \frac{Z_o + Z_i}{2Z_o} P_2 \tag{Formula 8A}$$

$$P_r = \frac{Z_o - Z_i}{2Z_o} P_2 \tag{Formula 8B}$$

Note that by defining the equivalent acoustic impedance Z_o as Formula 12, Formula 8 can be used to calculate pressures P_r and P_2 for the one-to-two pipe. The definition of this new acoustic impedance Z_o may be used in many outward pipes in a later section.

12.2   Power Transmission of a One-to-Two Pipe

12.2.1   Power Reflection and Transmission of a One-to-Two Pipe

The power reflection and transmission coefficients in a one-to-two pipe are defined as:

$$R_w \equiv \frac{w_r}{w_i} = \frac{Re\,(P_r)\,Re\left(\frac{P_r}{Z_r}\right)}{Re\,(P_i)\,Re\left(\frac{P_i}{Z_i}\right)}$$

$$T_{w1} \equiv \frac{w_1}{w_i} = \frac{Re\,(P_1)\,Re\left(\frac{P_1}{Z_1}\right)}{Re\,(P_i)\,Re\left(\frac{P_i}{Z_i}\right)}$$

$$T_{w2} \equiv \frac{w_2}{w_i} = \frac{Re\,(P_2)\,Re\left(\frac{P_2}{Z_2}\right)}{Re\,(P_i)\,Re\left(\frac{P_i}{Z_i}\right)}$$

where, for traveling plane waves, the following quantities are real numbers:

$$Z_i, Z_1, Z_2, P_i, P_r, P_1, P_2 \text{ real}$$

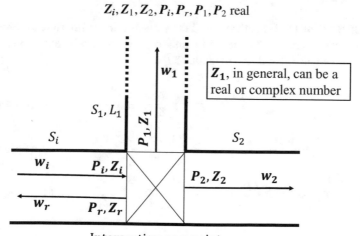

Intersection as a point

12.2.2   Special Case of Power Reflection and Transmission: A One-to-Two Pipe with No Returning Waves

The formulas (Formulas 13A–13E) of power reflection and transmission for a one-to-two pipe with no returning waves in the outlet pipes are similar to the formulas (Formulas 11C–11D) of power reflection and transmission for a one-to-one pipe with no returning waves in the outlet pipes:

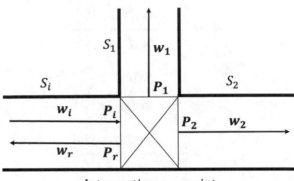

Intersection as a point

Formula 13 (Power Reflection and Transmission in a One-to-Two Pipe)
Power reflection and transmission in a one-to-two pipe can be formulated with real acoustic impedance Z and real magnitude P as:

$$R_w = \frac{w_r}{w_i} = -\left(\frac{P_r}{P_i}\right)^2 = -\left(\frac{Z_o - Z_i}{Z_o + Z_i}\right)^2 \qquad \text{(Formula 13A)}$$

$$T_{w2} = \frac{w_2}{w_i} = \left(\frac{P_2}{P_i}\right)^2 \frac{Z_i}{Z_2} = \left(\frac{2Z_o}{Z_o + Z_i}\right)^2 \frac{Z_i}{Z_2} \qquad \text{(Formula 13B)}$$

$$T_{w1} = \frac{w_1}{w_i} = \left(\frac{P_1}{P_i}\right)^2 \frac{Z_i}{Z_1} = \left(\frac{2Z_o}{Z_o + Z_i}\right)^2 \frac{Z_i}{Z_1} \qquad \text{(Formula 13C)}$$

$$w_i + w_r = w_1 + w_2 \qquad \text{(Formula 13D)}$$

where:

$$\frac{1}{Z_o} = \frac{1}{Z_1} + \frac{1}{Z_2} \qquad \text{(Formula 12E)}$$

And $Z_i, Z_1, Z_2, P_i, P_r, P_1,$ and P_2 are real numbers for traveling plane waves (no returning waves in Pipes 1 and 2).

Proof of Formulas 13 A–C

For traveling plane waves (forward and backward) in pipes:

- The acoustic impedance Z is a real number.
- The complex amplitude of pressure P is also a real number.

Therefore, Formula 13 for the power of acoustic waves can be formulated with real acoustic impedance Z and real magnitude P as:

$$w = Re\,(P)\,Re\left(\frac{P}{Z}\right) = P^2 \frac{1}{Z}$$

In the one-to-two pipe, the following quantities for traveling plane waves are real numbers:

$$Z_i, Z_r, \quad Z_1, Z_2, P_i, P_r, P_1, P_2 \text{ real}$$

and:

$$Z_r = -Z_i \qquad\qquad \text{(Formula 4)}$$

Therefore:

$$w_i = Re\,(P_i)\,Re\left(\frac{P_i}{Z_i}\right) = P_i^2 \frac{1}{Z_i}$$

$$w_r = Re\,(P_r)\,Re\left(\frac{P_r}{Z_r}\right) = P_r^2 \frac{1}{Z_r} = -P_r^2 \frac{1}{Z_i}$$

$$w_1 = Re\,(P_1)\,Re\left(\frac{P_1}{Z_1}\right) = P_1^2 \frac{1}{Z_1}$$

$$w_2 = Re\,(P_2)\,Re\left(\frac{P_2}{Z_2}\right) = P_2^2 \frac{1}{Z_2}$$

The power reflection and transmission coefficients are defined below and can be simplified as:

$$R_w \equiv \frac{w_r}{w_i} = \frac{Re\,(P_r)\,Re\left(\frac{P_r}{Z_r}\right)}{Re\,(P_i)\,Re\left(\frac{P_i}{Z_i}\right)} = \frac{\frac{P_r^2}{Z_r}}{\frac{P_i^2}{Z_i}} = -\left(\frac{P_r}{P_i}\right)^2$$

$$T_{w1} \equiv \frac{w_1}{w_i} = \frac{Re\,(P_1)\,Re\left(\frac{P_1}{Z_1}\right)}{Re\,(P_i)\,Re\left(\frac{i}{Z_i}\right)} = \frac{\frac{P_1^2}{Z_1}}{\frac{P_i^2}{Z_i}} = \left(\frac{P_1}{P_i}\right)^2 \frac{Z_i}{Z_1}$$

$$T_{w2} \equiv \frac{w_2}{w_i} = \frac{Re\,(P_2)\,Re\left(\frac{P_2}{Z_2}\right)}{Re\,(P_i)\,Re\left(\frac{i}{Z_i}\right)} = \frac{\frac{P_2^2}{Z_2}}{\frac{P_i^2}{Z_i}} = \left(\frac{P_2}{P_i}\right)^2 \frac{Z_i}{Z_2}$$

Based on Formula 8:

$$\frac{P_{b1L}}{P_{f1R}} = \frac{Z_{foL} - Z_{f1}}{Z_{foL} + Z_{f1}} \qquad \text{(Formula 8C)}$$

$$\frac{P_{foL}}{P_{f1R}} = \frac{2Z_{foL}}{Z_{foL} + Z_{f1}} \qquad \text{(Formula 8D)}$$

The power reflection and transmission coefficients can be formulated as:

$$R_w = \frac{\frac{P_r^2}{Z_r}}{\frac{P_i^2}{Z_i}} = -\left(\frac{P_r}{P_i}\right)^2 = -\left(\frac{Z_o - Z_i}{Z_o + Z_i}\right)^2 \qquad \text{(Formula 13A)}$$

$$T_{w1} = \frac{\frac{P_1^2}{Z_1}}{\frac{P_i^2}{Z_i}} = \left(\frac{P_1}{P_i}\right)^2 \frac{Z_i}{Z_1} = \left(\frac{2Z_o}{Z_o + Z_i}\right)^2 \frac{Z_i}{Z_1} \qquad \text{(Formula 13B)}$$

$$T_{w2} = \frac{\frac{P_2^2}{Z_2}}{\frac{P_i^2}{Z_i}} = \left(\frac{P_2}{P_i}\right)^2 \frac{Z_i}{Z_2} = \left(\frac{2Z_o}{Z_o + Z_i}\right)^2 \frac{Z_i}{Z_2} \qquad \text{(Formula 13C)}$$

Proof of Formula 13 D

Based on the conservation of energy, the power that goes into a system equals the power that comes out of it:

(POWER IN) = (POWER OUT)

Therefore, the summation of power on the left pipe is equal to the summation of power on the upper pipe (Pipe 1) and the right pipe (Pipe 2):

$$w_i + w_r = w_1 + w_2 \qquad \text{(Formula 13D)}$$

Note that w_i, w_1, and w_2 all do positive work, but w_r does negative work due to the negative impedance (180° phase difference between the pressure and velocity).

Dividing the equation above by w_i yields the relationship between the power ratios as shown below:

$$1 + \frac{w_r}{w_i} = \frac{w_1}{w_i} + \frac{w_2}{w_i}$$

$$\rightarrow 1 + R_w = T_{w1} + T_{w2}$$

Example 12.1: (Power in the One-to-Two Pipe)

The cross-section areas of Pipe i, Pipe 1, and Pipe 2 as shown below
are $S_i = 0.0830\ [m^2]$, $S_1 = 0.0415\ [m^2]$, and $S_2 = 0.0830\ [m^2]$, respectively:

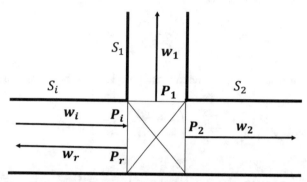

Intersection as a point

Use 415 rayls for the characteristic impedance ($\rho_o c$) of air.

(a) Determine the power reflection coefficient R_w.
(b) Determine the power transmission coefficients T_{w1}, T_{w2}.
(c) Validate your answers in Parts a and b using the conservation of energy.

Solution of Example 12.1

The acoustic impedances in Pipe i, Pipe 1, and Pipe 2 can be calculated based on
Formulas 3 and 4 as:

$$Z_i = \frac{\rho_o c}{S_i} = \frac{415}{0.083} = 50 \qquad \text{(Formula 3 : forward wave)}$$

$$Z_r = \frac{-\rho_o c}{S_i} = \frac{-415}{0.083} = -50 \qquad \text{(Formula 4 : backward wave)}$$

$$Z_1 = \frac{\rho_o c}{S_1} = \frac{415}{0.0415} = 100 \qquad \text{(Formula 3 : forward wave)}$$

$$Z_2 = \frac{\rho_o c}{S_2} = \frac{415}{0.083} = 50 \qquad \text{(Formula 3 : forward wave)}$$

Note that Z_i, Z_r, Z_1, and Z_2 are real numbers.
The equivalent acoustic impedance:

$$\frac{1}{Z_o} = \frac{1}{Z_1} + \frac{1}{Z_2}$$

$$\rightarrow Z_o = \frac{1}{\frac{1}{Z_1} + \frac{1}{Z_2}} = \frac{1}{\frac{1}{100} + \frac{1}{50}} = \frac{100}{3}$$

(a) The power reflection coefficients can be calculated based on Formula 13A as:

$$R_w = -\left(\frac{Z_o - Z_i}{Z_o + Z_i}\right)^2 = -\left(\frac{\frac{100}{3} - 50}{\frac{100}{3} + 50}\right)^2 = -0.04 \qquad \text{(Formula 13A)}$$

(b) The power transmission coefficients can be calculated based on Formula 13B as:

$$T_{w2} = \left(\frac{2Z_o}{Z_o + Z_i}\right)^2 \frac{Z_i}{Z_2} = \left(\frac{2 \times \frac{100}{3}}{\frac{100}{3} + 50}\right)^2 \frac{50}{50} = 0.64 \qquad \text{(Formula 13B)}$$

$$T_{w1} = \left(\frac{2Z_o}{Z_o + Z_i}\right)^2 \frac{Z_i}{Z_1} = \left(\frac{2 \times \frac{100}{3}}{\frac{100}{3} + 50}\right)^2 \frac{50}{100} = 0.32 \qquad \text{(Formula 13C)}$$

(c) Based on the conservation of energy (Formula 13D):

$$w_i + w_r = w_1 + w_2 \qquad \text{(Formula 13D)}$$

Divide the equation above by w_i to obtain the power ratio:

$$1 + \frac{w_r}{w_i} = \frac{w_1}{w_i} + \frac{w_2}{w_i}$$

$$\rightarrow 1 + R_w = T_{w1} + T_{w2}$$

Substituting the power reflection ($R_w = -0.04$) and power transmission ($T_{w1} = 0.32$ and $T_{w2} = 0.64$) into the conservation of energy ($1 + R_w = T_{w1} + T_{w2}$) gives:

$$1 + R_w = T_{w1} + T_{w2}$$

$$\rightarrow 1 - 0.04 = 0.32 + 0.64$$

The equation above validates the power reflection (R_w) and power transmission (T_{w1}) calculated in Parts a and b.

12.3   Low-Pass Filters

A low-pass filter can be constructed by connecting three pipes in series. The cross-sections of the pipes are S_i, S_2, and S_o as shown below. The length of the middle pipe is L:

$$\frac{P_r}{P_i} = \frac{\left(1-\frac{Z_i}{Z_o}\right)\cos(kL)+j\left(\frac{Z_2}{Z_o}-\frac{Z_i}{Z_2}\right)\sin(kL)}{\left(1+\frac{Z_i}{Z_o}\right)\cos(kL)+j\left(\frac{Z_2}{Z_o}+\frac{Z_i}{Z_2}\right)\sin(kL)}$$

$$\frac{P_o}{P_i} = \frac{2}{\left(1+\frac{Z_i}{Z_o}\right)\cos(kL)+j\left(\frac{Z_2}{Z_o}+\frac{Z_i}{Z_2}\right)\sin(kL)}$$

$x = 0 \qquad x = L$

Schematic of three sections in serial

Assumptions

1. The length L of the middle pipe (Pipe 2) is much less than the wavelength λ divided by 2π as:

$$L \ll \frac{\lambda}{2\pi} = \frac{1}{k}$$
$$\rightarrow kL \ll 1$$

2. The cross-section S_2 of the middle pipe (Pipe 2) is much greater than the cross-section S_i of the inlet pipe (Pipe i) and the cross-section S_o of the outlet pipe (Pipe o) as:

$$S_2 \gg S_i$$

$$S_2 \gg S_o$$

3. The cross-section S_i of the inlet pipe (Pipe i) is equal to the cross-section S_o of the outlet pipe (Pipe o) as:

$$S_i = S_o$$

Objective

Show that based on the three assumptions above, the power reflection R_w and transmission T_w coefficients of this low-pass filter can be formulated as:

$$T_w = \frac{1}{\left(1+\left(\frac{\omega}{\omega_c}\right)^2\right)^2}; \quad R_w = -\frac{\left(\frac{\omega}{\omega_c}\right)^4}{\left(1+\left(\frac{\omega}{\omega_c}\right)^2\right)^2} \text{ where } \omega_c \equiv \frac{2S_ic}{S_2L}.$$

Procedures
Step 1: Set up balancing equations of pressure and mass flow rate at the interfaces of the pipes:
Step 2: Calculate pressure ratios $\frac{P_r}{P_i}$ and $\frac{P_o}{P_i}$ of this low-pass filter.
Step 3: Calculate the power reflection coefficient R_w and the transmission coefficients T_w of this low-pass filter.

Step 1: Set Up Balancing Equations
The pressures and velocities in the low-pass filter are:

$$p_i = P_i e^{j(\omega t - kx)}, \qquad u_i = U_i e^{j(\omega t - kx)} = \frac{1}{Z_i} P_i e^{j(\omega t - kx)}$$

$$p_r = P_r e^{j(\omega t + kx)}, \qquad u_r = U_r e^{j(\omega t + kx)} = \frac{-1}{Z_i} P_r e^{j(\omega t + kx)}$$

$$p_f = P_f e^{j(\omega t - k(x-L))}, \quad u_f = U_f e^{j(\omega t - k(x-L))} = \frac{1}{Z_2} P_f e^{j(\omega t - k(x-L))}$$

$$p_b = P_b e^{j(\omega t + k(x-L))}, \quad u_b = U_b e^{j(\omega t + k(x-L))} = \frac{-1}{Z_2} P_b e^{j(\omega t + k(x-L))}$$

$$p_o = P_o e^{j(\omega t - k(x-L))}, \quad u_o = U_o e^{j(\omega t - k(x-L))} = \frac{1}{Z_o} P_o e^{j(\omega t - k(x-L))}$$

where the subscript i and r denote the incident and reflected waves, respectively. At the reflection point, there is also a transmitted wave, p_o. At either side of the reflection point, say $x = 0$, shown in the figure above, the pressures p must be the same, and the uS must also be the same.

Hence, the force balance and conservation of mass at $x = 0$ are:

$$P_i + P_r = P_f e^{jkL} + P_b e^{-jkL}$$

$$P_i - P_r = \frac{Z_i}{Z_2}\left(P_f e^{jkL} - P_b e^{-jkL}\right)$$

Also, the force balance and conservation of mass at $x = L$ are:

$$P_f + P_b = P_o$$

$$P_f - P_b = \frac{Z_2}{Z_o}P_o$$

Step 2: Calculate Pressure Ratios
Adding and subtracting the two conditions at $x = 0$ give:

$$P_i = \frac{1}{2}\left(1 + \frac{Z_i}{Z_2}\right)P_f e^{jkL} + \frac{1}{2}\left(1 - \frac{Z_i}{Z_2}\right)P_b e^{-jkL}$$

$$P_r = \frac{1}{2}\left(1 - \frac{Z_i}{Z_2}\right)P_f e^{jkL} + \frac{1}{2}\left(1 + \frac{Z_i}{Z_2}\right)P_b e^{-jkL}$$

Adding and subtracting the two conditions at $x = L$ yield:

$$P_f = \frac{1}{2}\left(1 + \frac{Z_2}{Z_o}\right)P_o$$

$$P_b = \frac{1}{2}\left(1 - \frac{Z_2}{Z_o}\right)P_o$$

Substitute the two equations above into the previous equation to get:

$$\frac{P_i}{P_o} = \frac{1}{4}\left(1 + \frac{Z_i}{Z_2}\right)\left(1 + \frac{Z_2}{Z_o}\right)e^{jkL} + \frac{1}{4}\left(1 - \frac{Z_i}{Z_2}\right)\left(1 - \frac{Z_2}{Z_o}\right)e^{-jkL}$$

$$= \frac{1}{4}\left(1 + \frac{Z_2}{Z_o} + \frac{Z_i}{Z_2} + \frac{Z_i}{Z_o}\right)e^{jkL} + \frac{1}{4}\left(1 - \frac{Z_2}{Z_o} - \frac{Z_i}{Z_2} + \frac{Z_i}{Z_o}\right)e^{-jkL}$$

$$= \frac{1}{2}\left(1 + \frac{Z_i}{Z_o}\right)\cos(kL) + j\frac{1}{2}\left(\frac{Z_2}{Z_o} + \frac{Z_i}{Z_2}\right)\sin(kL)$$

$$\frac{P_r}{P_o} = \frac{1}{4}\left(1 - \frac{Z_i}{Z_2}\right)\left(1 + \frac{Z_2}{Z_o}\right)e^{jkL} + \frac{1}{4}\left(1 + \frac{Z_i}{Z_2}\right)\left(1 - \frac{Z_2}{Z_o}\right)e^{-jkL}$$

$$= \frac{1}{4}\left(1 + \frac{Z_2}{Z_o} - \frac{Z_i}{Z_2} - \frac{Z_i}{Z_o}\right)e^{jkL} + \frac{1}{4}\left(1 - \frac{Z_2}{Z_o} + \frac{Z_i}{Z_2} - \frac{Z_i}{Z_o}\right)e^{-jkL}$$

$$= \frac{1}{2}\left(1 - \frac{Z_i}{Z_o}\right)\cos(kL) + \frac{1}{2}j\left(\frac{Z_2}{Z_o} - \frac{Z_i}{Z_2}\right)\sin(kL)$$

Using the two equations above yields:

$$\frac{P_r}{P_i} = \frac{\left(1 - \frac{Z_i}{Z_o}\right)\cos(kL) + j\left(\frac{Z_2}{Z_o} - \frac{Z_i}{Z_2}\right)\sin(kL)}{\left(1 + \frac{Z_i}{Z_o}\right)\cos(kL) + j\left(\frac{Z_2}{Z_o} + \frac{Z_i}{Z_2}\right)\sin(kL)}$$

$$\frac{P_o}{P_i} = \frac{2}{\left(1 + \frac{Z_i}{Z_o}\right)\cos(kL) + j\left(\frac{Z_2}{Z_o} + \frac{Z_i}{Z_2}\right)\sin(kL)}$$

where all the acoustic impedances are real numbers, as shown below:

$$Z_i = \frac{\rho_o c}{S_i}; Z_2 = \frac{\rho_o c}{S_2}; Z_o = \frac{\rho_o c}{S_o}$$

Step 3: Calculate Power Reflection and Transmission Coefficients

For plane waves in a pipe, Z_i, Z_2, and Z_o are real numbers, and also the incident wave P_i is a real number. However, from the equations above, P_r and P_o are complex numbers.

Then the power reflection and transmission coefficients can be formulated as:

$$R_w = \left(Re \left(\frac{P_r}{P_i} \right) \right)^2 \frac{Z_i}{Z_r} = - \left(Re \left(\frac{P_r}{P_i} \right) \right)^2 \qquad (Z_i, Z_r, P_i \text{ real})$$

$$T_w = \left(Re \left(\frac{P_o}{P_i} \right) \right)^2 \frac{Z_i}{Z_o} \qquad (Z_i, Z_o, P_i \text{ real})$$

where:

$$Re \left(\frac{P_o}{P_i} \right) = Re \left(\frac{2}{2 \cos (kL) + j \left(\frac{Z_2}{Z_o} + \frac{Z_i}{Z_2} \right) \sin (kL)} \right)$$

$$= \frac{4 \cos (kL)}{4 \cos^2 (kL) + \left(\frac{Z_2}{Z_o} + \frac{Z_i}{Z_2} \right)^2 \sin^2 (kL)}$$

$$Re \left(\frac{P_r}{P_i} \right) = Re \left(\frac{j \left(\frac{Z_2}{Z_o} - \frac{Z_i}{Z_2} \right) \sin (kL)}{2 \cos (kL) + j \left(\frac{Z_2}{Z_o} + \frac{Z_i}{Z_2} \right) \sin (kL)} \right) = \frac{\left(\left(\frac{Z_2}{Z_o} \right)^2 - \left(\frac{Z_i}{Z_2} \right)^2 \right) \sin^2 (kL)}{4 \cos^2 (kL) + \left(\frac{Z_2}{Z_o} + \frac{Z_i}{Z_2} \right)^2 \sin^2 (kL)}$$

where Z_i, Z_2, and Z_o are all real numbers.

Assumption #1:

$$kL \ll 1$$

Then:

$$\cos (kL) \cong 1,$$

$$\sin (kL) \cong kL, \quad (Z_i, Z_2, Z_o, P_i, P_r, P_o \text{ real})$$

Based on this assumption, $Re \left(\frac{P_o}{P_i} \right)$ and $Re \left(\frac{P_r}{P_i} \right)$ above can be simplified as:

$$Re\left(\frac{P_o}{P_i}\right) = \frac{4}{4 + \left(\frac{Z_2}{Z_o} + \frac{Z_i}{Z_2}\right)^2 (kL)^2} = \frac{1}{1 + \left(\left(\frac{Z_2}{Z_o} + \frac{Z_i}{Z_2}\right)\frac{kL}{2}\right)^2}$$

$$Re\left(\frac{P_r}{P_i}\right) = \frac{\left(\left(\frac{Z_2}{Z_o}\right)^2 - \left(\frac{Z_i}{Z_2}\right)^2\right)(kL)^2}{4 + \left(\frac{Z_2}{Z_o} + \frac{Z_i}{Z_2}\right)^2 (kL)^2} = \frac{\left(\left(\frac{Z_2}{Z_o}\right)^2 - \left(\frac{Z_i}{Z_2}\right)^2\right)\left(\frac{kL}{2}\right)^2}{1 + \left(\frac{Z_2}{Z_o} + \frac{Z_i}{Z_2}\right)^2 \left(\frac{kL}{2}\right)^2}$$

Assumption #2:

$$S_2 \gg S_i$$

$$S_2 \gg S_o$$

Then:

$$\frac{Z_i}{Z_2} = \frac{S_2}{S_i} \gg 1$$

$$\frac{Z_2}{Z_o} = \frac{S_o}{S_2} \cong 0$$

Based on this assumption, $Re\left(\frac{P_o}{P_i}\right)$ and $Re\left(\frac{P_r}{P_i}\right)$ become:

$$Re\left(\frac{P_o}{P_i}\right) = \frac{1}{1 + \left(\left(\frac{Z_2}{Z_o} + \frac{Z_i}{Z_2}\right)\frac{kL}{2}\right)^2} = \frac{1}{1 + \left(\frac{S_2 kL}{2S_i}\right)^2}$$

$$Re\left(\frac{P_r}{P_i}\right) = \frac{\left(\left(\frac{Z_2}{Z_o}\right)^2 - \left(\frac{Z_i}{Z_2}\right)^2\right)\left(\frac{kL}{2}\right)^2}{1 + \left(\frac{Z_2}{Z_o} + \frac{Z_i}{Z_2}\right)^2 \left(\frac{kL}{2}\right)^2} = \frac{-\left(\frac{S_2 kL}{2S_i}\right)^2}{1 + \left(\frac{S_2 kL}{2S_i}\right)^2}$$

Assumption #3:

$$S_i = S_o$$

Then:

$$Z_i = Z_o$$

Based on this assumption, T_w and R_w can be simplified as:

$$T_w = \left(Re\left(\frac{P_o}{P_i}\right) \right)^2 \frac{Z_i}{Z_o} = \frac{1}{\left(1 + \left(\frac{S_2 kL}{2S_i}\right)^2\right)^2}$$

$$R_w = -\left(Re\left(\frac{P_r}{P_i}\right) \right)^2 = -\left(Re\left(\frac{P_r}{P_i}\right) \right)^2 = -\frac{\left(\frac{S_2 kL}{2S_i}\right)^4}{\left(1 + \left(\frac{S_2 kL}{2S_i}\right)^2\right)^2}$$

Let:

$$\omega_c \equiv \frac{2S_i c}{S_2 L}.$$

Then:

$$\frac{S_2 kL}{2S_i} = \frac{S_2 \frac{\omega}{c} L}{2S_i} = \frac{\omega}{\omega_c}$$

Based on the definition of ω_c, the power transmission coefficient T_w and reflection coefficient R_w can be concluded as:

$$T_w = \frac{1}{\left(1 + \left(\frac{S_2 kL}{2S_i}\right)^2\right)^2} = \frac{1}{\left(1 + \left(\frac{\omega}{\omega_c}\right)^2\right)^2}$$

$$R_w = -\frac{\left(\frac{S_2 kL}{2S_i}\right)^4}{\left(1 + \left(\frac{S_2 kL}{2S_i}\right)^2\right)^2} = -\frac{\left(\frac{\omega}{\omega_c}\right)^4}{\left(1 + \left(\frac{\omega}{\omega_c}\right)^2\right)^2}$$

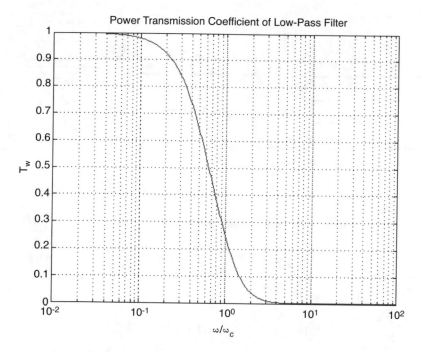

Power transmission coefficient as a function of the frequency ratio

Remarks

- The expanded middle section behaves like a low-pass filter. It filters out the wave for frequencies above $\omega_c = \frac{2Sc}{S_2 L}$.
- The impedance in a low-pass filter changes two times in series.

12.4 High-Pass Filters

A high-pass filter can be constructed by a pipe with a small hole. The cross-sections of the pipes are S_i, S_1, and S_2 as shown below. The length of outlet Pipe 1 (open side branch) is $L_1 = 0$. The radius of Pipe 1 is a and is related to the cross-section S_1 of Pipe 1 as $S_1 = \pi a^2$:

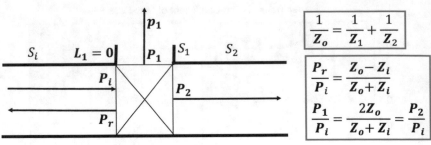

Intersection as a point

Assumptions
(1) Side pipe can be modeled as a point source:

$$p(r,t) = \frac{\rho_o ck}{4\pi r} Q_s e^{j\left(\omega t - kr + \theta_0 + \frac{\pi}{2}\right)}$$

$$v(r,t) = \frac{k}{4\pi r \cos(\phi)} Q_s e^{j\left(\omega t - kr - \phi + \theta_o + \frac{\pi}{2}\right)}$$

(2) The radius of Pipe 1, a, is much less than the wavelength λ divided by 2π as:

$$a \ll \frac{\lambda}{2\pi} = \frac{1}{k}$$

$$\rightarrow ka \ll 1$$

(3) The acoustic impedance Z_1 is calculated at:

$$r = \frac{8a}{3\pi}$$

Objective
Show that based on the three assumptions above, the power transmission T_{w2} coefficients to Pipe 2 of this high-pass filter can be formulated as:

$$T_w = \frac{1}{\left(1 + \left(\frac{\omega_c}{\omega}\right)^2\right)^2}$$

where $\omega_c = \frac{c}{2L'}$ and $L' = \frac{8S_i}{3\pi S_1} a$.

Procedures

Step 1: Estimate acoustic impedance Z_1 at the open side branch.
Step 2: Calculate pressure ratio using equivalent acoustic impedance.
Step 3: Calculate power transmission coefficient.

Step 1: Estimate Acoustic Impedance Z_1 at the Open Side Branch

The acoustic impedance Z_1 at the open side branch can be calculated from the pressure and velocity at the side open.

Assumption #1: Side pipe can be modeled as a point source.

Using the formulas of point sources, the pressure and the flow velocity are given by:

$$p(r,t) = \frac{\rho_o c k}{4\pi r} Q_s e^{j\left(\omega t - kr + \theta_0 + \frac{\pi}{2}\right)}$$

$$v(r,t) = \frac{k}{4\pi r \cos(\phi)} Q_s e^{j\left(\omega t - kr - \phi + \theta_o + \frac{\pi}{2}\right)}$$

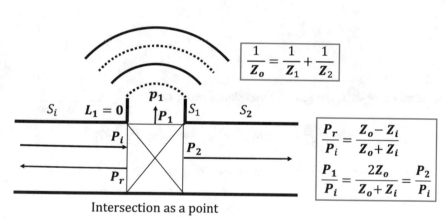

Intersection as a point

A pipe with a small hole as an open side branch

The acoustic impedance can be calculated by the above pressure and velocity:

$$Z_1 \equiv \frac{P_1}{S_1 v_1} = \frac{\frac{\rho_o c k}{4\pi r} Q_s e^{j\left(\omega t - kr + \theta_0 + \frac{\pi}{2}\right)}}{S_1 \frac{k}{4\pi r \cos(\phi)} Q_s e^{j\left(\omega t - kr - \phi + \theta_o + \frac{\pi}{2}\right)}}$$

Canceling out all common terms yields:

$$Z_1 = \frac{\rho_o c}{S_1} \cos(\phi) e^{j(\phi)}$$

where:

$$\cos(\phi) = \frac{kr}{\sqrt{1 + (kr)^2}}$$

Assumption #2: The radius a of Pipe 1 is much less than the wavelength λ divided by 2π as:

$$a \ll \frac{\lambda}{2\pi} = \frac{1}{k}$$

$$\rightarrow ka \ll 1$$

For $kr \ll 1$:

$$\cos(\phi) = \frac{kr}{\sqrt{1 + (kr)^2}} \cong kr$$

and:

$$\phi \cong \frac{\pi}{2}$$

Based on $kr \ll 1$, the acoustic impedance becomes:

$$\mathbf{Z}_1 = \frac{\rho_o c}{S_1} \cos(\phi) e^{j(\phi)} \cong \frac{\rho_o c}{S_1} kr e^{j\left(\frac{\pi}{2}\right)} = j \frac{\rho_o c}{S_1} kr$$

Assumption #3: The acoustic impedance Z_1 is calculated at:

$$r = \frac{8a}{3\pi}$$

Assume that the acoustic impedance is calculated at a distance $r = \frac{8a}{3\pi}$ from the point source, where a is the radius of opening. Then, the above acoustic impedance becomes:

$$\mathbf{Z}_1 \cong j \frac{\rho_o c}{S_1} kr \cong j \frac{\rho_o c}{S_1} k \frac{8a}{3\pi} \tag{6}$$

or:

$$\mathbf{Z}_1 \cong \frac{\rho_o c}{S_1} kr e^{j\left(\frac{\pi}{2}\right)} \cong \frac{\rho_o c}{S_1} k \frac{8a}{3\pi} e^{j\left(\frac{\pi}{2}\right)} \tag{7}$$

Note that the acoustic impedance is a complex number. More specifically speaking, this acoustic impedance is an imaginary number. This imaginary number has a phase separation of $\frac{\pi}{2}$ between the pressure and velocity:

$$Z_1 \equiv \frac{p_1}{S_1 v_1} = \frac{\rho_o c}{S_1} kre^{j\left(\frac{\pi}{2}\right)}$$

The $\frac{\pi}{2}$ phase angle indicates that there is a 90-degree phase difference between the pressure and velocity. The phase difference is similar to the phase delay of the damping of a structure. Also, the magnitude of the complex impedance is the ratio of magnitudes of pressure to velocity.

For the purpose of simplicity, the acoustic impedance in Eq. (6) can be condensed into:

$$Z_1 \cong j\frac{\rho_o c}{S_1} k\frac{8a}{3\pi} = j\frac{\rho_o c}{S_i} kL', \quad \text{where} \quad L' = \frac{8S_i}{3\pi S_1}a \tag{8}$$

In addition, at the intersection of the main pipe, the impedance becomes:

$$Z_2 = Z_i = \frac{\rho_o c}{S_i}$$

Therefore:

$$\frac{Z_2}{Z_i} = 1 \quad \text{and} \quad \frac{Z_1}{Z_i} = \frac{j\frac{\rho_o c}{S_i} kL'}{\frac{\rho_o c}{S_i}} = jkL'$$

Step 2: Calculate Pressure Ratio Using Equivalent Acoustic Impedance

A side branch on the pipe with a small hole to the outside medium behaves like a high-pass filter when the length of the side branch is very short.

The high-pass filter can be considered as a special case of the one-to-two pipe with the length of the side branch being zero, as analyzed in the previous section:

$$Z_o = \frac{Z_1 Z_2}{Z_1 + Z_2} = \frac{j\frac{\rho_o c}{S_i} kL' \frac{\rho_o c}{S_i}}{j\frac{\rho_o c}{S_i} kL' + \frac{\rho_o c}{S_i}} = \frac{jkL' Z_i}{jkL' + 1}$$

$$\rightarrow \quad \frac{Z_i}{Z_o} = \frac{jkL' + 1}{jkL'} = 1 - j\frac{1}{kL'}$$

Therefore, all formulas derived for a one-to-two pipe are valid for the high-pass filter. The pressure ratios for the one-to-two pipe are shown below:

$$\frac{P_r}{P_i} = \frac{Z_o - Z_i}{Z_o + Z_i}$$

$$\frac{P_1}{P_i} = \frac{P_2}{P_i} = \frac{2Z_o}{Z_o + Z_i}$$

Since we know all impedances of the high-pass filters, all pressures can be calculated using the above equations as the following:

$$\frac{P_r}{P_i} = \frac{Z_o - Z_i}{Z_o + Z_i} = \frac{1 - \frac{Z_i}{Z_o}}{1 + \frac{Z_i}{Z_o}} = \frac{1 - \frac{jkL'+1}{jkL'}}{1 + \frac{jkL'+1}{jkL'}} = \frac{j\frac{1}{2kL'}}{1 - j\frac{1}{2kL'}} = \frac{-\left(\frac{1}{2kL'}\right)^2 + j\frac{1}{2kL'}}{1 + \left(\frac{1}{2kL'}\right)^2}$$

$$\frac{P_2}{P_i} = \frac{2Z_o}{Z_o + Z_i} = \frac{2}{1 + \frac{Z_i}{Z_o}} = \frac{2}{1 + \frac{jkL'+1}{jkL'}} = \frac{1}{1 - j\frac{1}{2kL'}} = \frac{1 + j\frac{1}{2kL'}}{1 + \left(\frac{1}{2kL'}\right)^2}$$

$$\frac{P_1}{P_i} = \frac{P_2}{P_i} = \frac{1 + j\frac{1}{2kL'}}{1 + \left(\frac{1}{2kL'}\right)^2}$$

Step 3: Power Transmission Coefficient

The power transmission coefficient is defined as:

$$T_w = \frac{w_2}{w_i}$$

where:

$$w_i = \frac{1}{4}\left(P_i + P_i^*\right)\left(U_i + U_i^*\right)$$

$$w_2 = \frac{1}{4}\left(P_2 + P_2^*\right)\left(U_2 + U_2^*\right)$$

For the pipe with a side branch pipe as a small hole to the outside medium, Z_i and Z_2 are real numbers, but Z_1 and Z_o are complex numbers. In addition, P_i is real but P_2 is complex.

Hence, the power transmission through the main pipe is:

$$T_w = \left(Re\left(\frac{P_2}{P_i}\right)\right)^2 \frac{Z_i}{Z_2} = \frac{1}{\left(1 + \left(\frac{1}{2kL'}\right)^2\right)^2} = \frac{1}{\left(1 + \left(\frac{\omega_c}{\omega}\right)^2\right)^2} \qquad (Z_i, Z_2, P_i \text{ real})$$

The open side behaves like a high-pass filter; it filters out the wave for frequency $\omega = kc$ below $\omega_c = \frac{c}{2L'}$:

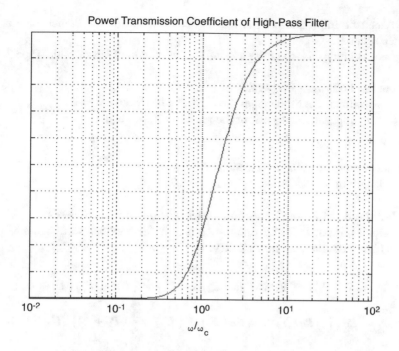

Power transmission coefficient as a function of the frequency ratio

Remarks

- Transmission pressures P_1 and P_2 are outward (no return) forward waves.
- Because pressures P_i, P_r, and P_2 are traveling plane waves, their acoustic imped-ances $Z_i(=\rho_0 c/S_i)$, $Z_r(=-\rho_0 c/S_i)$, and $Z_2(=\rho_0 c/S_2)$ are real numbers.
- The transmission pressure P_1 is a spherical wave, and the acoustic impedance Z_1 is a complex number.
- S_i: cross-section area of the inlet pipe.
- S_1: cross-section area of outlet Pipe 1 (side branch).
- S_2: cross-section area of outlet Pipe 2 (main pipe).
- L_1: length of outlet Pipe 1 (outlet side branch).
- p_i: pressure of the incident wave in the inlet pipe.
- p_r: pressure of the reflected wave in the inlet pipe.
- p_1: pressure of the transmitted wave in outlet Pipe 1 (outlet side branch).
- p_2: pressure of the transmitted wave in outlet Pipe 2 (outlet main branch).
- $P_i = P_{iR}$: complex pressure of the incident wave at the RHS of the inlet pipe.
- P_1: complex pressure of the transmitted wave in the side branch pipe at the intersection.
- $P_2 = P_{2L}$: complex pressure of the transmitted wave at the LHS of Pipe 2.
- Z_i: acoustic impedance of the incident wave at the RHS of the inlet pipe.

- Z_1: acoustic impedance of the transmitted wave inside the branch pipe at the intersection.
- Z_2: acoustic impedance of the transmitted wave at the LHS of outlet Pipe 2.

Assume the Following Properties:
- Cross-section areas of the pipes are S_i, S_1, and S_2 where $S_2 = S_i$.
- Acoustic impedances, $Z = z/S$, of the pipes are Z_i, Z_r, and Z_2

where $Z_r = -Z_i$; $Z_2 = Z_i$.

- Incident pressure at the intersection of the pipe is P_i.
- The length of the side branch pipe is $L_1 = 0$.

The Validation of Power Reflection and Transmission Coefficients
The formulas of power for the one-to-two pipe in the previous section can be used for the high-pass filter, as shown below:

$$w_i = Re\,(\,P_i)\,Re\,(\,V_i)S = Re\,(\,P_i)\,Re\,(\,U_i) = \frac{1}{4}\left(\,P_i + P_i^*\right)\left(\,U_i + U_i^*\right)$$

$$w_r = Re\,(\,P_r)\,Re\,(\,V_r)S = Re\,(\,P_r)\,Re\,(\,U_r) = \frac{1}{4}\left(\,P_r + P_r^*\right)\left(\,U_r + U_r^*\right)$$

$$w_1 = Re\,(\,P_1)\,Re\,(\,V_1)S = Re\,(\,P_1)\,Re\,(\,U_1) = \frac{1}{4}\left(\,P_1 + P_1^*\right)\left(\,U_1 + U_1^*\right)$$

$$w_2 = Re\,(\,P_2)\,Re\,(\,V_2)S = Re\,(\,P_2)\,Re\,(\,U_2) = \frac{1}{4}\left(\,P_2 + P_2^*\right)\left(\,U_2 + U_2^*\right)$$

For the pipe with a side branch pipe as a small hole to the outside medium, Z_i and Z_2 are real numbers, but Z_1 is a complex number. In addition, P_i is real but P_1 and P_2 are complex.

Therefore:

$$\frac{U_1}{U_i} = \frac{P_1}{P_i}\frac{Z_i}{Z_1} = \frac{1 + j\frac{1}{2kL'}}{1 + \left(\frac{1}{2kL'}\right)^2}\frac{1}{jkL'} = \frac{2\left(\frac{1}{2kL'}\right)^2 - j\frac{1}{kL'}}{1 + \left(\frac{1}{2kL'}\right)^2}$$

$$Re\left(\frac{P_r}{P_i}\right) = \frac{-\left(\frac{1}{2kL'}\right)^2}{1 + \left(\frac{1}{2kL'}\right)^2}, \qquad Re\left(\frac{P_2}{P_i}\right) = \frac{1}{1 + \left(\frac{1}{2kL'}\right)^2}$$

$$R_w = -\left(Re\left(\frac{P_r}{P_i}\right)\right)^2 = -\frac{\left(\frac{1}{2kL'}\right)^4}{\left(1 + \left(\frac{1}{2kL'}\right)^2\right)^2} = -\frac{\left(\frac{\omega_c}{\omega}\right)^4}{\left(1 + \left(\frac{\omega_c}{\omega}\right)^2\right)^2} \qquad (Z_i, Z_r, P_i \text{ real})$$

$$T_{w2} = \left(Re\left(\frac{P_2}{P_i}\right) \right)^2 \frac{Z_i}{Z_2} = \frac{1}{\left(1 + \left(\frac{1}{2kL'}\right)^2\right)^2} = \frac{1}{\left(1 + \left(\frac{\omega_c}{\omega}\right)^2\right)^2} \qquad (Z_i, Z_2, P_i \text{ real})$$

$$T_{w1} = \frac{(P_1 + P_1^*)(U_1 + U_1^*)}{(P_i + P_i^*)(U_i + U_i^*)} = \frac{1}{4}\frac{(P_1 + P_1^*)(U_1 + U_1^*)}{P_i U_i}$$

$$= \frac{1}{4}\left(\frac{P_1}{P_i} + \frac{P_1^*}{P_i^*}\right)\left(\frac{U_1}{U_i} + \frac{U_1^*}{U_i^*}\right)$$

$$= Re\left(\frac{P_1}{P_i}\right)\cdot Re\left(\frac{U_1}{U_i}\right) = \frac{2\left(\frac{1}{2kL'}\right)^2}{\left(1 + \left(\frac{1}{2kL'}\right)^2\right)^2} = \frac{2\left(\frac{\omega_c}{\omega}\right)^2}{\left(1 + \left(\frac{\omega_c}{\omega}\right)^2\right)^2} \qquad (P_i, U_i \text{ real})$$

$$-R_w + T_{w2} + T_{w1} = \frac{\left(\frac{\omega_c}{\omega}\right)^4}{\left(1 + \left(\frac{\omega_c}{\omega}\right)^2\right)^2} + \frac{1}{\left(1 + \left(\frac{\omega_c}{\omega}\right)^2\right)^2} + \frac{2\left(\frac{\omega_c}{\omega}\right)^2}{\left(1 + \left(\frac{\omega_c}{\omega}\right)^2\right)^2} = 1$$

12.5 Band-Stop Resonator

A general band-stop resonator can be constructed by a main pipe with a side branch of pipeline series. The resonant of the side branch will absorb the resonant from the main pipe at the resonant frequency of the side branch. The simple band-stop resonator and a Helmholtz resonator are shown in the figure below:

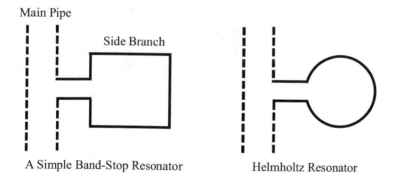

A Simple Band-Stop Resonator   Helmholtz Resonator

Objective

The Helmholtz resonator has a volume V connected to the remainder of the medium with a small neck of length L and an open area S, as shown in the figure below:

$$R_m = R_r + R_\omega, \quad R_r = \frac{1}{2}\rho_0 cS(ka)^2, \quad R_\omega = 2Mc\alpha_\omega$$

$$M\frac{d^2\xi}{d^2t} + R_m\frac{d\xi}{dt} + K\xi = Sp \;\Big|\; e^{\pm j\omega t}$$

$$Z = \frac{p}{uS} = \frac{Z_m}{S^2}$$

$$\xi = HSp, \quad H = (K - \omega^2 M \pm j\omega R_m)^{-1}$$

$$K = \rho_0 c^2 \frac{S^2}{V}$$

$$u = \frac{d\xi}{dt} = \pm j\omega H Sp$$

$$M = \rho_0 SL'$$

$$L' = L + 1.7a$$

$$Z_m = \frac{pS}{u} = \frac{1}{\pm j\omega H} = R_m \pm j\left(\omega M - \frac{K}{\omega}\right)$$

Schematic of a Helmholtz resonator

This passive acoustic system can be analyzed for a range of frequencies where $\lambda \gg L$, $\lambda \gg \sqrt{S}$, and $\lambda \gg V^{1/3}$.

Show that the power transmission T_w coefficients of the Helmholtz resonator can be formulated as:

$$T_w = \left(Re\left(\frac{P_2}{P_i}\right)\right)^2 \frac{Z_i}{Z_2} = \frac{1}{\left(1 + \left(\frac{kSV}{2S_2\left(k^2 L'V - S\right)}\right)^2\right)^2}$$

$$= \frac{1}{\left(1 + \left(\frac{D\left(\frac{k}{k_c}\right)}{\frac{k^2}{k_c^2} - 1}\right)^2\right)^2} \quad [Z_i, Z_2, P_i \text{ real}]$$

where:

$$k_c^2 = \frac{\omega_c^2}{c^2} = \frac{S}{L'V}, \quad D = \frac{1}{2}\frac{S}{S_2}\sqrt{\frac{V}{SL'}}$$

Step 1: Derivation of the Dynamic Equation of the Resonator

The volume of fluid in the neck area will act as a mass just as the open end of a flanged pipe for $\lambda \gg a$, where a is the radius of opening. The effective length of the neck is assumed to be $0.85a$ longer than the actual length of the pipe if the pipe is terminated with a wide flange at the opening. If the pipe has an un-flanged termination, it can be assumed that the effective length is longer only by $0.6a$. Since the interior end of the neck acts like a flanged termination, the effective length of the neck with flanged and un-flanged termination to the outside medium becomes:

$$L' = L + (0.85 + 0.85)a = L + 1.70a \qquad \text{(outer end flanged)}$$

$$L' = L + (0.85 + 0.60)a = L + 1.45a \qquad \text{(outer end un-flanged)}$$

Using the effective length, the mass of fluid moving in the neck is:

$$M = \rho_o S L'$$

The mass in the neck works against the stiffness of the fluid in volume V in addition to resistance from the frictional forces in the neck and the radiation impedance from the open end.

The stiffness of the fluid in volume V is calculated from the change of volume inside the container due to the motion of the mass in the neck. Hence, if ξ is the displacement of the neck mass, the change in volume V is then given by:

$$-S\xi = \Delta V$$

If V_o is the initial volume and V is the compressed volume, then $\Delta V = V - V_o$ and $\rho_o V_o = \rho V = M_V$, since the mass M_V of the volume V is preserved. Thus:

$$\Delta V = V - V_o = \frac{M_V}{\rho} - \frac{M_V}{\rho_o} = \frac{M_V}{\rho} \cdot \frac{\rho_o - \rho}{\rho_o} = -V \frac{\rho - \rho_o}{\rho_o} = -V \frac{\Delta \rho}{\rho_o}$$

Dividing both sides of this equation by the volume V and using the relationship above yield:

$$\frac{S\xi}{V} = -\frac{\Delta V}{V} = \frac{\Delta \rho}{\rho_o} = s_{ac}$$

where s_{ac} is the acoustic condensation. Now, using the relationship between acoustic pressure and condensation ($\frac{\Delta \rho}{\rho_o} = \frac{p}{\gamma P_0}$ and $c^2 = \frac{\gamma P_0}{\rho_0}$) from Chap. 2:

$$s_{ac} = \frac{p}{\rho_o c^2}$$

The acoustic pressure inside the given volume can thus be expressed as:

$$p = \rho_o c^2 s_{ac} = \rho_o c^2 \frac{S\xi}{V}$$

The force necessary to displace the fluid in the neck is simply $Sp = K\xi$ where K is the effective stiffness of the fluid inside V. Multiplying both sides of the pressure equation above by S yields:

$$Sp = \rho_o c^2 \frac{S^2}{V} \xi = K\xi$$

and the effective stiffness is:

$$K = \rho_o c^2 \frac{S^2}{V}$$

In addition to the inertial and stiffness forces acting on the fluid column in the neck, there are also resistive forces that are proportional to the fluid velocity. These forces are due to the radiation resistance R_r and the viscous resistance R_ω. These two constants provide damping to the mechanical system since the force they create is proportional to the velocity of the fluid in the neck. Hence, the total mechanical resistance coefficient is given by:

$$R_m = R_r + R_\omega$$

where:

$$R_r = \frac{1}{2} \rho_o cS(ka)^2 \qquad \text{(flanged opening)}$$

$$R_r = \frac{1}{4} \rho_o cS(ka)^2 \qquad \text{(un - flanged opening)}$$

and:

$$R_\omega = 2Mc\alpha_\omega$$

where α_ω is the absorption coefficient for wall losses.

The mass-spring system described here is forced by the pressure wave of p impinging on its opening. This force is simply the pressure of the wave multiplied by the neck cross-section area. That is:

$$f = Sp$$

Now the differential equation of the resonator can be written in terms of mass M, stiffness K, and the damping coefficient R_m given by the equations above as follows:

$$M\frac{d^2\xi}{d^2t} + R_m\frac{d\xi}{dt} + K\xi = Sp$$

Step 2: Mechanical Impedance of the Dynamic Equation

Let $\boldsymbol{\xi} = \boldsymbol{A}e^{j\omega t}$, where its complex conjugate is $\boldsymbol{\xi}^* = \boldsymbol{A}^*e^{-j\omega t}$, and we can then solve the following two equations step-by-step as follows for $\boldsymbol{p} = \boldsymbol{P}e^{j\omega t}$ and $\boldsymbol{p}^* = \boldsymbol{P}^*e^{-j\omega t}$. Note that $K, M, R_m, S,$ and ω are all real constants:

$$M\frac{d^2\boldsymbol{\xi}}{d^2t} + R_m\frac{d\boldsymbol{\xi}}{dt} + K\boldsymbol{\xi} = S\boldsymbol{p} = SPe^{j\omega t}; \qquad \boldsymbol{p} = \boldsymbol{P}e^{j\omega t}$$

$$M\frac{d^2\boldsymbol{\xi}^*}{d^2t} + R_m\frac{d\boldsymbol{\xi}^*}{dt} + K\boldsymbol{\xi}^* = S\boldsymbol{p}^* = SP^*e^{-j\omega t}; \qquad \boldsymbol{p}^* = \boldsymbol{P}^*e^{-j\omega t}$$

Let $\boldsymbol{\xi} = \boldsymbol{A}e^{j\omega t}$ to get $(K - \omega^2 M + j\omega R_m)\boldsymbol{A}e^{j\omega t} = SPe^{j\omega t}$.
Let $\boldsymbol{\xi}^* = \boldsymbol{A}^*e^{-j\omega t}$ to get $(K - \omega^2 M - j\omega R_m)\boldsymbol{A}^*e^{-j\omega t} = SP^*e^{-j\omega t}$.
Let:

$$H = \left(K - \omega^2 M + j\omega R_m\right)^{-1}$$

$$H^* = \left(K - \omega^2 M - j\omega R_m\right)^{-1}$$

to obtain:

$$A = \left(K - \omega^2 M + j\omega R_m\right)^{-1}SP = HSP$$

$$A^* = \left(K - \omega^2 M - j\omega R_m\right)^{-1}SP^* = H^*SP^*$$

Therefore:

$$\boldsymbol{\xi} = HSPe^{j\omega t}$$

$$\boldsymbol{\xi}^* = H^*SP^*e^{-j\omega t}$$

$$u = \frac{d\boldsymbol{\xi}}{dt} = j\omega HSPe^{j\omega t}$$

$$u^* = \frac{d\boldsymbol{\xi}^*}{dt} = -j\omega H^*SP^*e^{-j\omega t}$$

$$Z_m = \frac{PS}{u} = \frac{1}{j\omega H} = \frac{K - \omega^2 M + j\omega R_m}{j\omega} = R_m + j\left(\omega M - \frac{K}{\omega}\right)$$

$$Z_m^* = \frac{P^*S}{u^*} = \frac{1}{-j\omega H^*} = \frac{K - \omega^2 M - j\omega R_m}{-j\omega} = R_m - j\left(\omega M - \frac{K}{\omega}\right)$$

Note that the equations above show that all complex values, functions, or equations always have a complex conjugate part associated with them. Therefore, we can always hide the conjugate part, but when we need a real-world solution, we must show them.

Step 3: Transmission Coefficient Induced by a Side Resonator

Neglecting R_ω for wall losses, the input mechanical impedance of the resonator is:

$$Z_m = R_r + j\left(\omega M - \frac{K}{\omega}\right)$$

$$= \frac{1}{2}\rho_o cS(ka)^2 + j\left(ck\rho_o SL' - \frac{\rho_o c^2 S^2}{ckV}\right)$$

$$= \frac{1}{2}\rho_o cS\left((ka)^2 + j\frac{2(k^2 L'V - S)}{kV}\right)$$

$$Z_1 = \frac{Z_m}{S^2}, \qquad Z_2 = \frac{\rho_o c}{S_2}$$

$$\frac{P_2}{P_i} = \frac{2\,Z_1}{2Z_1 + Z_2} = \frac{(ka)^2 + j\frac{2(k^2 L'V - S)}{kV}}{(ka)^2 + j\frac{2(k^2 L'V - S)}{kV} + \frac{S}{S_2}}$$

For $ka \ll 1$, (Z_i, Z_2, P_i real) (Z_1, P_2 complex):

$$\frac{P_2}{P_i} = \frac{j\frac{2S_2(k^2 L'V - S)}{kSV}}{j\frac{2S_2(k^2 L'V - S)}{kSV} + 1} = \frac{\left(\frac{2S_2(k^2 L'V - S)}{kSV}\right)^2 + j\left(\frac{2S_2(k^2 L'V - S)}{kSV}\right)}{1 + \left(\frac{2S_2(k^2 L'V - S)}{kSV}\right)^2}$$

$$Re\left(\frac{P_2}{P_i}\right) = \frac{\left(\frac{2S_2(k^2 L'V - S)}{kSV}\right)^2}{1 + \left(\frac{2S_2(k^2 L'V - S)}{kSV}\right)^2} = \frac{1}{1 + \left(\frac{kSV}{2S_2(k^2 L'V - S)}\right)^2}$$

$$T_w = \left(Re\left(\frac{P_2}{P_i}\right)\right)^2 \frac{Z_i}{Z_2} = \frac{1}{\left(1 + \left(\frac{kSV}{2S_2(k^2 L'V - S)}\right)^2\right)^2}$$

$$= \cfrac{1}{\left(1 + \left(\cfrac{D\left(\frac{k}{k_c}\right)}{\frac{k^2}{k_c^2} - 1}\right)^2\right)^2} \qquad [\, Z_i, Z_2, \boldsymbol{P}_i \text{ real}]$$

where:

$$k_c^2 = \frac{\omega_c^2}{c^2} = \frac{S}{L'V}, \quad D = \frac{1}{2}\frac{S}{S_2}\sqrt{\frac{V}{SL'}}$$

The side Helmholtz resonator behaves like a band-stop filter or notch filter. The maximum attenuation of the filter occurs at the resonance frequency $\omega_c = c\sqrt{\frac{S}{L'V}}$ of the Helmholtz resonator. The bandwidth depends on the volume ratio $\frac{V}{SL'}$ and the area ratio of the pipes, $\frac{S}{S_2}$:

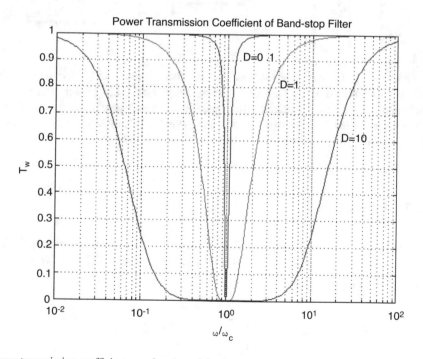

Power transmission coefficient as a function of the frequency ratio

12.6   Numerical Method for Modeling of Pipelines with Side Branches

Most acoustic filters can be modeled as a system of pipelines with side branches as shown in the figure below. The main pipe is the vertical pipe, and the side branch is the horizontal pipes in series as shown in the figure below:

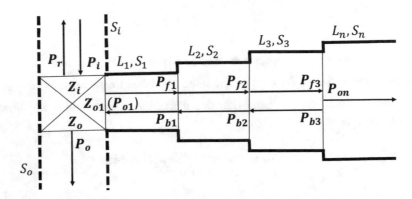

Hint: Doing this example by hand can be difficult because (1) the formula for acoustic impedance is in complex number format and (2) power transmission coefficients need to be calculated at multiple frequencies. You can use the MATLAB functions provided in Computer Code Section (Sect. 11.7) or any suitable programming language for this project.

12.7   Project

Use the MATLAB code provided in the Computer Program Section (Section 11.7) to calculate and plot the power transmission coefficient of the three basic filter designs:

(a) Low-pass filter as shown in the sketch below:

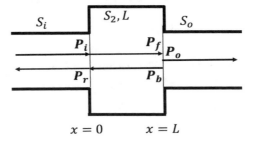

(b) High-pass filter:

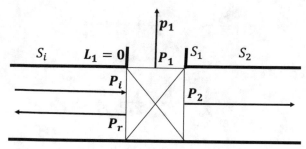

Intersection as a point

(c) Band-stop filter:

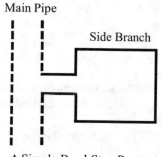

A Simple Band-Stop Resonator

12.8   Homework Exercises

Exercise 12.1

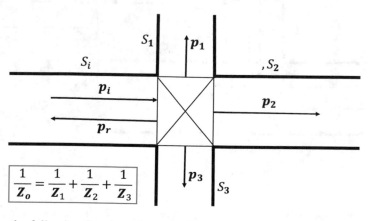

From the following four boundary conditions at the interface:

$$P_i + P_r = P_1 = P_2 = P_3 \tag{1}$$

$$\frac{P_i}{Z_i} - \frac{P_r}{Z_i} = \frac{P_1}{Z_1} + \frac{P_2}{Z_2} + \frac{P_3}{Z_3} \tag{2}$$

and:

$$\frac{1}{Z_o} = \frac{1}{Z_1} + \frac{1}{Z_2} + \frac{1}{Z_3}$$

Prove that:

$$\frac{P_r}{P_i} = \frac{Z_o - Z_i}{Z_o + Z_i}$$

$$\frac{P_1}{P_i} = \frac{P_2}{P_i} = \frac{P_3}{P_i} = \frac{2Z_o}{Z_o + Z_i}$$

Exercise 12.2

A side branch, length L, and a cross-section area S_{is} have an acoustic impedance at its termination given as $Z_{os} = \rho_o c / S_{is}$. The cross-section area of the main pipe is S_i ($=S_2$) as shown below.

The acoustic impedance of Z_1 at the intersection is:

$$Z_1 = Z_{is} \frac{Z_{os} + jZ_{is} \tan (kL)}{jZ_{os} \tan (kL) + Z_{is}}$$

Treat Z_1 and Z_2 as given values to solve the exercise:

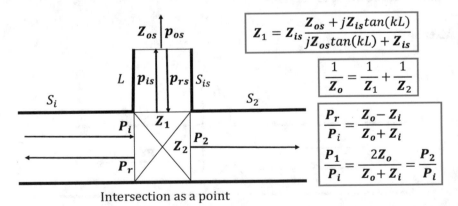

Intersection as a point

(a) Calculate the power reflection coefficient (R_w) from the cross-section at the intersection.

(b) Calculate the power transmission coefficient (T_{w2}) through the intersection of the main pipe.

(c) Calculate the power transmission coefficient (T_{w1}) through the side branch.

(Answers):

(a) $-\left(Re \left(\frac{Z_o - Z_i}{Z_o + Z_i} \right) \right)^2$; (b) $\left(Re \left(\frac{2Z_o}{Z_o + Z_i} \right) \right)^2$; (c) $Re \left(\frac{2Z_o}{Z_o + Z_i} \right) \, Re \left(\frac{2Z_o Z_i}{(Z_o + Z_i)Z_1} \right)$ where

$\frac{1}{Z_o} = \frac{1}{Z_1} + \frac{1}{Z_2}$

Nomenclature[1]

c	wave speed, the speed of sound
$\widehat{e}_x, \widehat{e}_y, \widehat{e}_z$	base vectors of Cartesian coordinates
$\widehat{e}_r, \widehat{e}_\theta, \widehat{e}_\phi$	base vectors of spherical coordinates
f	circular frequency
f_o	center frequency
f_{lower}	lower limit frequency
f_{upper}	upper limit frequency
I	acoustic intensity
I_a	moment of inertia of one particle
IL	insertion loss
K_a	kinetic energy of one particle
k	experimentally determined coefficient of air for excess absorption coefficient
k	wave number
k_x, k_y, k_z	components of wave number vector
L_p	pressure level
L_w	pressure level from direct source
L_x, L_y, L_z	dimensions of a rectangular cavity
m_a	mass of one particle
NR	noise reduction
n_x, n_y, n_z	components of a unit vector in wave propagation direction
PSIL	preferred speech interference level
P_o	averaged air pressure
P	absolute air pressure
p	pressure difference $(P - P_o)$
$p(x, t)$	one-dimensional pressure function
$p(x, y, z, t)$	three-dimensional pressure function

---

[1] Capital letter for amplitude and Bold facing for complex number.

© The Author(s), under exclusive license to Springer Nature Switzerland AG 2021
H. Lin et al., *Lecture Notes on Acoustics and Noise Control*,
https://doi.org/10.1007/978-3-030-88213-6

$p_+()$	forward traveling pressure wave
$p_-()$	backward traveling pressure wave
$p_s()$	standing pressure wave
Q	directivity of a source
Q_s	source strength
Q_ρ	mass flow rate
R	room constant
R	reflection coefficient; radius
Rm	mechanical resistance
RMS	root mean square
r, θ, φ	spherical coordinates
S	cross-section area
S	surface area of a room
SPL	sound pressure level
T	period
$T(t)$	time function
TL	transmission loss
T_{60}	reverberation time (decay to 60 dB)
U	flow rate amplitude (VS)
u	flow rate (vS)
V	velocity amplitude; volume of a room
V_a	volume occupied by one particle
v	flow velocity
v_{ca}	colliding speed of a particle
$v_f(x,t)$	one-dimensional flow velocity in x-direction
$\vec{v}_f(x,y,z,t)$	flow velocity
w	sound power
$X(x)$	space function
x, y, z	Cartesian coordinates
Z, Z_a	acoustic impedance ($p/u=p/(vS)=z/S$)
Z_m	mechanical impedance ($pS/u= zS$)
z	specific acoustic impedance (p/v)
α	sound absorption coefficient
$\bar{\alpha}$	average sound absorption coefficient
$\bar{\alpha}', \bar{\alpha}_{tot}$	total average absorption coefficient include excess absorption coefficient
$\bar{\alpha}_{ex}$	excess absorption coefficient
γ	ratio of specific heat
δ	total energy density
δ_d	direct energy density
δ_r	reverberant energy density
θ	phase angle
$\dot{\theta}_{ca}$	rotation speed of a particle
λ	wave length

ξ	location along the wave number vector
ρ	instantaneous mass density of the air
ρ_0	averaged mass density of the air
τ	transmission coefficient
ϕ	phase angle
ω	angular frequency

Appendices

Appendix 1: Discrete Fourier Transform

In Chap. 9, sound pressure functions expressed with frequency contents were given without explaining how to calculate them. Appendix 1 will demonstrate how to convert a time domain function to a frequency domain function using the discrete Fourier transform. This appendix can fill the gap between the time domain function and frequency domain function in Chap. 9.

Discrete Fourier Transform

Fourier Series for Periodical Time Function

The Fourier series for a periodical time function (T=period) can be formulated as:

$$p(t) = P_o + \sum_{k=1}^{\infty} \left(e^{j\omega_k t} P_k + e^{-j\omega_k t} P_k^* \right)$$

$$= A_o + \sum_{k=1}^{\infty} \left((A_k + jB_k) e^{j\omega_k t} + (A_k - jB_k) e^{-j\omega_k t} \right)$$

$$= A_o + 2 \sum_{k=1}^{\infty} \left[A_k \cos(\omega_k t) - B_k \sin(\omega_k t) \right]$$

where:

© The Author(s), under exclusive license to Springer Nature Switzerland AG 2021
H. Lin et al., *Lecture Notes on Acoustics and Noise Control*,
https://doi.org/10.1007/978-3-030-88213-6

$$P_k \equiv A_k + jB_k = \frac{1}{T}\int_0^T p(t)e^{-j\omega_k t}dt$$

$$\omega_k = k\left(\frac{2\pi}{T}\right) = k\Delta\omega = 2\pi\,k\Delta f, \qquad \Delta\omega = \frac{2\pi}{T}, \qquad \Delta f = \frac{1}{T}$$

Formulas of Discrete Fourier Series

In general application, $p(t)$ is given as $p_i = p(t_i) = p(i\Delta t)$ for $i = 0, 1, \ldots, N-1$, and $N = T/\Delta t$. Then, the integration for P_k can be changed to summation as follows:

$$P_k \equiv A_k + jB_k = \frac{1}{N\Delta t}\sum_{i=0}^{N-1}p_i e^{-j\omega_k t_i}\Delta t = \frac{1}{N}\sum_{i=0}^{N-1}p_i e^{-jk\left(\frac{2\pi}{N\Delta t}\right)i\Delta t}$$

$$= \frac{1}{N}\sum_{i=0}^{N-1}p_i\left[e^{-j\left(\frac{2\pi}{N}\right)}\right]^{ki} = \frac{1}{N}d_k$$

where:

$$d_k = \sum_{i=0}^{N-1}p_i\left[e^{-j\left(\frac{2\pi}{N}\right)}\right]^{ki}$$

is called the discrete Fourier transform (DFT) and can be computed by using the fft function in MATLAB as follows:

$$[d_k] = fft(p_i, N)$$

and therefore:

$$P_k \equiv A_k + jB_k = d_k/N = fft(p_i, N)/N$$

Example 1: The Cosine Function as Sound Pressure

A pure tone acoustic pressure $p(t)$ of frequency $f = 500$ [Hz] and amplitude $P = 3$ [Pa] is transformed into the frequency domain for calculating the A-weighted sound pressure level [dBA]:

$$p(t) = P\cos(2\pi ft)$$

Given:

(a) The number of discretized time domain data is $N = 16$.
(b) The time increment of time domain data is $\Delta t = 0.000125$ [s].
(c) The time domain pressure (n data) which is the input data for MATLAB fft:

Time domain pressure [Pa]

30	2.72.12	1.15	00	-1.15	-2.12	-2.77	-3-2.77	-2.12	-1.15	00	1.12.12	2.77

(d) The output from the MATLAB fft function is in the frequency domain, and there are N data as shown below:

Real part coefficients [Pa] of $P_k = fft(p_i, N)/N$

0	1.50	0	0	0	0	0	0	0	0	0	0	0	0	1.50

Imaginary part coefficients [Pa] of $P_k = fft(p_i, N)/N$

0	0	0	0	0	0	0	0	0	0	0	0	0	0	0

The P_k can be computed by using the fft function in MATLAB as follows:

$$P_k \equiv A_k + jB_k = d_k/N = fft(p_i, N)/N$$

Calculate:

The RMS pressure square of the given sound pressure $p(t)$
The A-weighted sound pressure level

Example 1: Solution
Procedures

Step 1 – Calculate the total time:

$$T = \Delta t \cdot N = 0.000125[s] \cdot 16 = 0.002 \ [s]$$

Step 2 – Calculate the frequency increment Δf:

$$\because \ N \cdot \Delta t \cdot \Delta f = 1$$

$$\rightarrow \Delta f = \frac{1}{N \cdot \Delta t} = \frac{1}{T} = \frac{1}{0.002} \ [Hz] = 500 \ [Hz]$$

Step 3 – Indicate frequency values to the frequency domain contents:
Real part coefficients [Pa]: $A_k = \text{Re} \ (P_k)$

0	1.50	0	0	0	0	0	0	0	0	0	0	0	0	1.50

Imaginary part coefficients [Pa]: $B_k = \text{Im} \ (P_k)$

0	0	0	0	0	0	0	0	0	0	0	0	0	0	0	0

Frequency [Hz]

0	500	1000	1500	2000	2500	3000	3500	4000	4500	5000	5500	6000	6500	7000	7500

Step 4 – Calculate the Nyquist frequency f_N:

$$f_N = \left(\frac{1}{\Delta t}\right)\frac{1}{2} = \frac{N}{T}\frac{1}{2} = \frac{16}{0.002} \cdot \frac{1}{2} = 4000 \ [Hz]$$

Step 5 – Replace frequencies after the Nyquist frequency with:

$$f_k = f_k - 2 * f_N = f_k - 8000$$

Real part coefficients [Pa]: $A_k = \text{Re}(P_k)$

0	1.50	0	0	0	0	0	0	0	0	0	0	0	0	0	1.50

Imaginary part coefficients [Pa]: $B_k = \text{Im}(P_k)$

0	0	0	0	0	0	0	0	0	0	0	0	0	0	0	0

Frequency [Hz]: $f_k = \Delta f \cdot k$

0	500	1000	1500	2000	2500	3000	3500	4000	-3500	-3000	-2500	-2000	-1500	-1000	-500

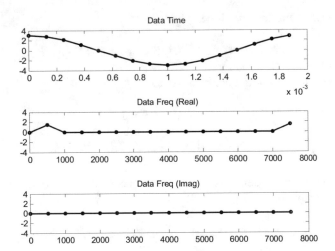

Remarks

Make sure that these are complex conjugate pairs.

Make sure that $A_{k\,=\,1}$ is the coefficient of the cosine function:

$$p(t) = A_o + 2 \sum_{k=1}^{\infty} [A_k \cos{(\omega_k t)} - B_k \sin{(\omega_k t)}] = 3 \cos{(2\pi \cdot 500 \cdot t)} [Pa]$$

Step 6 – Calculate the RMS pressure square:

$$p_{RMS}^2 = A_o^2 + 2 \sum_{k=1}^{\infty} (A_k^2 + B_k^2) = 2(1.5^2 + 0^2) = 4.5 \; [Pa^2]$$

Step 7 – Calculate the A-weighted sound pressure level:

$$L_p = 20 \log_{10} \left(\frac{P_{RMS,500Hz}}{P_r} \right) + 10 \log_{10} (W_{500Hz})$$

$$= 20 \log_{10} \left(\frac{\sqrt{2(1.5^2)}}{20E - 6} \right) [dB] - 3.2 \; [dB]$$

$$= 97.3 \; [dB]$$

FFT Using MATLAB fft Function

```
clear all
f=500; % Hz
P=3;   % Pa
T=1/f; % s
N=16; % number of data
%
dt=T/N; % time increment
df=1/T; % frequency increment
DataTime(:,1)=(0:1:N-1)*dt;
DataTime(:,2)=cos(2*pi*f*DataTime(:,1))*P;
% Discrete  FFT based on the Fourier series formula
DataFreq(:,1)=(0:1:N-1)*df;
DataFreq(:,2)=fft(DataTime(:,2),N)/N;
% get the real and imaginary parts
DataFreq(:,3)=real(DataFreq(:,2));   % real part
DataFreq(:,4)=imag(DataFreq(:,2)); % imaginary part
% plot results
subplot(3,1,1);
plot(DataTime(:,1),DataTime(:,2),'-ok','LineWidth',2,'MarkerSize',4);
title('Data Time'); xlabel('time [s]'); YLIM([-4 4]);
subplot(3,1,2);
plot(DataFreq(:,1),DataFreq(:,3),'-ok','LineWidth',2,'MarkerSize',4);
title('Data Freq (Real)'); xlabel('frequency [Hz]'); YLIM([-4 4])
subplot(3,1,3);
plot(DataFreq(:,1),DataFreq(:,4),'-ok','LineWidth',2,'MarkerSize',4);
title('Data Freq (Imag)'); xlabel('frequency [Hz]'); YLIM([-4 4])
% save the figure
saveas(gcf,'fig','emf')
```

Example 2: The Sine Function as Sound Pressure

A pure tone acoustic pressure $p(t)$ of frequency $f = 1000$ [Hz] and amplitude $P = 7$ [Pa] is transformed into the frequency domain for calculating the A-weighted sound pressure level [dBA]:

$$p(t) = P \sin(2\pi f t)$$

Given:

(a) The number of discretized time domain data is $N = 16$.

(b) The time increment of time domain data is $\Delta t = 0.000125$ [s].

(c) The time domain pressure (n data) which is the input data for MATLAB fft:

Time domain pressure [Pa]

3 0	4.95	7	4.95	0	-4.95	-7	-4.95	0	4.95	7	4.95	0	-4.95	-7	-4.95

(d) The frequency domain (N data) which is the output from the MATLAB fft function:

Real number part coefficients [Pa] of $P_k = fft(p_i, N)/N$

00	00	00	0	00	0	00	0	0	0	0	0	0	0	0	0

Imaginary number part coefficients [Pa] of $P_k = fft(p_i, N)/N$

00	0	0 -3.5	00	0	00	00	00	00	00	00	00	00	00	0 3.5	00

Calculate:

The RMS pressure square of the given sound pressure $p(t)$

Example 2: Solution
Procedures

Step 1 – Calculate the total time:

$$T = \Delta t \cdot N = 0.000125 \ [s] \cdot 16 = 0.002 \ [s]$$

Step 2 – Calculate the frequency increment Δf:

$$\because \ N \cdot \Delta t \cdot \Delta f = 1$$

$$\rightarrow \Delta f = \frac{1}{N \cdot \Delta t} = \frac{1}{T} = \frac{1}{0.002} \ [Hz] = 500 \ [Hz]$$

Step 3 – Indicate the frequency values to the frequency domain contents:

Real number part coefficients [Pa]: $A_k = \mathrm{Re} \ (P_k)$

0	0	0	0	0	0	0	0	0	0	0	0	0	0	0	0

Imaginary number part coefficients [Pa]: $B_k = \mathrm{Im} \ (P_k)$

0	0	-3.5	0	0	0	0	0	0	0	0	0	0	0	3.5	0

Frequency [Hz]

0	500	1000	1500	2000	2500	3000	3500	4000	4500	5000	5500	6000	6500	7000	7500

Step 4 – Calculate the Nyquist frequency, f_N:

$$f_N = \left(\frac{1}{\Delta t}\right)\frac{1}{2} = \frac{N}{T}\frac{1}{2} = \frac{N}{T}\frac{1}{2} = \frac{16}{0.002} \cdot \frac{1}{2} = 4000\ [Hz]$$

Step 5 – Replace frequencies after the Nyquist frequency with:

$$f_k = f_k - 2 * f_N = f_k - 8000$$

Real part coefficients [Pa]: $A_k = \text{Re}\ (P_k)$

0	0	0	0	0	0	0	0	0	0	0	0	0	0	0	0	0

Imaginary part coefficients [Pa]: $B_k = \text{Im}\ (P_k)$

0	0	-3.5	0	0	0	0	0	0	0	0	0	0	0	0	3.5	0

Frequency [Hz]: $f_k = \Delta f \cdot k$

0	500	1000	1500	2000	2500	3000	3500	4000	-3500	-3000	-2500	-2000	-1500	-1000	-500

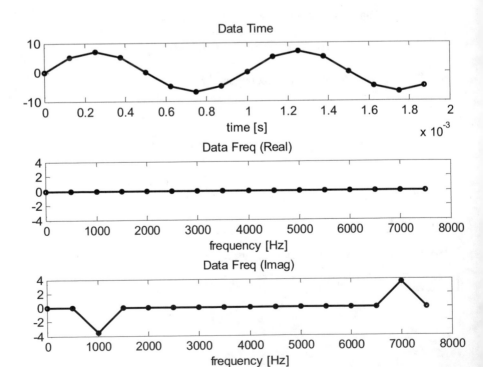

Remarks

Make sure that these are complex conjugate pairs.

Make sure that $B_{k=2}$ is the coefficient of the sine function:

$$p(t) = A_o + 2\sum_{k=1}^{\infty} [A_k \cos(\omega_k t) - B_k \sin(\omega_k t)] = 7\cos(2\pi \cdot 1000 \cdot t)[Pa]$$

Step 6 – Calculate the RMS pressure square:

$$p_{RMS}^2 = A_o^2 + 2\sum_{k=1}^{\infty} (A_k^2 + B_k^2) = 2(0^2 + (-3.5)^2)\ [Pa^2] = 24.5\ [Pa^2]$$

Step 7 – Calculate the A-weighted sound pressure level:

$$L_p = 20\log_{10}\left(\frac{P_{RMS,1000Hz}}{P_r}\right) + 10\log_{10}(W_{1000Hz})$$

$$= 20\log_{10}\left(\frac{\sqrt{2(3.5^2)}}{20E-6}\right)[dB] + 0\ [dB]$$

$$= 108\ [dB]$$

Example 3: Calculating Sound Pressure Level

Given:

The single-frequency pressure is:

$$p(x,t) = P_1 \cos(2\pi f_1 t)$$

where:

$$P_1 = 3\ [Pa]$$

$$f_1 = 500\ [Hz]$$

The RMS pressure square can be calculated in the time domain (Example 12.1) as:

$$p_{RMS}^2 = \frac{1}{T}\int_0^T p^2(t)dt = \frac{1}{T}\int_0^T P_1^2 \cos^2(2\pi f_1 t)dt = \frac{P_1^2}{2} = \frac{3^2}{2}\ [Pa^2]$$

or in the frequency domain (Example 12.3) as:

$$p^2_{RMS} = A^2_o + 2 \sum_{k=1}^{\infty} \left(A^2_k + B^2_k \right) = 2\left(1.5^2 + 0^2 \right) = 4.5 \left[Pa^2 \right]$$

Calculate:

The unweighted sound pressure level

Example 3: Solution

The unweighted sound pressure level is:

$$L_p = 10 \log_{10} \left(\frac{P^2_{RMS}}{P^2_r} \right) = 10 \log_{10} \left(\frac{P^2_{RMS}}{P^2_r} \right) = 10 \log_{10} \left(\frac{\left(\frac{3^2}{2} \right)}{\left(20 \times 10^{-6} \right)^2} \right)$$

$$= 100.5 [dB]$$

or equivalently:

$$L_p = 10 \log_{10} \left(\frac{P_{RMS}}{P_r} \right)^2 = 20 \log_{10} \left(\frac{P_{RMS}}{P_r} \right) = 20 \log_{10} \left(\frac{\frac{3}{\sqrt{2}}}{20 \times 10^{-6}} \right)$$

$$= 100.5 [dB]$$

Appendix 2: Power Spectral Density

In Sect. 9.3, the sound pressure level (SPL) spectrum of the octave band was obtained by manually adding the frequency contents to their corresponding bands. Practically, it is done using numerical methods. Appendix 2 will demonstrate how to compute the SPL spectrum numerically using accumulated energy. This appendix can be treated as a supplement to Sect. 9.3.

Power Spectral Density

The m-octave band spectrum is the power in each m-octave band and is typically represented in levels (dB) (i.e., the logarithmic scale) and should be operated on log level addition, subtraction, and averaging rules (see Sect. 9.1.3). The value is typically power-like (power, energy, or intensity of the wave), but it could also be non-power-like (RMS or peak amplitude of displacement, velocity, or acceleration of the wave). When the amplitudes of displacement are used, the amplitudes are usually squared, so that they are power-like and can be integrated or added among band frequencies. Whereas the amplitudes are squared, the spectrum is still called a *displacement*spectrum but not a *displacement square* spectrum. When adding octave band levels for specific physical quantities, care should be given to whether square quantities are being added.

The power spectral density, $S(f)$, is the power per frequency (usually in Hz), where the m-octave band spectrum is the power per m-octave band. Therefore, the total power is computed as $\int_0^\infty S(f)df$.

Accumulated Sound Pressure Square

From Section A.1.1.1, the total RMS pressure square is computed as follows:

$$p_{RMS}^2 = \frac{1}{T}\int_0^T p^2(t)dt = \frac{1}{T}\int_0^T \left(A_o + 2\sum_{k=1}^\infty [A_k \cos(\omega_k t) - B_k \sin(\omega_k t)]\right)^2 dt$$

$$= A_o^2 + 2\sum_{k=1}^\infty (A_k^2 + B_k^2) = \int_0^\infty S(f)df = \sum_{k=0}^\infty S(f_k)\Delta f$$

where $S(f_k) = 2(A_k^2 + B_k^2)/\Delta f = 2(A_k^2 + B_k^2)T$.

However, for the different weighting can be applied to each band, we need to compute the partial p_{RMS}^2 inside each band. One way to do that is first to compute the accumulated sound pressure square up to the frequency $f_n + \frac{1}{2}\Delta f$ as:

$$p_{RMS,Accu}^2\left(f_n + \frac{1}{2}\Delta f\right) = A_o^2 + 2\sum_{k=1}^n (A_k^2 + B_k^2) \quad \text{up to} \quad f_n + \frac{1}{2}\Delta f$$

Sound Pressure Level in Each Band

The partial p_{RMS}^2 inside the band center at f_o with the frequency limits $[f_{lower}, f_{upper}]$ can be obtained as:

$$p_{RMS,Band}^2 = p_{RMS,Accu}^2\left(f_{upper}\right) - p_{RMS,Accu}^2\left(f_{lower}\right)$$

where $p_{RMS,Accu}^2\left(f_{upper}\right)$ and $p_{RMS,Accu}^2\left(f_{lowper}\right)$ can be obtained by the linear interpolation from the accumulated values: $p_{RMS,Accu}^2\left(f_k + \frac{1}{2}\Delta f\right)$.

The sound pressure level in each band center at the frequency f_o is:

$$SPL_{f_o} = 10 * \log 10 \left(\frac{p^2_{RMS,Band}}{p^2_{ref}} \right) + \text{Aweighting}(f_o)$$

Example 1: Calculate Power Spectral Density

A discretized time ($\Delta t = 1/800$) domain data of the sound pressure [Pa] is given as:

[5,5.110,0.828,0.649,1,-2.945,-4.586,0.434,3,-2.281,-4.828,2.180,7,0.117,
-7.414,-3.262]

From $p(t) = \cos(2\pi \cdot 50t) + 2 \sin(2\pi \cdot 100t) + 3 \sin(2\pi \cdot 150t) + 4 \cos(2\pi \cdot 200t)$

Calculate:

(a) The RMS pressure square in the frequency domain
(b) The power spectral density of the given sound pressure $p(t)$

Example 1: Solution

The solution is obtained from the data set and plots of the MATLAB program. The step-by-step results are summarized as follows:

$\Delta t = 1/800$ [s], $N = 16$, $T = N\Delta t = 1/50$ [s], $\Delta f = 1/T = 50$ [Hz]

$$A_k = [0, 0.5, 0, 0, 2, 0, 0, 0, 0, 0, 0, 0, 2, 0, 0, 0.5]$$

$$B_k = [0, 0, -1, -1.5, 0, 0, 0, 0, 0, 0, 0, 0, 1.5, 1, 0, 0]$$

$$\sqrt{A_k^2 + B_k^2} = [0, 0.5, 1, 1.5, 2, 0, 0, 0, 0]$$

$$2(A_k^2 + B_k^2) = [0, 0.5, 2, 4.5, 8, 0, 0, 0, 0]$$

$$P^2_{RMS,Accu} = \sum 2(A_k^2 + B_k^2) = [0, 0.5, 2.5, 7, 15, 15, 15, 15, 15]$$

$$f_{Accu} = f_k + 25 = [25, 75, 125, 175, 225, 275, 325, 375, 425]$$

$$f_k = [31, 63, 125, 250]$$

$$f_{lower} = [22, 44, 88, 177, 354]$$

$$P^2_{RMS,lower} = [0, 0.1919, 1.0355, 7.2843, 15]$$

(a) $P^2_{RMS,Band} = [0.1919, 0.8436, 6.2487, 7.7157]$

$$SPL_{Band,unWeighting} = [112.8, 119.2, 127.9, 128.8]$$

$$aWeighting_{Band} = [-39.5, -26.2, -16.2, -8.7]$$

(b) $SPL_{Band,Weighted} = [73.3, 93.0, 111.7, 120.1]$

Exercise

An acoustic pressure $p(t)$ is transformed into the frequency domain for the calculation of the A-weighted sound pressure level [dBA]:

$$p(t) = 3\cos\left(2\pi \cdot 500 \cdot t\right) + 7\sin\left(2\pi \cdot 1000 \cdot t\right)[Pa]$$

Given:

(a) The number of discretized time domain data is $N = 16$.
(b) The time increment of time domain data is $\Delta t = 0.000125$ [s].
(c) The time domain pressure (N data) that is the input data for MATLAB fft:

Time domain pressure [Pa]

3	2.77	2.12	1.15	0	1.15	2.12	2.77	3	2.77	2.12	1.15	0	1.15	2.12	2.77

(d) The frequency domain (N data) which is the output from the MATLAB fft function:

Real number part coefficients [Pa] of $P_k = fft(p_i, N)/N$

0	1.5	0	0	0	0	0	0	0	0	0	0	0	0	0	1.5

Imaginary number part coefficients [Pa] of $P_k = fft(p_i, N)/N$

0	0	-3.5	0	0	0	0	0	0	0	0	0	0	0	3.5	0

Calculate:

The A-weighted sound pressure level
(Answers): 108.2 [dBA]

References

1. Blackstock, D. T. (2000). *Fundamentals of physical acoustics*. Wiley. ISBN:0-471-31979-1.
2. Harold, W., Lord, S., William, G., & Evensen Harold, A. (1987). *Noise control for engineers*. Krieger Publishing Company. ISBN:0-89464-255-3.
3. Harris, C. M. (1966). Absorption of sound in air. *Journal of the Acoustical Society of America, 40*, 148–159.
4. Irwin, J. D., & Graf, E. R. (1979). *Industrial noise and vibration control*. Prentice-Hall Inc.. ISBN:0-13-461574-3.
5. Kinsler, L. E., Frey Austin, R., Coppens Alan, B., & Sanders James, V. (1982). *Fundamentals of acoustics*. Wiley. ISBN:0-471-02933-5.
6. Magrab, E. B. (1975). *Environmental noise control*. Wiley. ISBN:0-471-56344-7.

Index

© The Author(s), under exclusive license to Springer Nature Switzerland AG 2021   373
H. Lin et al., *Lecture Notes on Acoustics and Noise Control*,
https://doi.org/10.1007/978-3-030-88213-6

Printed in the United States
by Baker & Taylor Publisher Services